Principles of Photochemistry

Principles of Photochemistry

Edited by
Nehemiah Wyatt

Larsen & Keller
www.larsen-keller.com

Principles of Photochemistry
Edited by Nehemiah Wyatt
ISBN: 978-1-63549-218-7 (Hardback)

© 2017 Larsen & Keller

▤ Larsen & Keller

Published by Larsen and Keller Education,
5 Penn Plaza,
19th Floor,
New York, NY 10001, USA

Cataloging-in-Publication Data

Principles of photochemistry / edited by Nehemiah Wyatt.
     p. cm.
Includes bibliographical references and index.
ISBN 978-1-63549-218-7
   1. Photochemistry. 2. Chemistry, Physical and theoretical.
3. Photoelectrochemistry. I. Wyatt, Nehemiah.
QD708.2 .P75 2017
541.35--dc23

The publisher's policy is to use permanent paper from mills that operate a sustainable forestry policy. Furthermore, the publisher ensures that the text paper and cover boards used have met acceptable environmental accreditation standards.

Printed and bound in the United States of America.

For more information regarding Larsen and Keller Education and its products, please visit the publisher's website www.larsen-keller.com

Table of Contents

Preface

This book provides comprehensive insights into the field of photochemistry. It elucidates the various concepts and theories related to this area. Photochemistry is the study of the chemical effects of light. This process is of significant importance as it enables photosynthesis in plants, helps in production of vitamin D and also, is crucial for vision. This book outlines the fundamental processes and applications of photochemistry in detail. It presents the complex subject of photochemistry in the most comprehensible and easy to understand language. The topics included in it are of utmost significance and are bound to provide incredible insights to readers. The textbook is appropriate for those seeking detailed information in this area.

A short introduction to every chapter is written below to provide an overview of the content of the book:

Chapter 1 - Photochemistry is a branch of chemistry; it studies the chemical effects of light. Photochemistry is very important in nature as it helps in photosynthesis. The chapter on photochemistry offers an insightful focus, keeping in mind the complex subject matter; **Chapter 2** - A number of fields are allied to photochemistry. Some of these fields are photoelectrochemistry and photogeochemistry. Photoelectrochemistry studies the interaction of light with electrochemical systems whereas photogeochemistry combines photochemistry and geochemistry into one subject. This text is a compilation of the various branches of photochemistry that form an integral part of the broader subject matter; **Chapter 3** - The main processes related to photochemistry are photochromism, bond softening, bond hardening, photocatalysis, scintillator and radiolysis. This chapter strategically encompasses and incorporates the major components and key concepts of photochemistry, providing a complete understanding; **Chapter 4** - The processes that are used in photoelectrochemistry are referred to as photoelectrochemical processes. It involves the transformation of light to other forms. Fluorescence, photoelectric effect, photoionization mode and absorption have been briefly explained in the following text; **Chapter 5** - The process used by plants to convert light into energy as a mechanism of feeding is known as photosynthesis. The major cause of the production of oxygen is photosynthesis. This section helps the readers in understanding all the topics related to photosynthesis; **Chapter 6** - Bioluminescence is the light emitted by a living organism. It occurs in a number of marine animals and as well as fungi. The light emitted by living organisms is emitted because of biophoton. The topics discussed in the chapter are of great importance to broaden the existing knowledge on bioluminescence; **Chapter 7** - The applications of photochemistry that have been discussed in the following section are ultraviolet germicidal irradiation, UV curing, light therapy, laser hair removal etc. The section serves as a source to understand the major applications related to photochemistry.

Finally, I would like to thank my fellow scholars who gave constructive feedback and my family members who supported me at every step.

Editor

Introduction to Photochemistry

Photochemistry is a branch of chemistry; it studies the chemical effects of light. Photochemistry is very important in nature as it helps in photosynthesis. The chapter on photochemistry offers an insightful focus, keeping in mind the complex subject matter.

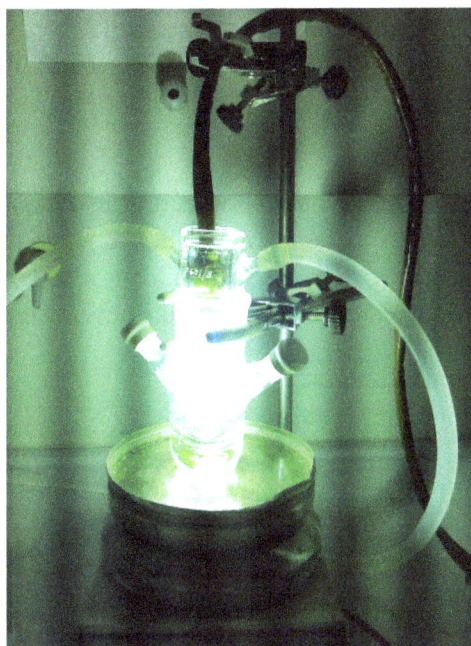

Photochemical immersion well reactor (50 mL) with a mercury-vapor lamp.

Photochemistry is the branch of chemistry concerned with the chemical effects of light. Generally, this term is used to describe a chemical reaction caused by absorption of ultraviolet (wavelength from 100 to 400 nm), visible light (400 – 750 nm) or infrared radiation (750 – 2500 nm).

In nature, photochemistry is of immense importance as it is the basis of photosynthesis, vision, and the formation of vitamin D with sunlight. Photochemical reactions proceed differently than thermal reactions. Photochemical paths access high energy intermediates that cannot be generated thermally, thereby overcoming large activation barriers in a short period of time, and allowing reactions otherwise inaccessible by thermal processes. Photochemistry is also destructive, as illustrated by the photodegradation of plastics.

Basics of Photochemistry

Grotthuss–Draper Law and Stark-einstein Law

Photoexcitation is the first step in a photochemical process where the reactant is elevated to a state of higher energy, an excited state. The first law of photochemistry, known as the Grotthuss–Draper law (for chemists Theodor Grotthuss and John W. Draper), states that light must be absorbed by a chemical substance in order for a photochemical reaction to take place. According to the second law of photochemistry, known as the Stark-Einstein law (for physicists Johannes Stark and Albert Einstein), for each photon of light absorbed by a chemical system, no more than one molecule is activated for a photochemical reaction, as defined by the quantum yield.

Fluorescence and Phosphorescence

When a molecule in the ground state (S_0) absorbs light, one electron is excited to a higher orbital level. This electron maintains its spin according to the spin selection rule; other transitions would violate the law of conservation of angular momentum. The excitation to a higher singlet state can be from HOMO to LUMO or to a higher orbital, so that singlet excitation states S_1, S_2, S_3... at different energies are possible.

Kasha's rule stipulates that higher singlet states would quickly relax by radiationless decay or internal conversion (IC) to S_1. Thus, S_1 is usually, but not always, the only relevant singlet excited state. This excited state S_1 can further relax to S_0 by IC, but also by an allowed radiative transition from S_1 to S_0 that emits a photon; this process is called fluorescence.

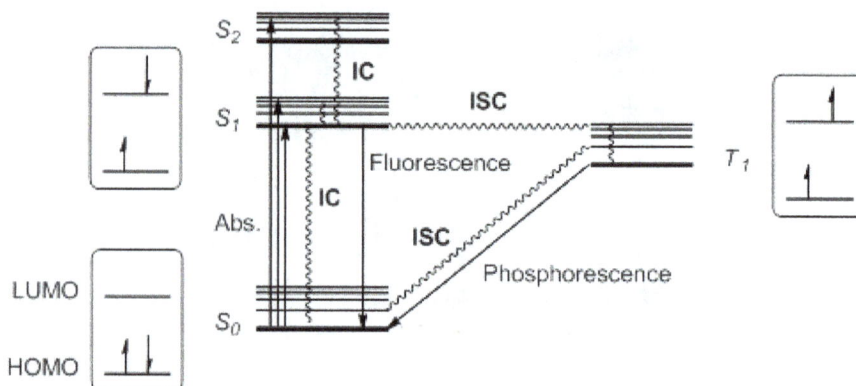

Jablonski diagram. Radiative paths are represented by straight arrows and non-radiative paths by curly lines.

Alternatively, it is possible for the excited state S_1 to undergo spin inversion and to generate a triplet excited state T_1 having two unpaired electrons with the same spin. This violation of the spin selection rule is possible by intersystem crossing (ISC) of the vibrational and electronic levels of S_1 and T_1. According to Hund's rule of maximum multiplicity, this T_1 state would be somewhat more stable than S_1.

This triplet state can relax to the ground state S_0 by radiationless IC or by a radiation pathway called phosphorescence. This process implies a change of electronic spin, which is forbidden by spin selection rules, making phosphorescence (from T_1 to S_0) much slower than fluorescence (from S_1 to S_0). Thus, triplet states generally have longer lifetimes than singlet states. These transitions are usually summarized in a state energy diagram or Jablonski diagram, the paradigm of molecular photochemistry.

These excited species, either S_1 or T_1, have a half empty low-energy orbital, and are consequently more oxidizing than the ground state. But at the same time, they have an electron in a high energy orbital, and are thus are more reducing. In general, excited species are prone to participate in electron transfer processes.

Experimental Set-up

Photochemical reactions require a light source that emits wavelengths corresponding to an electronic transition in the reactant. In the early experiments (and in everyday life), sunlight was the light source, although it is polychromatic. Mercury-vapor lamps are more common in the laboratory. Low pressure mercury vapor lamps mainly emit at 254 nm. For polychromatic sources, wavelength ranges can be selected using filters. Alternatively, laser beams are usually monochromatic (although two or more wavelengths can be obtained using nonlinear optics) and LEDs have a relatively narrowband that can be efficiently used, as well as Rayonet lamps, to get approximately monochromatic beams.

Photochemical immersion well reactor (750 mL) with a mercury-vapor lamp

Schlenk tube containing slurry of orange crystals of $Fe_2(CO)_9$ in acetic acid after its photochemical synthesis from $Fe(CO)_5$. The mercury lamp (connected to white power cords) can be seen on the left, set inside a water-jacketed quartz tube.

The emitted light must of course reach the targeted functional group without being blocked by the reactor, medium, or other functional groups present. For many applications, quartz is used for the reactors as well as to contain the lamp. Pyrex absorbs at wavelengths shorter than 275 nm. The solvent is an important experimental parameter. Solvents are potential reactants and for this reason, chlorinated solvents are avoided because the C-Cl bond can lead to chlorination of the substrate. Strongly absorbing solvents prevent photons from reaching the substrate. Hydrocarbon solvents absorb only at short wavelengths and are thus preferred for photochemical experiments requiring high energy photons. Solvents containing unsaturation absorb at longer wavelengths and can usefully filter out short wavelengths. For example, cyclohexane and acetone "cut off" (absorb strongly) at wavelengths shorter than 215 and 330 nm, respectively.

Photochemistry in Combination with Flow Chemistry

Continuous flow photochemistry offers multiple advantages over batch photochemistry. Photochemical reactions are driven by the number of photons that are able to activate molecules causing the desired reaction. The large surface area to volume ratio of a microreactor maximizes the illumination, and at the same time allows for efficient cooling, which decreases the thermal side products.

Principles

In the case of photochemical reactions, light provides the activation energy. Simplistically, light is one mechanism for providing the activation energy required for many reactions. If laser light is employed, it is possible to selectively excite a molecule so as to produce a desired electronic and vibrational state. Equally, the emission from a particular state may be selectively monitored, providing a measure of the population of that state. If the chemical system is at low pressure, this enables scientists to observe the energy distribution of the products of a chemical reaction before the differences in energy have been smeared out and averaged by repeated collisions.

The absorption of a photon of light by a reactant molecule may also permit a reaction to occur not just by bringing the molecule to the necessary activation energy, but also by changing the symmetry of the molecule's electronic configuration, enabling an otherwise inaccessible reaction path, as described by the Woodward–Hoffmann selection rules. A 2+2 cycloaddition reaction is one example of a pericyclic reaction that can be analyzed using these rules or by the related frontier molecular orbital theory.

Some photochemical reactions are several orders of magnitude faster than thermal reactions; reactions as fast as 10^{-9} seconds and associated processes as fast as 10^{-15} seconds are often observed.

The photon can be absorbed directly by the reactant or by a photosensitizer, which absorbs the photon and transfers the energy to the reactant. The opposite process is called quenching when a photoexited state is deactivated by a chemical reagent.

Most photochemical transformations occur through a series of simple steps known as primary photochemical processes. One common example of these processes is the excited state proton transfer.

Photochemical Reactions

Examples of Photochemical Reactions

- Photosynthesis: plants use solar energy to convert carbon dioxide and water into glucose and oxygen.

- Human formation of vitamin D by exposure to sunlight.

- Bioluminescence: *e.g.* In fireflies, an enzyme in the abdomen catalyzes a reaction that produced light.

- Polymerizations started by photoinitiators, which decompose upon absorbing light to produce the free radicals for radical polymerization.

- Photodegradation of many substances, e.g. polyvinyl chloride and Fp. Medicine bottles are often made with darkened glass to prevent the drugs from photodegradation.

- Photodynamic therapy: light is used to destroy tumors by the action of singlet oxygen generated by photosensitized reactions of triplet oxygen. Typical photosensitizers include tetraphenylporphyrin and methylene blue. The resulting singlet oxygen is an aggressive oxidant, capable of converting C-H bonds into C-OH groups.

- Photoresist technology, used in the production of microelectronic components.

- Vision is initiated by a photochemical reaction of rhodopsin.

- Toray photochemical production of ε-caprolactame.

- Photochemical production of artemisinin, anti-malaria drug.

Organic Photochemistry

Examples of photochemical organic reactions are electrocyclic reactions, radical reactions, photoisomerization and Norrish reactions.

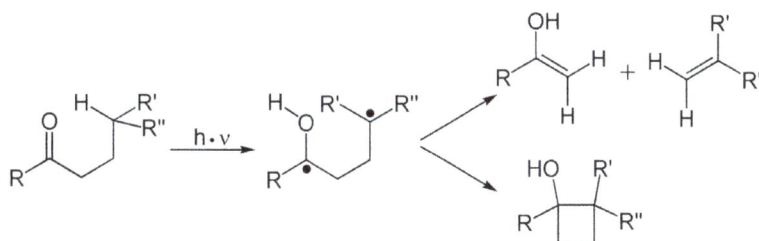

Alkenes undergo many important reactions that proceed via a photon-induced π to π^* transition. The first electronic excited state of an alkene lack the π-bond, so that rotation about the C-C bond is rapid and the molecule engages in reactions not observed thermally. These reactions include cis-trans isomerization, cycloaddition to other (ground state) alkene to give cyclobutane derivatives. The cis-trans isomerization of a (poly)alkene is involved in retinal, a component of the machinery of vision. The dimerization of alkenes is relevant to the photodamage of DNA, where thymine dimers are observed upon illuminating DNA to UV radiation. Such dimers interfere with transcription. The beneficial effects of sunlight are associated with the photochemically induced retro-cyclization (decyclization) reaction of ergosterol to give vitamin D. In the DeMayo reaction, an alkene reacts with a 1,3-diketone reacts via its enol to yield a 1,5-diketone. Still another common photochemical reaction is Zimmerman's Di-pi-methane rearrangement.

In an industrial application, about 100,000 tonnes of benzyl chloride are prepared annually by the gas-phase photochemical reaction of toluene with chlorine. The light is absorbed by chlorine molecule, the low energy of this transition being indicated by the yellowish color of the gas. The photon induces homolysis of the Cl-Cl bond, and the resulting chlorine radical converts toluene to the benzyl radical:

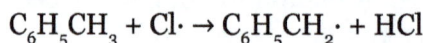

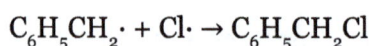

$$Cl_2 + h\nu \rightarrow 2\ Cl\cdot$$

$$C_6H_5CH_3 + Cl\cdot \rightarrow C_6H_5CH_2\cdot + HCl$$

$$C_6H_5CH_2\cdot + Cl\cdot \rightarrow C_6H_5CH_2Cl$$

Mercaptans can be produced by photochemical addition of hydrogen sulfide (H_2S) to alpha olefins.

Inorganic and Organometallic Photochemistry

Coordination complexes and organometallic compounds are also photoreactive. These reactions can entail cis-trans isomerization. More commonly photoreactions result in dissociation of ligands, since the photon excites an electron on the metal to an orbital that is antibonding with respect to the ligands. Thus, metal carbonyls that resist thermal substitution undergo decarbonylation upon irradiation with UV light. UV-irradiation of a THF solution of molybdenum hexacarbonyl gives the THF complex, which is synthetically useful:

$$Mo(CO)_6 + THF \rightarrow Mo(CO)_5(THF) + CO$$

In a related reaction, photolysis of iron pentacarbonyl affords diiron nonacarbonyl (see figure):

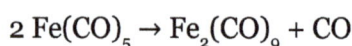

$$2\ Fe(CO)_5 \rightarrow Fe_2(CO)_9 + CO$$

Historical

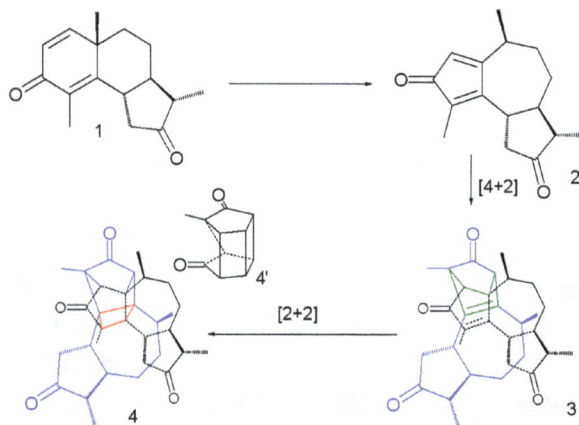

The first step is a rearrangement reaction to a cyclopentadienone intermediate **2**, the second one a dimerization in a Diels-Alder reaction (**3**) and the third one an intramolecular [2+2]cycloaddition (**4**). The bursting effect is attributed to a large change in crystal volume on dimerization.

Although bleaching has long been practiced, the first photochemical reaction was described by Trommsdorf in 1834. He observed that crystals of the compound α-santonin when exposed to sunlight turned yellow and burst. In a 2007 study the reaction was described as a succession of three steps taking place within a single crystal.

References

- Wayne, C. E.; Wayne, R. P. Photochemistry, 1st ed.; Oxford University Press: Oxford, United Kingdom, reprinted 2005. ISBN 0-19-855886-4.

- P. Klán, J. Wirz Photochemistry of Organic Compounds: From Concepts to Practice. Wiley, Chichester, 2009, ISBN 978-1405190886.

- N. J. Turro, V. Ramamurthy, J. C. Scaiano Modern Molecular Photochemistry of Organic MoKsenija Glusac "What has light ever done for chemistry?" Nature Chemistry 2016, volume 8, 734–73. doi:10.1038/nchem.2582

- Peplow, M. Chemistry World (Updated: 17th Apr 2013) <http://www.rsc.org/chemistryworld/2013/04/sanofi-launches-malaria-drug-production>

2

Allied Fields of Photochemistry

A number of fields are allied to photochemistry. Some of these fields are photoelectrochemistry and photogeochemistry. Photoelectrochemistry studies the interaction of light with electrochemical systems whereas photogeochemistry combines photochemistry and geochemistry into one subject. This text is a compilation of the various branches of photochemistry that form an integral part of the broader subject matter.

Photoelectrochemistry

Photoelectrochemistry is a subfield of study within physical chemistry concerned with the interaction of light with electrochemical systems. It is an active domain of investigation. One of the pioneers of this field of electrochemistry was the German electrochemist Heinz Gerischer. The interest in this domain is high in the context of development of renewable energy conversion and storage technology.

Historical Approach

Photoelectrochemistry has been intensively studied in the 70-80s because of the first peak oil crisis. Because fossil fuels are non-renewable, it is necessary to develop processes to obtain renewable ressources and use clean energy. Artificial photosynthesis, photoelectrochemical water splitting and regenerative solar cells are of special interest in this context.

H. Gerischer, H. Tributsch, AJ. Nozik, AJ. Bard, A. Fujishima, K. Honda, PE. Laibinis, K. Rajeshwar, TJ Meyer, PV. Kamat, N.S. Lewis, R. Memming, JOM. Bockris are researchers which have contributed a lot to the field of photoelectrochemistry.

Semiconductor's Electrochemistry

Introduction

Semiconductor material has a band gap and generates a pair of electron and hole per absorbed photon if the energy of the photon is higher than the band gap of the semiconductor. This property of semiconductor materials has been successfully used to convert solar energy into electrical energy by photovoltaic devices.

In photocatalysis the electron-hole pair is immediately used to drive a redox reaction but the problem is that the electron-hole pair suffer from fast recombinations. In photoelectrocatalysis, a differential potential is applied to diminish the number of recombinations between the electrons and the holes. This allows an increase in the yield of light's conversion into chemical energy.

Semiconductor-electrolyte Interface

When a semiconductor comes into contact with a liquid (redox species), to maintain electrostatic equilibrium, there will be a charge transfer between the semiconductor and liquid phase if formal redox potential of redox species lies inside semiconductor band gap. At thermodynamic equilibrium, the Fermi level of semiconductor and the formal redox potential of redox species are aligned at the interface between semiconductor and redox species. This introduces a downward band bending in a n-type semiconductor for n-type semiconductor/liquid junction (Figure 1(a)) and an upward band bending in a p-type semiconductor for a p-type semiconductor/liquid junction (Figure 1(b)). This characteristic of semiconductor/liquid junctions is similar to a rectifying semiconductor/metal junction or Schottky junction. Ideally to get a good rectifying characteristics at the semiconductor/liquid interface, the formal redox potential must be close to the valence band of the semiconductor for a n-type semiconductor and close to the conduction band of the semiconductor for a p-type semiconductor. The semiconductor/liquid junction has one advantage over the rectifying semiconductor/metal junction in that the light is able to travel through to the semiconductor surface without much reflection; whereas most of the light is reflected back from the metal surface at a semiconductor/metal junction. Therefore, semiconductor/liquid junctions can also be used as photovoltaic devices similar to solid state p–n junction devices. Both n-type and p-type semiconductor/liquid junctions can be used as photovoltaic devices to convert solar energy into electrical energy and are called photoelectrochemical cells. In addition, a semiconductor/liquid junction could also be used to directly convert solar energy into chemical energy by virtue of photoelectrolysis at the semiconductor/liquid junction.

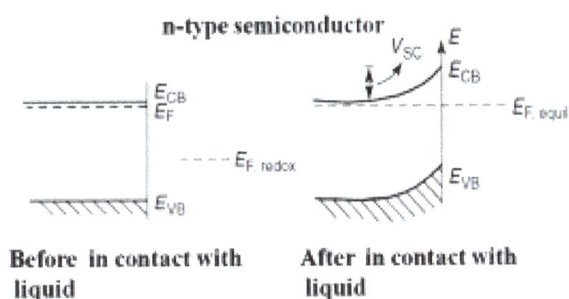

Figure 1(a) band diagram of n-type semiconductor/liquid junction

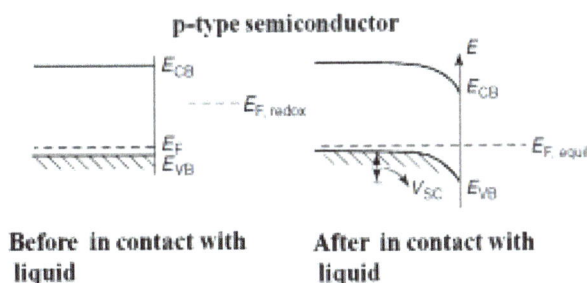

Figure 1(b) band diagram of p-type semiconductor/liquid junction

Experimental Setup

Semiconductors are usually studied in a photoelectrochemical cell. Different configurations exist with a three electrode device. The phenomenon to study happens at the working electrode WE while the differential potential is applied between the WE and a reference electrode RE (saturated calomel, Ag/AgCl). The current is measured between the WE and the counter electrode CE (carbon vitreous, platinum gauze). The working electrode is the semiconductor material and the electrolyte is composed of a solvent, an electrolyte and a redox specie.

A UV-vis lamp is usually used to illuminate the working electrode. The photoelectrochemical cell is usually made with a quartz window because it does not absorb the light. A monochromator can be used to control the wavelength sent to the WE.

Main Absorbers Used in Photoelectrochemistry

Semiconductor IV

C(diamond), Si, Ge, SiC, SiGe

Semiconductor III-V

BN, BP, BAs, AlN, AlP, AlAs, GaN, GaP, GaAs, InN, InP, InAs...

Semiconductor II-VI

CdS, CdSe, CdTe, ZnO, ZnS, ZnSe, ZnTe, MoS_2, $MoSe_2$, $MoTe_2$, WS_2, WSe_2

Metal Oxides

TiO_2, Fe_2O_3, Cu_2O

Organic Dyes

Methylene blue

Applications

Photoelectrochemical Splitting of Water

Photoelectrochemistry has been intensively studied in the field of hydrogen production from water and solar energy. The photoelectrochemical splitting of water was historically discovered by Fujishima and Honda in 1972 onto TiO_2 electrodes. Recently many materials have shown promising properties to split efficiently water but TiO_2 remains cheap, abundant, stable against photo-corrosion. The main problem of TiO_2 is its bandgap which is 3 or 3.2 eV according to its crystallinity (anatase or rutile). These values are too high and only the wavelength in the UV region can be absorbed. To increase the performances of this material to split water with solar wavelength, it is necessary to sensitize the TiO_2. Currently Quantum Dots sensitization is very promising but more research is needed to find new materials able to absorb the light efficiently. Recently water a splitting membrane concept has been developed. This method to split water is very similar to the principle of fuel cells but in a reverse way.

Artificial Photosynthesis

Photosynthesis is the natural process that converts CO_2 using light to produce hydrocarbon compounds such as sugar. The depletion of fossil fuels encourages scientists to find alternatives to produce hydrocarbon compounds. Artificial photosynthesis is a promising method mimicking the natural photosynthesis to produce such compounds. The photoelectrochemical reduction of CO2 is much studied because of its worldwide impact. Many researchers aim to find new semiconductors to develop stable and efficient photo-anodes and photo-cathodes.

Regenerative Cells or Dye-sensitized Solar Cell (Graetzel Cell)

Dye-sensitized solar cells or DSSCs use TiO_2 and dyes to absorb the light. This absorption induces the formation of electron-hole pairs which are used to oxidize and reduce the same redox couple, usually I^-/I_3^-. Consequently, a differential potential is created which induces a current.

Photogeochemistry

Photogeochemistry merges photochemistry and geochemistry into the study of light-induced chemical reactions that occur or may occur among natural components of Earth's surface. Photogeochemistry has been recently defined as the photochemistry of Earth-abundant minerals in shaping the biogeochemistry of Earth; this indeed describes the core of photogeochemical study, although other facets may be admitted into the definition.

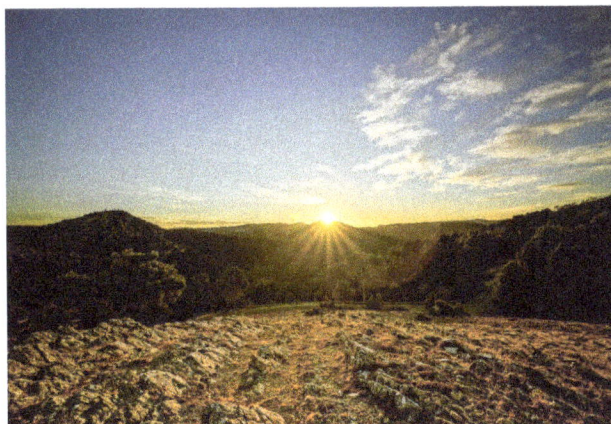

Sunlight facilitates chemical reactions among components of Earth's surface.

The Domain of Photogeochemistry

The context of a photogeochemical reaction is implicitly the surface of Earth, since that is where sunlight is available (although other sources of light such as chemiluminescence would not be strictly excluded from photogeochemical study). Reactions may occur among components of land such as rocks, soil and detritus; components of surface water such as sediment and dissolved organic matter; and components of the atmospheric boundary layer directly influenced by contact with land or water, such as mineral aerosols and gases. Visible and medium- to long-wave ultraviolet radiation is the main source of energy for photogeochemical reactions; wavelengths of light shorter than about 290 nm are completely absorbed by the present atmosphere, and are therefore practically irrelevant, except in consideration of atmospheres different from that of Earth today.

Photogeochemical reactions are limited to chemical reactions not facilitated by living organisms. The reactions comprising photosynthesis in plants and other organisms, for example, are not considered photogeochemistry, since the physiochemical context for these reactions is installed by the organism, and must be maintained in order for these reactions to continue (i.e. the reactions cease if the organism dies). In contrast, if a certain compound is produced by an organism, and the organism dies but the compound remains, this compound may still participate independently in a photogeochemical reaction even though its origin is biological (e.g. biogenic mineral precipitates or organic compounds released from plants into water).

The study of photogeochemistry is primarily concerned with naturally occurring materials, but may extend to include other materials, inasmuch as they are representative of, or bear some relation to, those found on Earth. For example, many inorganic compounds have been synthesized in the laboratory to study photocatalytic reactions. Although these studies are usually not undertaken in the context of environmental or Earth sciences, the study of such reactions is relevant to photogeochemistry if there is a geochemical implication (i.e. similar reactants or reaction mechanisms occur naturally). Similarly, photogeochemistry may also include photochemical reactions of naturally occurring materials that are not touched by sunlight, if there is the possibility that these materials may become exposed (e.g. deep soil layers uncovered by mining).

Iron(III) oxides and oxyhydroxides, such as these cliffs of ochre, are common catalysts in photogeochemical reactions.

Except for several isolated instances, studies that fit the definition of photogeochemistry have not been explicitly specified as such, but have been traditionally categorized as photochemistry, especially at the time when photochemistry was an emerging field or new facets of photochemistry were being explored. Photogeochemical research, however, may be set apart in light of its specific context and implications, thereby bringing more exposure to this "poorly explored area of experimental geochemistry". Past studies that fit the definition of photogeochemistry may be designated retroactively as such.

Early Photogeochemistry

The first efforts that can be considered photogeochemical research can be traced to the "form-aldehyde hypothesis" of Adolf von Baeyer in 1870, in which formaldehyde was proposed to be the initial product of plant photosynthesis, formed from carbon dioxide and water through the action of light on a green leaf. This suggestion inspired numerous attempts to obtain formaldehyde *in vitro*, which can retroactively be considered photogeochemical studies. Detection of organic compounds such as formaldehyde and sugars was reported by many workers, usually by exposure of a solution of carbon dioxide to light, typically a mercury lamp or sunlight itself. At the same time, many other workers reported negative results. One of the pioneer experiments was that of Bach in 1893, who observed the formation of lower uranium oxides upon irradiation of a solution of uranium acetate and carbon dioxide, implying the formation of formaldehyde. Some experiments included reducing agents such as hydrogen gas, and others detected formaldeyhde or other products in the absence of any additives, although the possibility was admitted that reducing power may have been produced from the decomposition of water during the experiment. In addition to the main focus on synthesis of formaldehyde and simple sugars, other light-assisted reactions were occasionally reported, such as the decomposition of formaldehyde and subsequent release of methane, or the formation of formamide from carbon monoxide and ammonia.

In 1912 Benjamin Moore summarized the main facet of photogeochemistry, that of inorganic photocatalysis: "the inorganic colloid must possess the property of transforming sunlight, or some other form of radiant energy, into chemical energy." Many experiments, still focused on how plants assimilate carbon, did indeed explore the effect of a "transformer" (catalyst); some effective "transformers" were similar to naturally occurring minerals, including iron(III) oxide or colloidal iron hydroxide; cobalt carbonate, copper carbonate, nickel carbonate; and iron(II) carbonate. Working with an iron oxide catalyst, Baly concluded in 1930 that "the analogy between the laboratory process and that in the living plant seems therefore to be complete," referring to his observation that in both cases, a photochemical reaction takes place on a surface, the activation energy is supplied in part by the surface and in part by light, efficiency decreases when the light intensity is too great, the optimal temperature of the reaction is similar to that of living plants, and efficiency increases from the blue to the red end of the light spectrum.

At this time, however, the intricate details of plant photosynthesis were still obscure, and the nature of photocatalysis in general was still actively being discovered; Mackinney in 1932 stated that "the status of this problem [photochemical CO_2 reduction] is extraordinarily involved." As in many emerging fields, experiments were largely empirical, but the enthusiasm surrounding this early work did lead to significant advances in photochemistry. The simple but challenging principle of transforming solar energy into chemical energy capable of performing a desired reaction remains the basis of application-based photocatalysis, most notably artificial photosynthesis (production of solar fuels).

After several decades of experiments centered around the reduction of carbon dioxide, interest began to spread to other light-induced reactions involving naturally occurring materials. These experiments usually focused on reactions analogous to known biological processes, such as soil nitrification, for which the photochemical counterpart "photonitrification" was first reported in 1930.

Classifying Photogeochemical Reactions

Photogeochemical reactions may be classified based on thermodynamics and/or the nature of the materials involved. In addition, when ambiguity exists regarding an analogous reaction involving light and living organisms (phototrophy), the term "photochemical" may be used to distinguish a particular abiotic reaction from the corresponding photobiological reaction. For example, "photo-oxidation of iron(II)" can refer to either a biological process driven by light (phototrophic or photobiological iron oxidation) or a strictly chemical, abiotic process (photochemical iron oxidation). Similarly, an abiotic process that converts water to O_2 under the action of light may be designated "photochemical oxidation of water" rather than simply "photooxidation of water", in order to distinguish it from photobiological oxidation of water potentially occurring in the same environment (by algae, for example).

Thermodynamics

Photogeochemical reactions are described by the same principles used to describe photochemical reactions in general, and may be classified similarly:

1. Photosynthesis: in the most general sense, photosynthesis refers to any light-activated reaction for which the change in free energy (ΔG°) is positive for the reaction itself (without considering the presence of a catalyst or light). The products have higher energy than the reactants, and therefore the reaction is thermodynamically unfavorable, except through the action of light in conjunction with a catalyst. Examples of photosynthetic reactions include the splitting of water to form H_2 and O_2, the reaction of CO_2 and water to form O_2 and reduced carbon compounds such as methanol and methane, and the reaction of N_2 with water to yield NH_3 and O_2.

2. Photocatalysis: this refers to reactions that are accelerated by the presence of a catalyst (the light itself is not the catalyst as may be erroneously implied). The overall reaction has a negative change in free energy, and is therefore thermodynamically favored. Examples of photocatalytic reactions include the reaction of organic compounds with O_2 to form CO_2 and water, and the reaction of organic compounds with water to give H_2 and CO_2.

3. Uncatalyzed photoreactions: photoinduced or photoactivated reactions proceed via the action of light alone. For example, photodegradation of organic compounds often proceeds without a catalyst if the reactants themselves absorb light.

Nature of Reactants

Any reaction in the domain of photogeochemistry, either observed in the environment or studied in the laboratory, may be broadly classified according to the nature of the materials involved.

1. Reactions among naturally occurring compounds. Photogeochemistry, both observational and exploratory, is concerned with reactions among materials known to occur naturally, as this reflects what happens or may happen on Earth.

2. Reactions in which one or more of the reactants are not known to occur naturally. Studies of reactions among materials related to naturally occurring materials may contribute to

understanding of natural processes. These complementary studies are relevant to photo-geochemistry in that they illustrate reactions that may have a natural counterpart. For example, it has been shown that soils, when irradiated, can generate reactive oxygen species and that clay minerals present in soils can accelerate the degradation of synthetic chemicals; it may therefore be postulated that naturally occurring compounds are similarly affected by sunlight acting on soil. The conversion of N_2 to NH_3 has been observed upon irradiation in the presence of the iron titanate $Fe_2Ti_2O_7$. While such a compound is not known to occur naturally, it is related to ilmenite ($FeTiO_3$) and pseudobrookite (Fe_2TiO_5), and can form upon heating of ilmenite; this may imply a similar reaction with N_2 for the naturally occurring minerals.

Photogeochemical Catalysts

Direct Catalysts

Direct photogeochemical catalysts act by absorbing light and subsequently transferring energy to reactants.

Semiconducting Minerals

The majority of observed photogeochemical reactions involve a mineral catalyst. Many naturally occurring minerals are semiconductors that absorb some portion of solar radiation. These semiconducting minerals are frequently transition metal oxides and sulfides and include abundant, well-known minerals such as hematite (Fe_2O_3), magnetite (Fe_3O_4), goethite and lepidocrocite (FeOOH), and pyrolusite (MnO_2). Radiation of energy equal to or greater than the band gap of a semiconductor is sufficient to excite an electron from the valence band to a higher energy level in the conduction band, leaving behind an electron hole (h^+); the resulting electron-hole pair is called an exciton. The excited electron and hole can reduce and oxidize, respectively, species having suitable redox potentials relative to the potentials of the valence and conduction bands. Semiconducting minerals with appropriate band gaps and appropriate band energy levels can catalyze a vast array of reactions, most commonly at mineral-water or mineral-gas interfaces.

Organic Compounds

Organic compounds such as "bio-organic substances" and humic substances are also able to absorb light and act as catalysts or sensitizers, accelerating photoreactions that normally occur slowly or facilitating reactions that might not normally occur at all.

Indirect Catalysts

Some materials, such as certain silicate minerals, absorb little or no solar radiation, but may still participate in light-driven reactions by mechanisms other than direct transfer of energy to reactants.

Production of Reactive Species

Indirect photocatalysis may occur via the production of a reactive species which then participates in another reaction. For example, photodegradation of certain compounds has been observed in the presence of kaolinite and montmorillonite, and this may proceed via the formation of reactive oxygen species at the surface of these clay minerals. Indeed, reactive oxygen species have been observed when soil surfaces are exposed to sunlight. The ability of irradiated soil to generate singlet oxygen was found to be independent of the organic matter content, and both the mineral and organic components of soil appear to contribute to this process. Indirect photolysis in soil has been observed to occur at depths of up to 2 mm due to migration of reactive species; in contrast, direct photolysis (in which the degraded compound itself absorbs light) was restricted to a "photic depth" of 0.2 to 0.4 mm. Like certain minerals, organic matter in solution, as well as particulate organic matter, may act as an indirect catalyst via formation of singlet oxygen which then reacts with other compounds.

Surface Sensitization

Indirect catalysts may also act through surface sensitization of reactants, by which species sorbed to a surface become more susceptible to photodegradation.

True Catalysis

Strictly speaking, the term "catalysis" should not be used unless it can be shown that the number of product molecules produced per number of active sites is greater than one; this is difficult to do in practice, although it is often assumed to be true if there is no loss in the photoactivity of the catalyst for an extended period of time. Reactions that are not strictly catalytic may be designated "assisted photoreactions". Furthermore, phenomena that involve complex mixtures of compounds (e.g. soil) may be hard to classify unless complete reactions (not just individual reactants or products) can be identified.

Experimental Approaches

The great majority of photogeochemical research is performed in the laboratory, as it is easier to demonstrate and observe a particular reaction under controlled conditions. This includes confirming the identity of materials, designing reaction vessels, controlling light sources, and adjusting the reaction atmosphere. However, observation of natural phenomena often provides initial inspiration for further study. For example, during the 1970s it was generally agreed that nitrous oxide (N_2O) has a short residence time in the troposphere, although the actual explanation for its removal was unknown. Since N_2O does not absorb light at wavelengths greater than 280 nm, direct photolysis had been discarded as a possible explanation. It was then observed that light would decompose chloromethanes when they were absorbed on silica sand, and this occurred at wavelengths far above the absorption spectra for these compounds. The same phenomenon was observed for N_2O, leading to the conclusion that particulate matter in the atmosphere is responsible for the destruction of N_2O via surface-sensitized photolysis. Indeed, the idea of such a sink for atmospheric N_2O was supported by several reports of low concentrations of N_2O in the air above deserts, where there is a high amount of suspended particulate matter. As another example, the observation that the amount of nitrous acid in the atmosphere greatly increases during the day lead to insight into the surface photochemistry of humic acids and soils and an explanation for the original observation.

Photogeochemical Reactions

The following table lists some reported reactions that are relevant to photogeochemical study, including reactions that involve only naturally occurring compounds as well as complementary reactions that involve synthetic but related compounds. The selection of reactions and references given is merely illustrative and may not exhaustively reflect current knowledge, especially in the case of popular reactions such as nitrogen photofixation for which there is a large body of literature. Furthermore, although these reactions have natural counterparts, the probability of encountering optimal reaction conditions may be low in some cases; for example, most experimental work concerning CO_2 photoreduction is intentionally performed in the absence of O_2, since O_2 almost always suppresses the reduction of CO_2. In natural systems, however, it is uncommon to find an analogous context where CO_2 and a catalyst are reached by light but there is no O_2 present.

References

- Kisch, Horst (2015). Semiconductor photocatalysis: principles and applications. Wiley. ISBN 978-3-527-33553-4.

- Georgiou, CD; et al. (2015). "Evidence for photochemical production of reactive oxygen species in desert soils". Nature Communications. 6: 7100. doi:10.1038/ncomms8100.

- Appiani, E; McNeill, K (2015). "Photochemical production of singlet oxygen from particulate organic matter". Environmental Science and Technology. 49: 3514–3522. doi:10.1021/es505712e.

- Kisch H. 2015. Semiconductor photocatalysis for atom-economic reactions. In: Bahnemann DW and Robertson PKJ (Eds). Environmental Photochemistry III. Springer. p. 186.

- Glaeser, SP; Berghoff, BA; Stratmann, V; Grossart, HP; Glaeser, J (2014). "Contrasting effects of singlet oxygen and hydrogen peroxide on bacterial community composition in a humic lake". PLOS ONE. 9: e92518. doi:10.1371/journal.pone.0092518.

- Gankanda, A; Grassian, VH (2014). "Nitrate photochemistry on laboratory proxies of mineral dust aerosol: wavelength dependence and action spectra". Journal of Physical Chemistry C. 118: 29117–29125. doi:10.1021/jp504399a.

- Li, K; An, X; Park, KH; Khraisheh, M; Tng, J (2014). "A critical review of CO2 photoconversion: catalysts and reactors". Catalysis Today. 224: 3–12. doi:10.1016/j.cattod.2013.12.006.

- Ismail, AA; Bahnemann, DW (2014). "Photochemical splitting of water for hydrogen production by photocatalysis: a review". Solar Energy Materials and Solar Cells. 128: 85–101. doi:10.1016/j.solmat.2014.04.037.

3

Processes of Photochemistry

The main processes related to photochemistry are photochromism, bond softening, bond hardening, photocatalysis, scintillator and radiolysis. This chapter strategically encompasses and incorporates the major components and key concepts of photochemistry, providing a complete understanding.

Photochromism

Photochromism is the reversible transformation of a chemical species between two forms by the absorption of electromagnetic radiation, where the two forms have different absorption spectra. Trivially, this can be described as a reversible change of colour upon exposure to light. The phenomenon was discovered in the late 1880s, including work by Markwald, who studied the reversible change of color of 2,3,4,4-tetrachloronaphthalen-1(4H)-one in the solid state. He labeled this phenomenon "phototropy", and this name was used until the 1950s when Yehuda Hirshberg, of the Weizmann Institute of Science in Israel proposed the term "photochromism". Photochromism can take place in both organic and inorganic compounds, and also has its place in biological systems (for example retinal in the vision process).

A photochromic eyeglass lens, after exposure to sunlight while part of the lens remained covered by paper.

Overview

Photochromism does not have a rigorous definition, but is usually used to describe compounds that undergo a reversible photochemical reaction where an absorption band in the visible part of the electromagnetic spectrum changes dramatically in strength or wavelength. In many cases, an absorbance band is present in only one form. The degree of change required for a photochemical reaction to be dubbed "photochromic" is that which appears dramatic by eye, but in essence there is no dividing line between photochromic reactions and other photochemistry. Therefore, while the trans-cis isomerization of azobenzene is considered a photochromic reaction, the analogous reaction of stilbene is not. Since photochromism is just a special case of a photochemical reaction, almost any photochemical reaction type may be used to produce photochromism with appropriate molecular design. Some of the most common processes involved in photochromism are pericyclic reactions, cis-trans isomerizations, intramolecular hydrogen transfer, intramolecular group transfers, dissociation processes and electron transfers (oxidation-reduction).

Another requirement of photochromism is two states of the molecule should be thermally stable under ambient conditions for a reasonable time. All the same, nitrospiropyran (which back-isomerizes in the dark over ~10 minutes at room temperature) is considered photochromic. All photochromic molecules back-isomerize to their more stable form at some rate, and this back-isomerization is accelerated by heating. There is therefore a close relationship between photochromic and thermochromic compounds. The timescale of thermal back-isomerization is important for applications, and may be molecularly engineered. Photochromic compounds considered to be "thermally stable" include some diarylethenes, which do not back isomerize even after heating at 80 C for 3 months.

Since photochromic chromophores are dyes, and operate according to well-known reactions, their molecular engineering to fine-tune their properties can be achieved relatively easily using known design models, quantum mechanics calculations, and experimentation. In particular, the tuning of absorbance bands to particular parts of the spectrum and the engineering of thermal stability have received much attention.

Sometimes, and particularly in the dye industry, the term "irreversible photochromic" is used to describe materials that undergo a permanent color change upon exposure to ultraviolet or visible light radiation. Because by definition photochromics are reversible, there is technically no such thing as an "irreversible photochromic"—this is loose usage, and these compounds are better referred to as "photochangable" or "photoreactive" dyes.

Apart from the qualities already mentioned, several other properties of photochromics are important for their use. These include

- Quantum yield of the photochemical reaction. This determined the efficiency of the photochromic change with respect to the amount of light absorbed. The quantum yield of isomerization can be strongly dependent on conditions.

- Fatigue resistance. In photochromic materials, fatigue refers to the loss of reversibility by processes such as photodegradation, photobleaching, photooxidation, and other side reactions. All photochromics suffer fatigue to some extent, and its rate is strongly dependent on the activating light and the conditions of the sample.

- Photostationary state. Photochromic materials have two states, and their interconversion can be controlled using different wavelengths of light. Excitation with any given wavelength of light will result in a mixture of the two states at a particular ratio, called the "photostationary state". In a perfect system, there would exist wavelengths that can be used to provide 1:0 and 0:1 ratios of the isomers, but in real systems this is not possible, since the active absorbance bands always overlap to some extent.

- Polarity and solubility. In order to incorporate photochromics in working systems, they suffer the same issues as other dyes. They are often charged in one or more state, leading to very high polarity and possible large changes in polarity. They also often contain large conjugated systems that limit their solubility.

Photochromic Complexes

A *photochromic complex* is a kind of chemical compound that has photoresponsive parts on its ligand. These complexes have a specific structure: photoswitchable organic compounds are attached to metal complexes. For the photocontrollable parts, thermally and photochemically stable chromophores (azobenzene, diarylethene, spiropyran, etc.) are usually used. And for the metal complexes, a wide variety of compounds that have various functions (redox response, luminescence, magnetism, etc.) are applied.

The photochromic parts and metal parts are so close that they can affect each other's molecular orbitals. The physical properties of these compounds shown by parts of them (i.e., chromophores or metals) thus can be controlled by switching their other sites by external stimuli. For example, photoisomerization behaviors of some complexes can be switched by oxidation and reduction of their metal parts. Some other compounds can be changed in their luminescence behavior, magnetic interaction of metal sites, or stability of metal-to-ligand coordination by photoisomerization of their photochromic parts.

Classes of Photochromic Materials

Photochromic molecules can belong to various classes: triarylmethanes, stilbenes, azastilbenes, nitrones, fulgides, spiropyrans, naphthopyrans, spiro-oxazines, quinones and others.

Spiropyrans and Spirooxazines

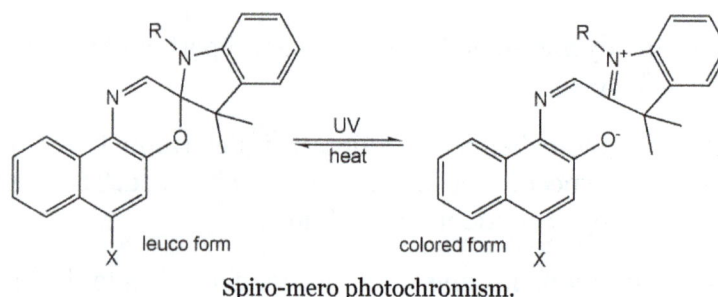

Spiro-mero photochromism.

One of the oldest, and perhaps the most studied, families of photochromes are the spiropyrans. Very closely related to these are the spirooxazines. For example, the spiro form of an oxazine is a colorless leuco dye; the conjugated system of the oxazine and another aromatic part of the molecule is separated by a sp³-hybridized "spiro" carbon. After irradiation with UV light, the bond between the spiro-carbon and the oxazine breaks, the ring opens, the spiro carbon achieves sp² hybridization and becomes planar, the aromatic group rotates, aligns its π-orbitals with the rest of the molecule, and a conjugated system forms with ability to absorb photons of visible light, and therefore appear colorful. When the UV source is removed, the molecules gradually relax to their ground state, the carbon-oxygen bond reforms, the spiro-carbon becomes sp^3 hybridized again, and the molecule returns to its colorless state.

This class of photochromes in particular are thermodynamically unstable in one form and revert to the stable form in the dark unless cooled to low temperatures. Their lifetime can also be affected by exposure to UV light. Like most organic dyes they are susceptible to degradation by oxygen and free radicals. Incorporation of the dyes into a polymer matrix, adding a stabilizer, or providing a barrier to oxygen and chemicals by other means prolongs their lifetime

Diarylethenes

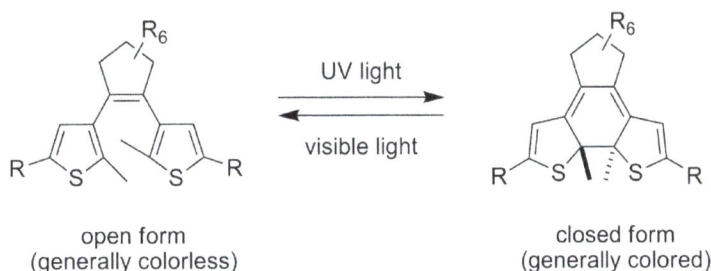

open form
(generally colorless)

closed form
(generally colored)

Dithienylethene photochemistry.

The "diarylethenes" were first introduced by Irie and have since gained widespread interest, largely on account of their high thermodynamic stability. They operate by means of a 6-pi electrocyclic reaction, the thermal analog of which is impossible due to steric hindrance. Pure photochromic dyes usually have the appearance of a crystalline powder, and in order to achieve the color change, they usually have to be dissolved in a solvent or dispersed in a suitable matrix. However, some diarylethenes have so little shape change upon isomerization that they can be converted while remaining in crystalline form.

Azobenzenes

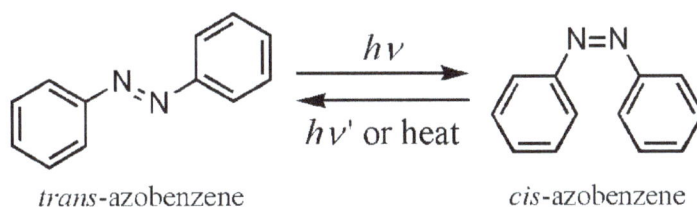

trans-azobenzene

cis-azobenzene

Azobenzene photoisomerization.

The photochromic trans-cis isomerization of azobenzenes has been used extensively in molecular switches, often taking advantage of its shape change upon isomerization to produce a supramolecular result. In particular, azobenzenes incorporated into crown ethers give switchable receptors and azobenzenes in monolayers can provide light-controlled changes in surface properties.

Photochromic Quinones

Some quinones, and phenoxynaphthacene quinone in particular, have photochromicity resulting from the ability of the phenyl group to migrate from one oxygen atom to another. Quinones with good thermal stability have been prepared, and they also have the additional feature of redox activity, leading to the construction of many-state molecular switches that operate by a mixture of photonic and electronic stimuli.

Inorganic Photochromics

Many inorganic substances also exhibit photochromic properties, often with much better resistance to fatigue than organic photochromics. In particular, silver chloride is extensively used in the manufacture of photochromic lenses. Other silver and zinc halides are also photochromic. Yttrium hydride is another inorganic material with photochromic properties.

Photochromic Coordination Compounds

Photochromic coordination complexes are relatively rare in comparison to the organic compounds listed above. There are two major classes of photochromic coordination compounds. Those based on sodium nitroprusside and the ruthenium sulfoxide compounds. The ruthenium sulfoxide complexes were created and developed by Rack and coworkers. The mode of action is an excited state isomerization of a sulfoxide ligand on a ruthenium polypyridine fragment from S to O or O to S. The difference in bonding from between Ru and S or O leads to the dramatic color change and change in Ru(III/II) reduction potential. The groundstate is always S-bonded and the metastable state is always O-bonded.Typically, absorption maxima changes of nearly 100 nm are observed. The metastable states (O-bonded isomers) of this class often revert thermally to their respective ground states (S-bonded isomers), although a number of examples exhibit two-color reversible photochromism. Ultrafast spectroscopy of these compounds has revealed exceptionally fast isomerization lifetimes ranging from 1.5 nanoseconds to 48 picoseconds.

Applications

Sunglasses

One of the most famous reversible photochromic applications is color changing lenses for sunglasses, as found in eyeglasses. The largest limitation in using PC technology is that the materials cannot be made stable enough to withstand thousands of hours of outdoor exposure so long-term outdoor applications are not appropriate at this time.

The switching speed of photochromic dyes is highly sensitive to the rigidity of the environment around the dye. As result, they switch most rapidly in solution and slowest in the rigid environment like a polymer lens. In 2005 it was reported that attaching flexible polymers with low glass transition temperature (for example siloxanes or poly(butyl acrylate)) to the dyes allow them to switch much more rapidly in a rigid lens. Some spirooxazines with siloxane polymers attached switch at near solution-like speeds even though they are in a rigid lens matrix.

Supramolecular Chemistry

Photochromic units have been employed extensively in supramolecular chemistry. Their ability to give a light-controlled reversible shape change means that they can be used to make or break molecular recognition motifs, or to cause a consequent shape change in their surroundings. Thus, photochromic units have been demonstrated as components of molecular switches. The coupling of photochromic units to enzymes or enzyme cofactors even provides the ability to reversibly turn enzymes "on" and "off", by altering their shape or orientation in such a way that their functions are either "working" or "broken".

Data Storage

The possibility of using photochromic compounds for data storage was first suggested in 1956 by Yehuda Hirshberg. Since that time, there have been many investigations by various academic and commercial groups, particularly in the area of 3D optical data storage which promises discs that can hold a terabyte of data. Initially, issues with thermal back-reactions and destructive reading dogged these studies, but more recently more-stable systems have been developed.

Novelty Items

Reversible photochromics are also found in applications such as toys, cosmetics, clothing and industrial applications. If necessary, they can be made to change between desired colors by combination with a permanent pigment.

Solar Energy Storage

Researchers at the Center for Exploitation of Solar Energy at the University of Copenhagen Department of Chemistry are studying, the Photochromic Dihydroazulene–Vinylheptafulvene System, for possible application to harvest solar energy and store it for significant amounts of time. Although storage lifetimes are attractive, for a real device it must of course be possible to trigger the back-reaction, which calls for further iterations in the future.

Photoluminescence

Fluorescent solutions under UV-light. Absorbed photons are rapidly re-emitted under longer wavelength.

Photoluminescence (abbreviated as PL) is light emission from any form of matter after the absorption of photons (electromagnetic radiation). It is one of many forms of luminescence (light emission) and is initiated by photoexcitation (excitation by photons), hence the prefix *photo-*. Following excitation various relaxation processes typically occur in which other photons are re-radiated.

Time periods between absorption and emission may vary: ranging from short femtosecond-regime for emission involving free-carrier plasma in inorganic semiconductors up to milliseconds for phosphorescent processes in molecular systems; and under special circumstances delay of emission may even span to minutes or hours.

Observation of photoluminescence at a certain energy can be viewed as indication that excitation populated an excited state associated with this transition energy.

While this is generally true in atoms and similar systems, correlations and other more complex phenomena also act as sources for photoluminescence in many-body systems such as semiconductors. A theoretical approach to handle this is given by the semiconductor luminescence equations.

Forms of Photoluminescence

Photoluminescence processes can be classified by various parameters such as the energy of the exciting photon with respect to the emission. Resonant excitation describes a situation in which photons of a particular wavelength are absorbed and equivalent photons are very rapidly re-emitted. This is often referred to as resonance fluorescence. For materials in solution or in the gas phase, this process involves electrons but no significant internal energy transitions involving molecular features of the chemical substance between absorption and emission. In crystalline inorganic semiconductors where an electronic band structure is formed, secondary emission can be more complicated as events may contain both coherent such as resonant Rayleigh scattering where a fixed phase relation with the driving light field is maintained (i.e. energetically elastic processes where no losses are involved) and incoherent contributions (or inelastic modes where some energy channels into an auxiliary loss mode),

The latter originate, e.g., from the radiative recombination of excitons, Coulomb-bound electron-hole pair states in solids. Resonance fluorescence may also show significant quantum optical correlations.

More processes may occur when a substance undergoes internal energy transitions before re-emitting the energy from the absorption event. Electrons change energy states by either resonantly gaining energy from absorption of a photon or losing energy by emitting photons. In chemistry-related disciplines, one often distinguishes between fluorescence and phosphorescence. The prior is typically a fast process, yet some amount of the original energy is dissipated so that re-emitted light photons will have lower energy than did the absorbed excitation photons. The re-emitted photon in this case is said to be red shifted, referring to the reduced energy it carries following this loss. For phosphorescence, absorbed photons undergo intersystem crossing where they enter into a state with altered spin multiplicity, usually a triplet state. Once energy from this absorbed electron is transferred in this triplet state, electron transition back to the lower singlet energy states is quantum mechanically forbidden, meaning that it happens much more slowly than other transitions. The result is a slow process of radiative transition back to the singlet state, sometimes lasting minutes or hours. This is the basis for "glow in the dark" substances.

Photoluminescence is an important technique for measuring the purity and crystalline quality of semiconductors such as GaAs and InP and for quantification of the amount of disorder present in a system. Several variations of photoluminescence exist, including photoluminescence excitation (PLE) spectroscopy.

Time-resolved photoluminescence (TRPL) is a method where the sample is excited with a light pulse and then the decay in photoluminescence with respect to time is measured. This technique is useful for measuring the minority carrier lifetime of III-V semiconductors like gallium arsenide (GaAs).

Photoluminescence Properties of Direct-gap Semiconductors

In a typical PL experiment, a semiconductor is excited with a light-source that provides photons with an energy larger than the bandgap energy. The incoming light excites a polarization that can be described with the semiconductor Bloch equations. Once the photons are absorbed, electrons and holes are formed with finite momenta $L+O2+\mathbf{k}$ in the conduction and valence bands, respectively. The excitations then undergo energy and momentum relaxation towards the band gap minimum. Typical mechanisms are Coulomb scattering and the interaction with phonons. Finally, the electrons recombine with holes under emission of photons.

Ideal, defect-free semiconductors are many-body systems where the interactions of charge-carriers and lattice vibrations have to be considered in addition to the light-matter coupling. In general, the PL properties are also extremely sensitive to internal electric fields and to the dielectric environment (such as in photonic crystals) which impose further degrees of complexity. A precise microscopic description is provided by the semiconductor luminescence equations.

Ideal Quantum-well Structures

An ideal, defect-free semiconductor quantum well structure is a useful model system to illustrate the fundamental processes in typical PL experiments. The discussion is based on results published in Klingshirn (2012) and Balkan (1998).

The fictive model structure for this discussion has two confined quantized electronic and two hole subbands, e1, e2 and h1,h2, respectively. The linear absorption spectrum of such a structure shows the exciton resonances of the first (e1h1) and the second quantum well subbands (e2h2), as well as the absorption from the corresponding continuum states and from the barrier.

Photoexcitation

In general, three different excitation conditions are distinguished: resonant, quasi-resonant, and non-resonant. For the resonant excitation, the central energy of the laser corresponds to the lowest exciton resonance of the quantum well. No or only a negligible amount of the excess energy is injected to the carrier system. For these conditions, coherent processes contribute significantly to the spontaneous emission. The decay of polarization creates excitons directly. The detection of PL is challenging for resonant excitation as it is difficult to discriminate contributions from the excitation, i.e., stray-light and diffuse scattering from surface roughness. Thus, speckle and resonant Rayleigh-scattering are always superimposed to the incoherent emission.

In case of the non-resonant excitation, the structure is excited with some excess energy. This is the typical situation used in most PL experiments as the excitation energy can be discriminated using a spectrometer or an optical filter. One has to distinguish between quasi-resonant excitation and barrier excitation.

For quasi-resonant conditions, the energy of the excitation is tuned above the ground state but still below the barrier absorption edge, for example, into the continuum of the first subband. The polarization decay for these conditions is much faster than for resonant excitation and coherent contributions to the quantum well emission are negligible. The initial temperature of the carrier system is significantly higher than the lattice temperature due to the surplus energy of the injected carriers. Finally, only the electron-hole plasma is initially created. It is then followed by the formation of excitons.

In case of barrier excitation, the initial carrier distribution in the quantum well strongly depends on the carrier scattering between barrier and the well.

Relaxation

Initially, the laser light induces coherent polarization in the sample, i.e., the transitions between electron and hole states oscillate with the laser frequency and a fixed phase. The polarization dephases typically on a sub-100 fs time-scale in case of nonresonant excitation due to ultra-fast Coulomb- and phonon-scattering.

The dephasing of the polarization leads to creation of populations of electrons and holes in the conduction and the valence bands, respectively. The lifetime of the carrier populations is rather long, limited by radiative and non-radiative recombination such as Auger recombination. During this lifetime a fraction of electrons and holes may form excitons, this topic is still controversially discussed in the literature. The formation rate depends on the experimental conditions such as lattice temperature, excitation density, as well as on the general material parameters, e.g., the strength of the Coulomb-interaction or the exciton binding energy.

The characteristic time-scales are in the range of hundreds of picoseconds in GaAs; they appear to be much shorter in wide-gap semiconductors.

Directly after the excitation with short (femtosecond) pulses and the quasi-instantaneous decay of the polarization, the carrier distribution is mainly determined by the spectral width of the excitation, e.g., a laser pulse. The distribution is thus highly non-thermal and resembles a Gaussian distribution, centered at a finite momentum. In the first hundreds of femtoseconds, the carriers are scattered by phonons, or at elevated carrier densities via Coulomb-interaction. The carrier system successively relaxes to the Fermi–Dirac distribution typically within the first picosecond. Finally, the carrier system cools down under the emission of phonons. This can take up to several nanoseconds, depending on the material system, the lattice temperature, and the excitation conditions such as the surplus energy.

Initially, the carrier temperature decreases fast via emission of optical phonons. This is quite efficient due to the comparatively large energy associated with optical phonons, (36meV or 420K in GaAs) and their rather flat dispersion, allowing for a wide range of scattering processes under conservation of energy and momentum. Once the carrier temperature decreases below the value corresponding to the optical phonon energy, acoustic phonons dominate the relaxation. Here, cooling is less efficient due their dispersion and small energies and the temperature decreases much slower beyond the first tens of picoseconds. At elevated excitation densities, the carrier cooling is further inhibited by the so-called hot-phonon effect. The relaxation of a large number of hot carriers leads to a high generation rate of optical phonons which exceeds the decay rate into acoustic phonons. This creates a non-equilibrium "over-population" of optical phonons and thus causes their increased reabsorption by the charge-carriers significantly suppressing any cooling. A system thus cools slower, the higher the carrier density is.

Radiative Recombination

The emission directly after the excitation is spectrally very broad, yet still centered in the vicinity of the strongest exciton resonance. As the carrier distribution relaxes and cools, the width of the PL peak decreases and the emission energy shifts to match the ground state of the exciton for ideal samples without disorder. The PL spectrum approaches its quasi-steady-state shape defined by the distribution of electrons and holes. Increasing the excitation density will change the emission spectra. They are dominated by the excitonic ground state for low densities. Additional peaks from higher subband transitions appear as the carrier density or lattice temperature are increased as these states get more and more populated. Also, the width of the main PL peak increases significantly with rising excitation due to excitation-induced dephasing and the emission peak experiences a small shift in energy due to the Coulomb-renormalization and phase-filling.

In general, both exciton populations and plasma, uncorrelated electrons and holes, can act as sources for photoluminescence as described in the semiconductor-luminescence equations. Both yield very similar spectral features which are difficult to distinguish; their emission dynamics, however, vary significantly. The decay of excitons yields a single-exponential decay function since the probability of their radiative recombination does not depend on the carrier density. The probability of spontaneous emission for uncorrelated electrons and holes, is approximately proportional to the product of electron and hole populations eventually leading to a non-single-exponential decay described by a hyperbolic function.

Effects of Disorder

Real material systems always incorporate disorder. Examples are structural defects in the lattice or disorder due to variations of the chemical composition. Their treatment is extremely challenging for microscopic theories due to the lack of detailed knowledge about perturbations of the ideal structure. Thus, the influence of the extrinsic effects on the PL is usually addressed phenomenologically. In experiments, disorder can lead to localization of carriers and hence drastically increase the photoluminescence life times as localized carriers cannot as easily find nonradiative recombination centers as can free ones.

Photoluminescent Material for Temperature Detection

In phosphor thermometry, the temperature dependence of the photoluminescence process is exploited to measure temperature.

Experimental Methods

Photoluminescence spectroscopy is a widely used technique for characterisation of the optical and electronic properties of semiconductors and molecules. In chemistry, it is more often referred to as fluorescence spectroscopy, but the instrumentation is the same. The relaxation processes can be studied using Time-resolved fluorescence spectroscopy to find the decay lifetime of the photoluminescence. These techniques can be combined with microscopy, to map the intensity (Confocal microscopy) or the lifetime (Fluorescence-lifetime imaging microscopy) of the photoluminescence across a sample (e.g. a semiconducting wafer, or a biological sample that has been marked with fluorescent molecules).

Bond Softening

Bond softening is an effect of reducing the strength of a chemical bond by strong laser fields. To make this effect significant, the strength of the electric field in the laser light has to be comparable with the electric field the bonding electron "feels" from the nuclei of the molecule. Such fields are typically in the range of 1–10 V/Å, which corresponds to laser intensities 10^{13}–10^{15} W/cm². Nowadays, these intensities are routinely achievable from table-top Ti:Sapphire lasers.

Theory

Theoretical description of bond softening can be traced back to early work on dissociation of diatomic molecules in intense laser fields. While the quantitative description of this process requires quantum mechanics, it can be understood qualitatively using quite simple models.

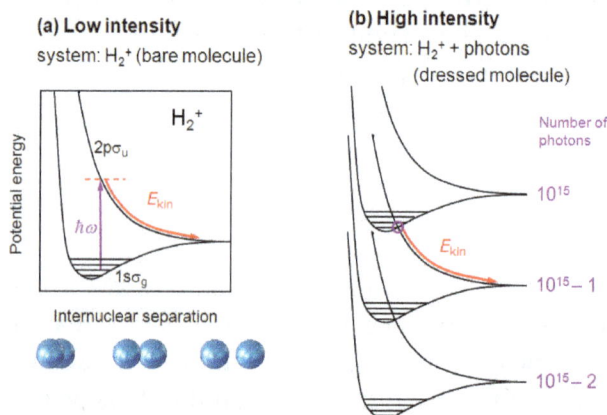

Two theoretical models of a molecule interacting with laser field. At low intensity (a) it is convenient to plot molecular energy curves and indicate photon transitions with vertical arrows. At high intensity (b) it is more appropriate to "dress" the molecular curves in photons and consider photon transitions at the curve crossings.

Low-intensity Description

Consider the simplest diatomic molecule, the H_2^+ ion. The ground state of this molecule is bonding and the first excited state is antibonding. This means that when we plot the potential energy of the molecule (i.e. the average electrostatic energy of the two protons and the electron plus the kinetic energy of the latter) as the function of proton-proton separation, the ground state has a minimum but the excited state is repulsive. Normally, the molecule is in the ground state, in one of the lowest vibrational levels (marked by horizontal lines).

In the presence of light, the molecule may absorb a photon (violet arrow), provided its frequency matches the energy difference between the ground and the excited states. The excited state is unstable and the molecule dissociates within femtoseconds into hydrogen atom and a proton releasing kinetic energy (red arrow). This is the usual description of photon absorption, which works well at low intensity. At high intensity, however, the interaction of the light with the molecule is so strong that the potential energy curves become distorted. To take this distortion into account requires "dressing" the molecule in photons.

Dressing in Photons at High Intensity

At high laser intensity absorptions and stimulated emissions of photons are so frequent that the molecule cannot be regarded as a system separate from the laser field; the molecule is "dressed" in photons forming a single system. However, the number of photons in this system varies when photons are absorbed and emitted. Therefore, to plot the energy diagram of the dressed molecule, we need to repeat the energy curves at each number of photons. The number of photons is very large but only a few curve repetitions need to be considered in this very tall ladder, as shown in Fig. 1b.

In the dressed model, photon absorption (and emission) is no longer represented by vertical transitions. As the energy must be conserved, photon absorption occurs at the curve crossings. For example, if the molecule is in the ground electronic state with 10^{15} photons present, it can jump to the repulsive state absorbing a photon at the curve crossing (violet circle) and dissociate to the $10^{15}-1$ photon limit (red arrow). This "curve jumping" is in fact continuous and can be explained in terms of avoided crossings.

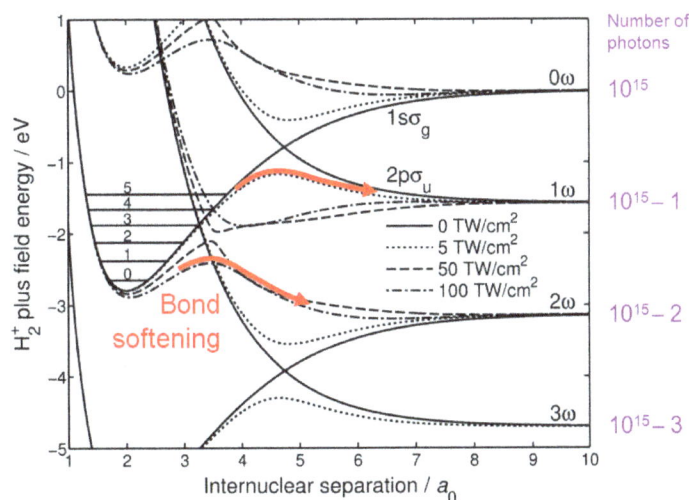

Figure 2: Distortion of molecular energy curves dressed in photons for increasing laser intensity. Curve crossings become anticrossings, which induces bond softening. The distorted curves have been calculated from undistorted ones in Matlab using Hamiltonian diagonalisation.

Energy Curve Distortion

When strong laser field perturbs the molecule, its energy levels are no longer the same as in the absence of the field. To calculate the new energy levels, the perturbation must be included as off-diagonal elements of the Hamiltonian, which has to be diagonalised. In consequence, the crossings turn into anticrossings and the higher the laser intensity, the larger the gap of the anticrossing as shown in Fig. 2. The molecule can dissociate along the lower branch of the anticrossings as indicated by the red arrows.

The top arrow represents one photon absorption, which is a continuous process. In the region of the anticrossing the molecule is in a superposition of the ground and the excited states, continuously exchanging energy with the laser field. As the internuclear separation increases, the molecule absorbs energy and the electronic wavefunction evolves to the antibonding state on the femtosecond timescale. The H_2^+ ion dissociates to the 1ω limit.

The bottom arrow represents a process initiated at the 3-photon gap. As the system passes through this gap, the 1-photon gap is wide open and the system slides along the top branch of the 1-photon anticrossing. The molecule dissociates to the 2ω limit via absorption of 3 photons followed by re-emission of 1 photon. (One-step even-photon absorptions and emissions are forbidden by the symmetry of the system.)

The anticrossing curves are adiabatic, i.e. they are accurate only for infinitely slow transitions. When the dissociation is fast and the gap is small, a diabatic transition may occur where the system ends up on the other branch of the anticrossing. The probability of such a transition is described by the Landau–Zener formula. When applied to the dissociation through the 3-photon gap, the formula gives a small probability of the H_2^+ molecular ion ending up in the 3ω dissociation limit without emitting any photons.

Experimental Confirmation

The "bond softening" phrase was coined by Phil Bucksbaum in 1990 at the time of its experimental observation. A Nd:YAG laser was used to generate intense pulses of about 80 ps duration at the second harmonic of 532 nm. In a vacuum chamber, the pulses were focused on molecular hydrogen under low pressure (about 10^{-6} mbar) inducing ionization and dissociation. The kinetic energy of protons was measured in a time-of-flight (TOF) spectrometer. The proton TOF spectra revealed three peaks of kinetic energy spaced by a half of the photon energy. As the neutral H atom was taking the other half of the photon energy, this was an unambiguous confirmation of the bond softening process leading to the 1ω, 2ω and 3ω dissociation limits. Such a process which absorbs more than the minimum number of photons is known as above-threshold dissociation.

A comprehensive review puts the mechanism of bond softening in a broader research context. Anticrossings of diatomic energy curves have many similarities to the conical intersections of energy surfaces in polyatomic molecules.

Bond Hardening

Bond hardening is a process of creating a new chemical bond by strong laser fields—an effect opposite to bond softening. However, it is not opposite in the sense that the bond becomes stronger, but in the sense that the molecule enters a state that is diametrically opposite to the bond-softened state. Such states require laser pulses of high intensity, in the range of 10^{13}–10^{15} W/cm², and they disappear once the pulse is gone.

Theory

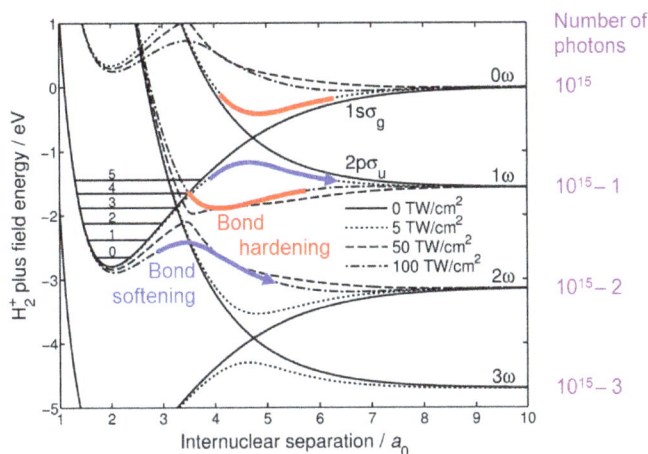

Energy curves of the H_2^+ ion dressed in photons for a few laser intensities. Bond hardening creates a new bound state on the upper branch of anticrossings. Bond softening dissociates the molecule along the lower branches.

Bond hardening and bond softening share the same theoretical basis, which is described under the latter entry. Briefly, the ground and the first excited energy curves of the H_2^+ ion are dressed in photons. The laser field perturbs the curves and turns their crossings into anticrossings. Bond softening occurs on the lower branches of the anticrossings and bond hardening happens if the molecule is excited to the upper branches –

To trap the molecule in the bond-hardened state, the anticrossing gap cannot be too small or too large. If it is too small, the system can undergo a diabatic transition to the lower branch of the anticrossing and dissociate via bond softening. If the gap is too large, the upper branch becomes shallow or even repulsive, and the system can also dissociate. This means that bound bond-hardened states can exist only in relatively narrow range of laser intensities, which makes them difficult to observe.

Experimental Search for Bond Hardening

When the existence of bond *softening* was experimentally verified in 1990, the attention turned to bond hardening. Rather noisy photoelectron spectra reported in the early 1990s implied bond hardening occurring at the 1-photon and 3-photon anticrossings. These reports were received with great interest because bond hardening could explain apparent stabilization of the molecular bond in strong laser fields accompanied by a collective ejection of several electrons. However, instead of more convincing evidence, new negative results relegated bond hardening to a remote theoretical possibility. Only at the end of the decade, the reality of bond hardening was established in an experiment where the laser pulse duration was varied by chirping.

Conclusive Evidence

The results of the chirp experiment are shown in Fig. 2 in the form of a map. The central "crater" of the map is a signature of bond hardening. To appreciate the uniqueness of this signature requires explaining other features on the map.

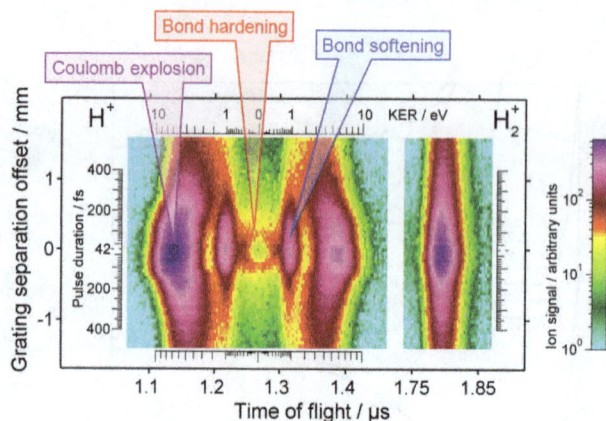

Figure 2: Signature of bond hardening in proton time-of-flight spectra with varying laser pulse duration. The kinetic energy release (KER) varies with pulse duration for bond hardening, unlike in bond softening where it is constant.

The horizontal axis of the map gives the time-of-flight (TOF) of ions produced in ionization and fragmentation of molecular hydrogen exposed to intense laser pulses. The left panel reveals several proton peaks; the right panel shows relatively uninteresting, single peak of molecular hydrogen ion.

The vertical axis gives grating position of the compressor in a chirped pulse amplifier of the Ti:Sapphire laser used in the experiment. The grating position controls the pulse duration, which is shortest (42 fs) for the zero position and increases in both directions. While the stretched pulses are also chirped, it is not the chirp but the pulse duration that matters in this experiment, as corroborated by the symmetry of the map in respect to the zero position line. The pulse energy is kept constant, therefore the shortest pulses are also most intense producing most ions at the zero position.

Kinetic Energy Variation

The proton TOF spectra allow one to measure the kinetic energy release (KER) in the dissociation process. Protons ejected towards the detector have shorter TOFs than protons ejected away from the detector because the latter have to be turned back by an external electric field applied to the interaction region. This forward-backward symmetry is reflected in the symmetry of the proton map in respect to zero KER ($1.27\,\mu s$ TOF).

The most energetic protons come from the Coulomb explosion of the molecule, where laser field completely strips H_2 from electrons and the two bare protons repel each other with strong Coulombic force, unimpeded by any chemical bond. The stripping process it not instantaneous but occurs in a stepwise fashion, on the rising edge of the laser pulse. The shorter the laser pulse, the quicker the stripping process and there is less time for the molecule to dissociate before the Coulomb force attains its full strength. Therefore, the KER is highest for the shortest pulses, as demonstrated by the outer curving "lobes" in Fig. 2.

The second pair of proton peaks (1 eV KER) comes from bond softening of the H_2^+ ion, which dissociates into a proton and a neutral hydrogen atom (undetected). The dissociation starts at the 3-photon gap and proceeds to the 2ω limit (the lower blue arrow in Fig. 1). Since both the initial and the final energies of this process are fixed by the 1.55 eV photon energy, the KER is also constant producing the two vertical lines in Fig. 2.

The lowest energy protons are produced by the bond hardening process, which also starts at the 3-photon gap but proceeds to the 1ω limit (the lower red trough in Fig. 1). Since the initial and the final energies are also fixed here, the KER should also be constant but clearly it is not, as the round shape of the central "crater" demonstrates it in Fig. 2. To explain this variation, the dynamics of the H_2^+ states needs to be considered.

Dynamics of Bond Hardening

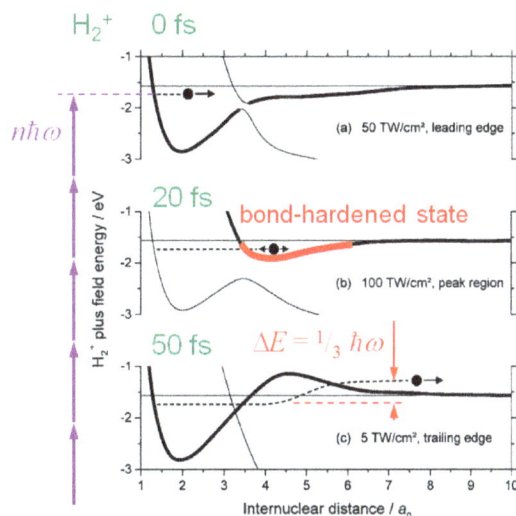

Figure 3: Evolution of bond hardening in laser field. An H_2^+ wave packet is created by absorption of n photons on the leading edge of the laser pulse (a). Trapping occurs near peak intensity (b). The wave packet is lifted up and released with some kinetic energy by the trailing edge of the pulse (c). A fraction of the photon energy is absorbed from the field.

The H_2^+ ion is created on the leading edge of the laser pulse in the multiphoton ionization process. Since the equilibrium internuclear separation for the neutral molecule is smaller than for the ionized one, the ionic nuclear wave packet finds itself on the repulsive side of the ground state potential well and starts to cross it.

In a few femtoseconds it takes the wave packet to cross the potential well, the laser intensity is still modest and the 3-photon gap is small allowing the wave packet to cross it diabatically. At large internuclear separations, the gentle slope of the potential well slowly turns the wave packet back, so when the packet returns to the 3-photon gap, the laser intensity is significantly higher and the gap is wide open trapping the wave packet in a bond-hardened state, which lasts throughout the highest intensities.

When the laser intensity falls, the bond-hardened energy curve returns to the original shape, flex-ing up, lifting the wave packet and releasing about a half of it to the 1ω limit. The faster intensity falls, the higher the wave packet is lifted and more energy it gains, which explains why the KER of the "crater" in Fig. 1 is highest at the shortest laser pulse. This energy gain, however, is not induced by the rising edge of the laser pulse as one would naively expect, but by the falling edge.

A Fraction of a Photon?

Note that the maximum energy gain of the nuclear wave packet is about $\frac{1}{3}\hbar\omega$ and continuously decreases with the pulse duration. Does it mean we can have a fraction of a photon? There are two valid answers to this puzzling proposition.

Breakdown of the Photon Model

One can say that the photon is not a particle but as a mere quantum of energy that is usually exchanged in integer multiples of $\hbar\omega$, but not always, as it is the case in the above experiment. From this point of view, photons are quasiparticles, akin to phonons and plasmons, in a sense less "real" than electrons and protons. Before dismissing this view as unscientific, its worth recalling the words of Willis Lamb, who won a Nobel prize in the area of quantum electrodynamics:

There is no such thing as a photon. Only a comedy of errors and historical accidents led to its popularity among physicists and optical scientists.

Dynamic Raman Effect

Alternatively, one can save the photon concept by recalling that the laser field is very strong and the pulse is very short. Indeed, the electric field in the laser pulse is so strong that during the process depicted in Fig. 3 about a hundred of photon absorptions and stimulated emissions can take place. And since the pulse is short, it has sufficiently wide bandwidth to accommodate absorption of photons that are more energetic than the re-emitted ones, giving the net result of a fraction of $\hbar\omega$. Effectively, we have a kind of dynamic Raman effect.

Zero-photon Dissociation

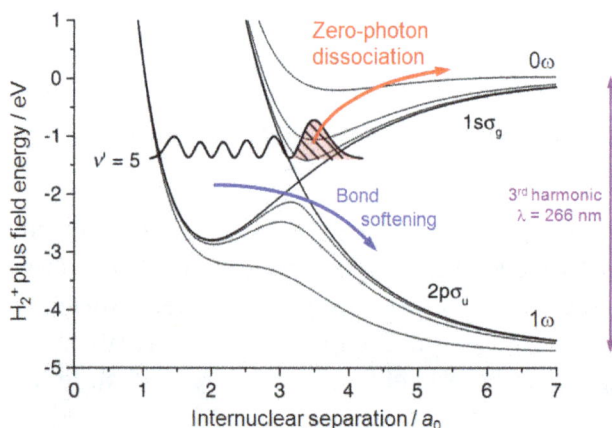

Figure 4: Zero-photon dissociation. The 3rd harmonic of the Ti:Sapphire laser can lift the trapped wave packet all the way to the 0ω limit. The molecule is dissociated without absorption of a net number of photons.

Even more striking challenge to the photon concept comes from the zero-photon dissociation process (ZPD), where nominally no photons are absorbed but some energy is still extracted from the laser field. To demonstrate this process, molecular hydrogen was exposed to 250 fs pulses of the 3rd harmonic of a Ti:Sapphire laser. Since the photon energy was 3 times higher, the spacing of the energy curves shown in Fig. 2 was 3 times larger, replacing the 3-photon crossing with a 1-photon one, as shown in Fig. 4. As before, the laser field changed the crossing to anticrossing, bond softening was induced on its lower branch and bond hardening trapped a part of the vibrational wave packet on the upper branch. In increasing laser intensity the anticrossing gap was getting wider, lifting the wave packet to the 0ω limit and dissociating the molecule with very small KER.

The experimental signature of the ZPD was a proton peak at zero KER. Moreover, the probability of a proton being promoted to this peak was found to be independent of the laser intensity, which confirms that it is induced by a zero-photon process because the probability of multiphoton processes is proportional to the intensity, I, raised to the number of photons absorbed, giving $I^0 =$ const.

Photocatalysis

In the experiment above, photons from a light source (out of frame on the right hand side) are absorbed by the surface of the titanium dioxide disc, exciting electrons within the material. These then react with the water molecules, splitting it into its constituents of hydrogen and oxygen. In this experiment, chemicals dissolved in the water prevent the formation of oxygen, which would otherwise recombine with the hydrogen.

In chemistry, photocatalysis is the acceleration of a photoreaction in the presence of a catalyst. In catalysed photolysis, light is absorbed by an adsorbed substrate. In photogenerated catalysis, the photocatalytic activity (PCA) depends on the ability of the catalyst to create electron–hole pairs, which generate free radicals (e.g. hydroxyl radicals: •OH) able to undergo secondary reactions. Its practical application was made possible by the discovery of water electrolysis by means of titanium dioxide. The commercially used process is called the advanced oxidation process (AOP). There are several ways the AOP can be carried out; these may (but do not necessarily) involve TiO_2 or even the use of UV light. Generally the defining factor is the production and use of the hydroxyl radical.

Types of Photocatalysis

Homogeneous Photocatalysis

In homogeneous photocatalysis, the reactants and the photocatalysts exist in the same phase. The most commonly used homogeneous photocatalysts include ozone and photo-Fenton systems (Fe^+ and Fe^+/H_2O_2). The reactive species is the $\cdot OH$ which is used for different purposes. The mechanism of hydroxyl radical production by ozone can follow two paths.

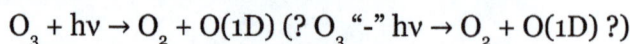

$$O_3 + hv \rightarrow O_2 + O(1D)\ (?\ O_3\ \text{"-"}\ hv \rightarrow O_2 + O(1D)\ ?)$$

$$O(1D) + H_2O \rightarrow \cdot OH + \cdot OH$$

$$O(1D) + H_2O \rightarrow H_2O_2$$

$$H_2O_2 + hv \rightarrow \cdot OH + \cdot OH$$

Similarly, the Fenton system produces hydroxyl radicals by the following mechanism

$$Fe^{2+} + H_2O_2 \rightarrow HO\cdot + Fe^{3+} + OH^-$$

$$Fe^{3+} + H_2O_2 \rightarrow Fe^{2+} + HO\cdot 2 + H^+$$

$$Fe^{2+} + HO\cdot \rightarrow Fe^{3+} + OH^-$$

In photo-Fenton type processes, additional sources of OH radicals should be considered: through photolysis of H_2O_2, and through reduction of Fe^{3+} ions under UV light:

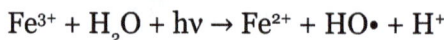

$$H_2O_2 + hv \rightarrow HO\cdot + HO\cdot$$

$$Fe^{3+} + H_2O + hv \rightarrow Fe^{2+} + HO\cdot + H^+$$

The efficiency of Fenton type processes is influenced by several operating parameters like concentration of hydrogen peroxide, pH and intensity of UV. The main advantage of this process is the ability of using sunlight with light sensitivity up to 450 nm, thus avoiding the high costs of UV lamps and electrical energy. These reactions have been proven more efficient than the other photocatalysis but the disadvantages of the process are the low pH values which are required, since iron precipitates at higher pH values and the fact that iron has to be removed after treatment.

Heterogeneous Photocatalysis

Heterogeneous catalysis has the catalyst in a different phase from the reactants. Heterogeneous photocatalysis is a discipline which includes a large variety of reactions: mild or total oxidations, dehydrogenation, hydrogen transfer, $^{18}O_2-^{16}O_2$ and deuterium-alkane isotopic exchange, metal deposition, water detoxification, gaseous pollutant removal, etc.

Most common heterogeneous photocatalysts are transition metal oxides and semiconductors, which have unique characteristics. Unlike the metals which have a continuum of electronic states, semiconductors possess a void energy region where no energy levels are available to promote recombination of an electron and hole produced by photoactivation in the solid. The void region, which extends from the top of the filled valence band to the bottom of the vacant conduction band, is called the band gap. When a photon with energy equal to or greater than the materials band gap is absorbed by the semiconductor, an electron is excited from the valence band to the conduction band, generating a positive hole in the valence band. The excited electron and hole can recombine and release the energy gained from the excitation of the electron as heat. Recombination is undesirable and leads to an inefficient photocatalyst. The ultimate goal of the process is to have a reaction between the excited electrons with an oxidant to produce a reduced product, and also a reaction between the generated holes with a reductant to produce an oxidized product. Due to the generation of positive holes and electrons, oxidation-reduction reactions take place at the surface of semiconductors. In the oxidative reaction, the positive holes react with the moisture present on the surface and produce a hydroxyl radical.

Oxidative reactions due to photocatalytic effect:

$$UV + MO \rightarrow MO (h + e^-)$$

Here MO stands for metal oxide ---

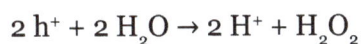

$$h^+ + H_2O \rightarrow H^+ + \cdot OH$$

$$2\,h^+ + 2\,H_2O \rightarrow 2\,H^+ + H_2O_2$$

$$H_2O_2 \rightarrow 2\,\cdot OH$$

The reductive reaction due to photocatalytic effect:

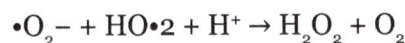

$$e^- + O_2 \rightarrow \cdot O_2^-$$

$$\cdot O_2^- + HO\cdot 2 + H^+ \rightarrow H_2O_2 + O_2$$

$$HOOH \rightarrow HO\cdot + \cdot OH$$

Ultimately, the hydroxyl radicals are generated in both the reactions. These hydroxyl radicals are very oxidative in nature and non selective with redox potential of (E_0 = +3.06 V)

Applications

- Conversion of water to hydrogen gas by photocatalytic water splitting. An efficient photocatalyst in the UV range is based on a sodium tantalite ($NaTaO_3$) doped with La and loaded with a cocatalyst nickel oxide. The surface of the sodium tantalite crystals is grooved with so called nanosteps that is a result of doping with lanthanum (3–15 nm range). The NiO particles which facilitate hydrogen gas evolution are present on the edges, with the oxygen gas evolving from the grooves.

- Use of titanium dioxide in self-cleaning glass. Free radicals generated from TiO_2 oxidize organic matter.

- Disinfection of water by supported titanium dioxide photocatalysts, a form of solar water disinfection (SODIS).

- Use of titanium dioxide in self-sterilizing photocatalytic coatings (for application to food contact surfaces and in other environments where microbial pathogens spread by indirect contact).

- Oxidation of organic contaminants using magnetic particles that are coated with titanium dioxide nanoparticles and agitated using a magnetic field while being exposed to UV light.

- Conversion of carbon dioxide into gaseous hydrocarbons using titanium dioxide in the presence of water. As an efficient absorber in the UV range, titanium dioxide nanoparticles in the anatase and rutile phases are able to generate excitons by promoting electrons across the band gap. The electrons and holes react with the surrounding water vapor to produce hydroxyl radicals and protons. At present, proposed reaction mechanisms usually suggest the creation of a highly reactive carbon radical from carbon monoxide and carbon dioxide which then reacts with the photogenerated protons to ultimately form methane. Although the efficiencies of present titanium dioxide based photocatalysts are low, the incorporation of carbon based nanostructures such as carbon nanotubes and metallic nanoparticles have been shown to enhance the efficiency of these photocatalysts.

- Sterilization of surgical instruments and removal of unwanted fingerprints from sensitive electrical and optical components.

- A less-toxic alternative to tin and copper-based antifouling marine paints, ePaint, generates hydrogen peroxide by photocatalysis.

- Decomposition of crude oil with TiO_2 nanoparticles: by using titanium dioxide photocatalysts and UV-A radiation from the sun, the hydrocarbons found in crude oil can be turned into H_2O and CO_2. Higher amounts of oxygen and UV radiation increased the degradation of the model organics. These particles can be placed on floating substrates, making it easier to recover and catalyze the reaction. This is relevant since oil slicks float on top of the ocean and photons from the sun target the surface more than the inner depth of the ocean. By covering floating substrates like woodchips with epoxy adhesives, water logging can be prevented and TiO_2 particles can stick to the substrates. With more research, this method should be applicable to other organics.

- Decontamination of water with photocatalysis and adsorption: the removal and destruction of organic contaminants in groundwater can be addressed through the impregnation of adsorbents with photoactive catalysts. These adsorbents attract contaminating organic atoms/molecules like tetrachloroethylene to them. The photoactive catalysts impregnated inside speed up the degradation of the organics. Adsorbents are placed in packed beds for 18 hours, which would attract and degrade the organic compounds. The spent adsorbents would then be placed in regeneration fluid, essentially taking away all organics still attached by passing hot water counter-current to the flow of water during the adsorption process to speed up the reaction. The regeneration fluid then gets passed through the fixed beds of silica gel photocatalysts to remove and decompose the rest of the organics left. Through the use of fixed bed reactors, the regeneration of adsorbents can help increase the efficiency.

- Decomposition of polyaromatic hydrocarbons (PAHs). Triethylamine (TEA) was utilized to solvate and extract the polyaromatic hydrocarbons (PAHs) found in crude oil. By solvating these PAHs, TEA can attract the PAHs to itself. Once removed, TiO_2 slurries and UV light can photocatalytically degrade the PAHs. The figure shows the high success rate of this experiment. With high yielding of recoveries of 93–99% of these contaminants, this process has become an innovative idea that can be finalized for actual environmental usage. This procedure demonstrates the ability to develop photocatalysts that would be performed at ambient pressure, ambient temperature, and at a cheaper cost.

Quantification of Photocatalytic Activity

ISO 22197 -1:2007 specifies a test method for the determination of the nitric oxide removal performance of materials that contain a photocatalyst or have photocatalytic films on the surface.

Specific FTIR systems are used to characterise photocatalytic activity and/or passivity especially with respect to Volatile Organic Compounds VOCs and representative matrices of the binders applied.

Recent studies show that mass spectrometry can be a powerful tool to determine photocatalytic activity of certain materials by following the decomposition of gaseous pollutants such as nitrogen oxides or carbon dioxide

Photoredox Catalysis

Photoredox catalysis is a branch of catalysis that harnesses the energy of visible light to accelerate a chemical reaction via a single-electron transfer. This area is named as a combination of "photo-" referring to light and redox, a condensed expression for the chemical processes of reduction and oxidation. In particular, photoredox catalysis employs small quantities of a light-sensitive compound that, when excited by light, can mediate the transfer of electrons between chemical compounds that otherwise would not react. Photoredox catalysts are generally drawn from three classes of materials: transition-metal complexes, organic dyes and semiconductors. While each class of materials has advantages, soluble transition-metal complexes are used most often.

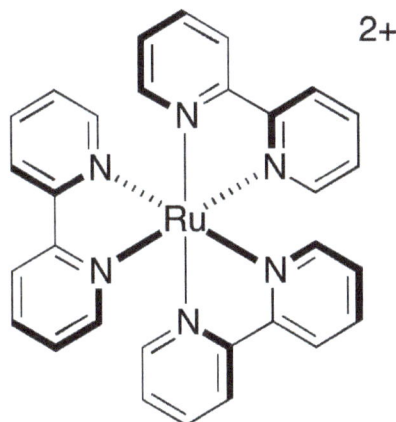

Study of this branch of catalysis led to the development of new methods to accomplish known and new chemical transformations. One attraction to the area is that photoredox catalysts are often less toxic than other reagents often used to generate free radicals, such as organotin reagents. Furthermore, while photoredox catalysts generate potent redox agents while exposed to light, they are innocuous under ordinary conditions Thus transition-metal complex photoredox catalysts are in some ways more attractive than stoichiometric redox agents such as quinones. The properties of photoredox catalysts can be modified by changing ligands and the metal, reflecting the somewhat modular nature of the catalyst.

While photoredox catalysis has most often been applied to generate known reactive intermediates in a novel way, the study of this mode of catalysis led to the discovery of new organic reactions, such as the first direct functionalization of the β-arylation of saturated aldehydes. Although the D_3-symmetric transition-metal complexes used in many photoredox-catalyzed reactions are chiral, the use of enantioenriched photoredox catalysts led to low levels of enantioselectivity in a photoredox-catalyzed aryl-aryl coupling reaction, suggesting that the chiral nature of these catalysts is not yet a highly effective means of transmitting stereochemical information in photoredox reactions. However, while synthetically useful levels of enantioselectivity have not been achieved using chiral photoredox catalysts alone, optically-active products have been obtained through the synergistic combination of photoredox catalysis with chiral organocatalysts such as secondary amines and Brønsted acids.

Photochemistry of Transition Metal Photocatalysts

The activity of a photoredox catalyst can be described in three steps. First, a molecule of the catalyst in its ground state, where the electrons are distributed among the lowest-energy combination of states, interacts with light and moves into a long-lived excited state, where the electrons are not distributed among the lowest-energy combination of available states. Second, the photoexcited catalyst interacts by an outer sphere electron transfer process to "quench" the excited state and to activate one of the other components of the chemical reaction. Finally, a second single electron transfer occurs to return the catalyst to its original oxidation state and electron arrangement.

Photoexcitation

The first step of this process, photoexcitation, is initiated by absorption of a photon, promoting an electron from the highest occupied molecular orbital (HOMO) of the photocatalyst to another spin-allowed state. Thus, since the ground state of the photocatalyst is a singlet state (a state with no total electron spin), the photon absorption excites the catalyst to another singlet state of higher energy. This excitation is realized as a metal-to-ligand charge transfer, where the electron moves from an orbital centered on the metal (e.g. a d orbital) to an orbital localized on the ligands (e.g. the π* orbital of an aromatic ligand).

While absorption of a photon can occur for any energy corresponding to the difference between two singlet states, the excited electronic state relaxes to the lowest energy singlet excited state through internal conversion, a process whereby energy is dissipated as vibrational energy rather than as electromagnetic radiation. This singlet excited state can relax further by two distinct processes: the catalyst may fluoresce, radiating a photon and returning to the singlet ground state, or it can move to the lowest energy triplet excited state (a state where two unpaired electrons have the same spin) by a second non-radiative process termed intersystem crossing.

Direct relaxation of the excited triplet to the ground state, termed phosphorescence, requires both emission of a photon and inversion of the spin of the excited electron. The spin-forbidden nature of this pathway means that it is a slow process and therefore that the triplet excited state has a substantial average lifetime. For the common photocatalyst tris-(2,2'-bipyridyl)ruthenium chloride (also known as Ru(bipy)$_3{}^{2+}$), the lifetime of the triplet excited state has been measured to be approximately 1100 ns. This span of time is long enough that other relaxation pathways, specifically electron-transfer pathways, can occur more rapidly than decay of the catalyst to its ground state.

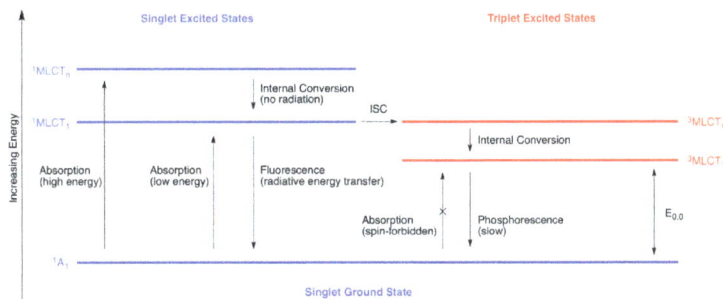

The long-lived triplet excited state accessible by photoexcitation is both a more potent reducing agent and a more potent oxidizing agent than the ground state of the catalyst molecule. In other words, the catalyst can more readily give up one its electrons or accept an electron from an external source. The catalyst excited state is a stronger reductant because its highest energy electron has been excited to an even higher energy state through the photoexcitation process. Similarly, the excited catalyst is a stronger oxidant because one of the catalyst's lowest energy orbitals, which is fully occupied in the ground state, is only singly occupied after photoexcitation and is therefore available for an external electron to occupy. Since organometallic photocatalysts consist of a coordinatively saturated metal complex, i.e. a structure that cannot form any additional bonds, electron transfer cannot take place by an inner sphere mechanism through a direct bond of the metal complex to another reagent. Instead, electron transfer must take place via an outer sphere process, where the electron tunnels between the catalyst and another molecule.

Outer Sphere Electron Transfer

Marcus' theory of outer sphere electron transfer predicts that such a tunneling process will occur most quickly in systems where the electron transfer is thermodynamically favorable (i.e. between strong reductants and oxidants) and where the electron transfer has a low intrinsic barrier.

The intrinsic barrier of electron transfer derives from the Franck–Condon principle, stating that electronic transition takes place more quickly given greater overlap between the initial and final electronic states. Interpreted loosely, this principle suggests that the barrier of an electronic transition is related to the degree to which the system seeks to reorganize. For an electronic transition with a system, the barrier is therefore related to the "overlap" between the initial and final wave functions of the excited electron–i.e. the degree to which the electron needs to "move" in the transition.

In an intermolecular electron transfer, a similar role is played by the degree to which the nuclei seek to move in response to the change in their new electronic environment. Immediately after electron transfer, the nuclear arrangement of the molecule, previously an equilibrium, now represents a vibrationally excited state and must relax to its new equilibrium geometry. Rigid systems, whose geometry is not greatly dependent on oxidation state, therefore experience less vibrational excitation during electron transfer, and have a lower intrinsic barrier. Photocatalysts such as Ru(bipy)3, are held in a rigid arrangement by flat, bidentate ligands arranged in an octahedral geometry around the metal center. Therefore, the complex does not undergo much reorganization during electron transfer. Since electron transfer of these complexes is fast, it is likely to take place within the duration of the catalyst's active state, i.e. during the lifetime of the triplet excited state.

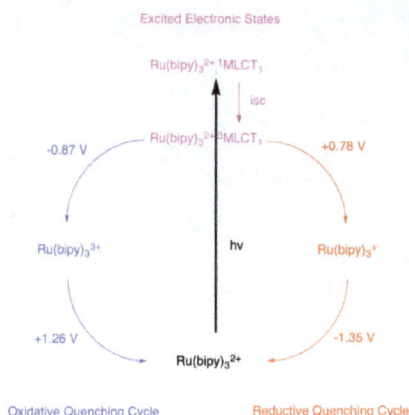

Catalyst Regeneration

The final step in the photocatalytic cycle is the regeneration of the photocatalyst in its ground state. At this stage, the catalyst exists as the ground state of either its oxidized or reduced forms, depending on whether it received or gave up an electron. These oxidation states have a strong driving force to return to their equilibrium oxidation state and act as a potent single-electron reductant or oxidant to satisfy that driving force.

To regenerate the original ground state, the catalyst must participate in a second outer-sphere electron transfer. In many cases, this electron transfer takes place with a stoichiometric two-electron reductant or oxidant, although in some cases this step has involves a second reagent. The case where the excited state catalyst is first reduced and then oxidized to return to its resting state is known as a reductive quenching cycle. Conversely, the case where the excited state catalyst is first oxidized and then reduced to return to its resting state is known as an oxidative quenching cycle. These two possible cycles can be distinguished by a Stern–Volmer experiment.

Since the electron transfer step of the catalytic cycle takes place from the triplet excited state, it competes with phosphorescence as a relaxation pathway. The Stern–Volmer experiment measures the intensity of phosphorescence while varying the concentration of each possible quenching agent. When the concentration of the actual quenching agent is varied, the rate of electron transfer and the degree of phosphorescence is affected. This relationship is modeled by the equation:

$$\left(\frac{I_0}{I}\right) = 1 + k_q * \tau_0 \times [Q]$$

Here, I_0 and I denote the emission intensity with and without quenching agent present, k_q the rate constant of the quenching process, τ_0 the excited-state lifetime in the absence of quenching agent and $[Q]$ the concentration of quenching agent. Thus, if the excited-state lifetime of the photoredox catalyst is known from other experiments, the rate constant of quenching in the presence of a single reaction component can be determined by measuring the change in emission intensity as the concentration of quenching agent changes.

Photophysical Properties

Redox Potentials

The redox potentials of photoredox catalysts must be matched to the reaction's other components. While ground state redox potentials are easily measured by cyclic voltammetry or other electrochemical methods, measuring the redox potential of an electronically excited state cannot be accomplished directly by these methods. However, two methods exist that allow estimation of the excited-state redox potentials and one method exists for the direct measurement of these potentials. To estimate the excited-state redox potentials, one method is to compare the rates of electron transfer from the excited state to a series of ground-state reactants whose redox potentials are known. A more common method to estimate these potentials is to use an equation developed by Rehm and Weller that describes the excited-state potentials as a correction of the ground-state potentials:

$$E^*_{1/2}{}^{ox} = E_{1/2}{}^{ox} - E_{0,0} + w$$

$$E^*_{1/2}{}^{red} = E_{1/2}{}^{red} + E_{0,0} + w_r$$

In these formulas, $E^*_{1/2}$ represents the reduction or oxidation potential of the excited state, $E_{1/2}$ represents the reduction or oxidation potential of the ground state, $E_{0,0}$ represents the difference in energy between the zeroth vibrational states of the ground and excited states and w_r represents the work function, an electrostatic interaction that arises due to the separation of charges that occurs during electron-transfer between two chemical species. The zero-zero excitation energy, $E_{0,0}$ is usually approximated by the corresponding transition in the fluorescence spectrum. This method allows calculation of approximate excited-state redox potentials from more easily measured ground-state redox potentials and spectroscopic data.

Direct measurement of the excited-state redox potentials is possible by applying a method known as phase-modulated voltammetry. This method works by shining light onto an electrochemical cell in order to generate the desired excited-state species, but to modulate the intensity of the light sinusoidally, so that the concentration of the excited-state species is not constant. In fact, the concentration of excited-state species in the cell should change exactly in phase with the intensity of light incident on the electrochemical cell. If the potential applied to the cell is strong enough for electron transfer to occur, the change in concentration of the redox-competent excited state can be measured as an alternating current (AC). Furthermore, the phase shift of the AC current relative to the intensity of the incident light corresponds to the average lifetime of an excited-state species before it engages in electron transfer.

In the following table, redox potentials are quoted as reduction potentials measured in comparison to the saturated calomel electrode (SCE). Thus, the oxidation potential of a chemical species is listed as the reduction potential of its oxidized form. More positive numbers indicate that the catalyst is reduced more readily, i.e. that it is a more strongly oxidizing species. Negative numbers indicate that the electrochemical reaction does not occur spontaneously in the direction written. In such a case, a voltage of the written magnitude must be applied to the catalyst in order to induce reduction. Thus, more negative reduction potentials indicate that the chemical species is more easily oxidized, and is a more strongly reducing species.

Photocatalyst	Structure	$E_{1/2}$(C⁺/C) (V vs SCE)	$E_{1/2}$(C/C⁻) (V vs SCE)	$E_{1/2}$(C⁺/C*) (V vs SCE)	$E_{1/2}$(C*/C⁻) (V vs SCE)	Excited-State Lifetime (ns)	Peak Excitation Wavelength (nm)	Peak Emission Wavelength (nm)	Reference
tris-(2,2'-bipyrimidine) ruthenium²⁺ (Ru(bpm)₃²⁺)	Schematic of tris-(2,2'-bipyrimidyl)ruthenium(II)	1.69	-0.91	-0.21	0.99	131	454	639	
tris-(2,2'-bipyrazine) ruthenium²⁺ (Ru(bpz)₃²⁺)	Schematic of tris-(2,2-bipyrazyl)ruthenium(II)	1.86	-0.80	-0.26	1.45	740	443	591	
tris-(2,2'-bipyridine) ruthenium²⁺ (Ru(bipy)₃²⁺)	Schematic diagram of Ru(bipy)₃²⁺, a typical photoredox catalyst	1.29	-1.33	-0.81	0.77	1100	452	615	
tris-(1,10-phenanthroline) ruthenium²⁺ (Ru(phen)₃²⁺)	Schematic of tris-(1,10-phenanthroline)ruthenium(II)	1.26	-1.36	-0.87	0.82	500	422	610	

Compound							
bis-(2-(2',4'-difluorophenyl)-5-trifluoromethylpyridine)(di-tert-butylbipyridine)iridium⁺ (Ir(dF(CF₃)ppy)₂(dtbbpy)⁺) — Schematic of bis-(2-(2',4'-difluorophenyl)-5-trifluoromethylpyridine)(di-tert-butylbipyridine)iridium(III)	470	380	2300	1.21	-0.89	-1.37	1.69
bis-(2-phenylpyridine)(di-tert-butylbipyridine)iridium⁺ (Ir(ppy)₂(dtbbpy)⁺) — Schematic of bis-(2-phenylpyridine)(di-tert-butylbipyridine)iridium(III)	581	378	557	0.66	-0.96	-1.51	1.21
fac-(tris-(2,2'-phenylpyridine))iridium (Ir(ppy)₃) — Schematic of fac-(tris-(2,2'-phenylpyridine))iridium(III)	494	375	1900	0.31	-1.73	-2.19	0.77

Schematic of bis-(2-(2',4'-difluorophenyl)-5-trifluoromethylpyridine)(di-tert-butylbipyridine)iridium(III)

Schematic of bis-(2-phenylpyridine)(di-tert-butylbipyridine)iridium(III)

Schematic of fac-(tris-(2,2'-phenylpyridine))iridium(III)

Ligand Electronegativity

The relative reducing and oxidizing natures of these photocatalysts can be understood intuitively by considering the ligands' electronegativity and the catalyst complex's metal center. More electronegative metals and ligands tend to stabilize electrons better than their less electronegative counterparts. Therefore, complexes with more electronegative complexes are more easily reduced (i.e. be more powerfully oxidizing) than more electropositive complexes. For example, the ligands 2,2'-bipyridine and 2,2'-phenylpyridine are isoelectronic structures, containing the same number and arrangement of electrons.

However, phenylpyridine replaces one of the nitrogen atoms in bipyridine with a carbon atom. Carbon is less electronegative than nitrogen is, so it less tightly holds electrons. Since the remainder of the ligand molecule is identical, this effect is transferred to the structure as a whole: phenylpyridine holds electrons less tightly than bipyridine, i.e. it is more strongly electron-donating and less electronegative as a ligand. Complexes with phenylpyridine ligands are therefore more strongly reducing and less strongly oxidizing than complexes with bipyridine ligands because the phenylpyridine complexes are more electron-rich and hold their electrons with slightly less strength.

In the same way, a fluorinated derivative of phenylpyridine is more electronegative than the corresponding simple phenylpyridine and complexes with fluorine-containing ligands are more strongly oxidizing and less strongly reducing than the unsubstituted phenylpyridine complex. The metal center's electronic influence on the complex is somewhat more complex than the ligand effect. According to the Pauling scale of electronegativity, both ruthenium and iridium have an electronegativity of 2.2. If this was the sole factor relevant to redox potentials, then complexes of ruthenium and iridium with the same ligands should be equally powerful photoredox catalysts.

However, considering the Rehm-Weller equation, the spectroscopic properties of the metal also play a role in determining the redox character of the excited state. In particular, the parameter $E_{0,0}$ is related to the emission wavelength of the complex and therefore to the size of the Stokes shift - the difference in energy between the maximum absorption and emission of a molecule.

Ruthenium complexes typically have large Stokes shifts, and therefore have low energy emission wavelengths and small zero-zero excitation energies, when compared to iridium complexes. In effect, this means that although ground-state ruthenium complexes can be potent reductants, the excited-state complex is a much less potent reductant or oxidant than a comparable iridium complex. This makes iridium photocatalysts more favorable for the development of general organic transformations because the stronger redox potentials of the excited catalyst allows the use of weaker stoichiometric reductants and oxidants or the use of less reactive substrates.

Applications

Reductive Dehalogenation

Reductive dehalogenation is a convenient method for removing halogen atoms that are introduced into a molecule by a method such as halolactonization. However, the standard method for removing halogen atoms requires the use of stoichiometric organotin reagents, such as tributyltin hydride. While this reaction is very powerful and orthogonal to other functional groups commonly present in organic molecules, organotin reagents are highly toxic. The cleavage of activated and reductively labile functional groups including sulfoniums and halogens is the earliest application of photoredox catalysis to organic synthesis, but early attempts were hampered by the need for specific or uncommon substrates or by the preferential formation of dimeric coupling products. More general methods are known. One method employs Ru(bipy)$_3^{2+}$ as the photocatalyst and a stoichiometric amine reductant to reduce "activated" carbon-halogen bonds, such as those with an adjacent carbonyl group or arene. These bonds are considered to be activated, because the radical they produce upon fragmentation is stabilized by conjugation with the carbonyl group or arene, respectively. The stoichiometric reductant present in this reaction transfers an electron to reduce the excited-state catalyst to the Ru(I) oxidation state. The reduced catalyst can then shuttle the transferred electron to the halogenated substrate, reducing the weak C-X bond and inducing fragmentation.

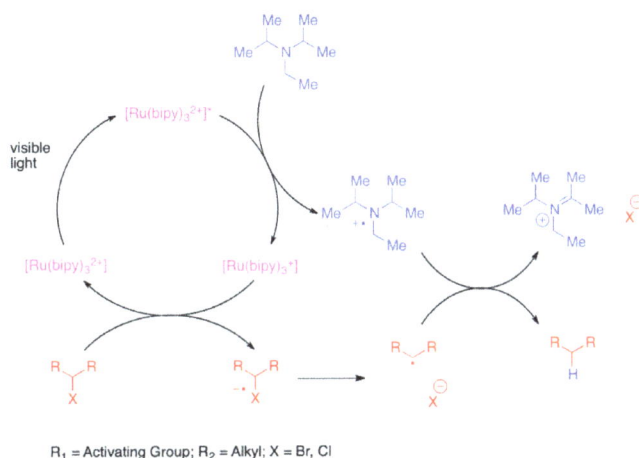

R$_1$ = Activating Group; R$_2$ = Alkyl; X = Br, Cl

Unactivated carbon-iodine bonds can be reduced using the strongly reducing photocatalyst tris-(2,2'-phenylpyridine)iridium (Ir(ppy)$_3^{2+}$). This updated reaction is mechanistically distinct from the previous transformation of activated bromides and chlorides. The increased reduction potential of this catalyst compared to Ru(bipy)$_3^{2+}$ allows direct reduction of the carbon-iodine bond without first interacting with a stoichiometric reductant. Thus, the iridium complex transfers an electron to the substrate, causing fragmentation of the substrate and oxidizing the catalyst to the Ir(IV) oxidation state. The oxidized photocatalyst is then easily returned to its original oxidation state through interaction with one of the reaction additives.

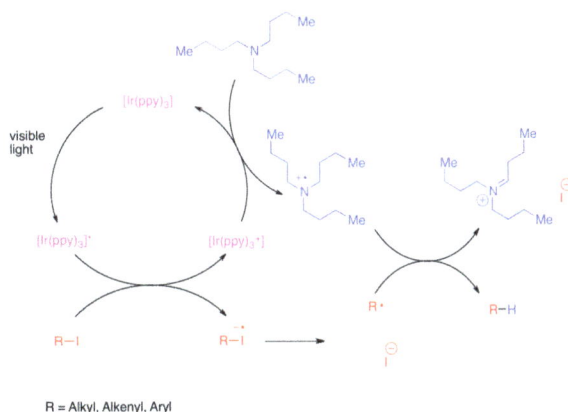

R = Alkyl, Alkenyl, Aryl

Just as tin-mediated radical dehalogenation reactions can be used to initiate cascade cyclizations to rapidly generate molecular complexity, the photocatlytic reductive dehalogenation allows access to the same types of complex products. In this work, a radical cascade cyclization that closed two five-membered rings and formed two new stereocenters, with good yield. This reductive dehalogenation protocol is a key step in a total synthesis of the natural product (+)-Gliocladin C.

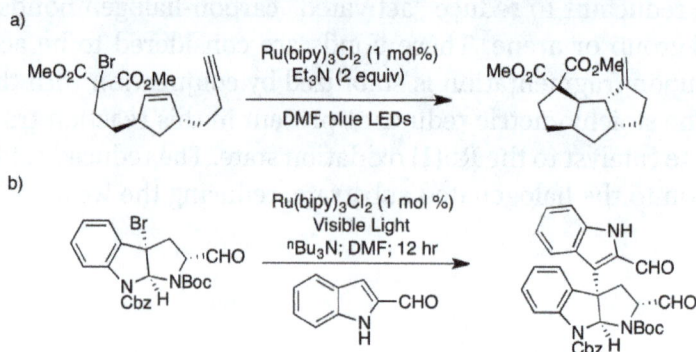

Oxidative Generation of Iminium Ions

Iminium ions are potent electrophiles useful for generating C-N bonds in complex molecules. However, the condensation of amines with carbonyl compounds to form iminium ions is often an unfavorable process, sometimes requiring harsh dehydration conditions. For this reason, alternative methods for iminium ion generation, particularly by oxidation from the corresponding amine, are a valuable synthesis tool. Iminium ions can be generated from activated amines by the use of Ir(dtbbpy)(ppy)$_2$PF$_6$ as a photoredox catalyst. This transformation is proposed to occur by oxidation of the amine to the aminium radical cation by the excited photocatalyst. The photocatalyst is strongly oxidizing due to the electrophilicity of iridium and the electron-poor nature of the fluorinated ligands. This is followed by H-atom transfer to a superstoichimetric oxidant, such as nitromethane (also functioning in the reaction as a nucleophile and as the solvent) or molecular oxygen. Finally, the reactive iminium ion formed by the H-atom transfer is quenched by reaction with nitromethane. Related transformations of amines with a wide variety of other nucleophiles have been investigated, such as cyanide (Strecker reaction), silyl enol ethers (Mannich reaction), dialkylphosphates, allyl silanes (aza-Sakurai reaction), indoles (Friedel-Crafts reaction), and copper acetylides.

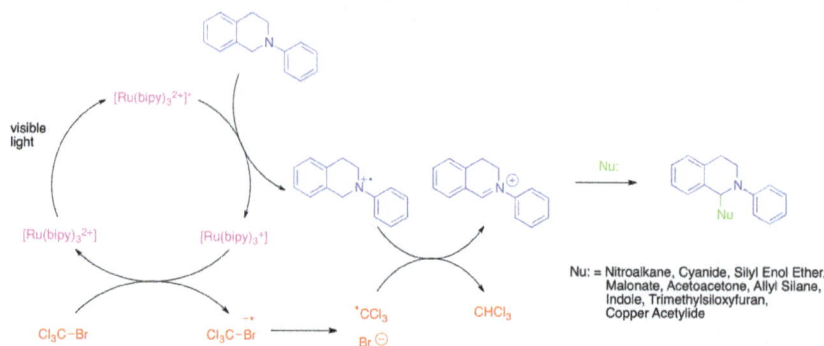

Similar photoredox generation of iminium ions has furthermore been achieved using purely organic photoredox catalysts, such as Rose Bengal and Eosin Y.

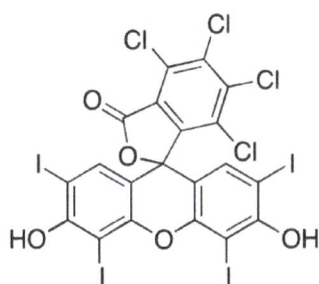

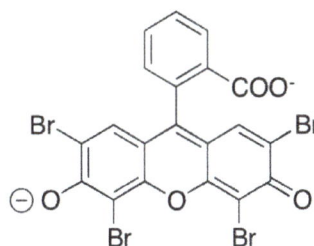

Rose Bengal   Eosin Y

An asymmetric variant of this reaction utilizes acyl nucleophile equivalents generated by N-heterocyclic carbene catalysis. This reaction method sidesteps the problem of poor enantioinduction from chiral photoredox catalysts by moving the source of enantioselectivity to the N-heterocyclic carbene.

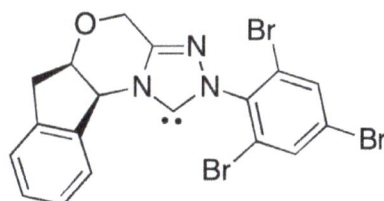

Oxidative Generation of Oxocarbenium Ions

The development of orthogonal protecting group chemistry is a crucial problem in organic synthesis because these protecting groups allow each instance of a common functional group, such as the hydroxyl group, to be distinguished during the synthesis of a complex molecule. One very common protecting group for the hydroxyl functional group is the *para*-methoxy benzyl (PMB) ether. This protecting group is chemically very similar to the less electron-rich benzyl ether. The usual method for selective cleavage of a PMB ether in the presence of a benzyl ether is through the use of strong stoichiometric oxidants such as 2,3-dichloro-5,6-dicyano-1,4-benzoquinone (DDQ) or ceric ammonium nitrate (CAN). PMB ethers are far more susceptible to oxidation than benzyl ethers because they are more electron-rich. The selective deprotection of PMB ethers can be achieved through the use of bis-(2-(2',4'-difluorophenyl)-5-trifluoromethylpyridine)-(4,4'-ditertbutylbipyridine)iridium(III) hexafluorophosphate (Ir[dF(CF$_3$) ppy]$_2$(dtbbpy)PF$_6$) and a mild stoichiometric oxidant such as bromotrichloromethane, CBrCl$_3$. The photoexcited iridium catalyst is reducing enough to fragment the polyhalomethane compound to form trichloromethyl radical, bromide anion and the Ir(IV) oxidation state of the catalyst. The electron-poor nature of the fluorinated ligands means that this iridium complex can be readily reduced: in particular, by an electron-rich arene such as a para-methoxy benzyl ether. After the arene is oxidized, it will readily participate in H-atom transfer with trichloromethyl radical to form chloroform and an oxocarbenium ion, which is readily hydrolyzed to reveal the free hydroxide. This reaction was demonstrated to be orthogonal to many common protecting groups, especially with the addition of a base to counteract the buildup of HBr during the reaction.

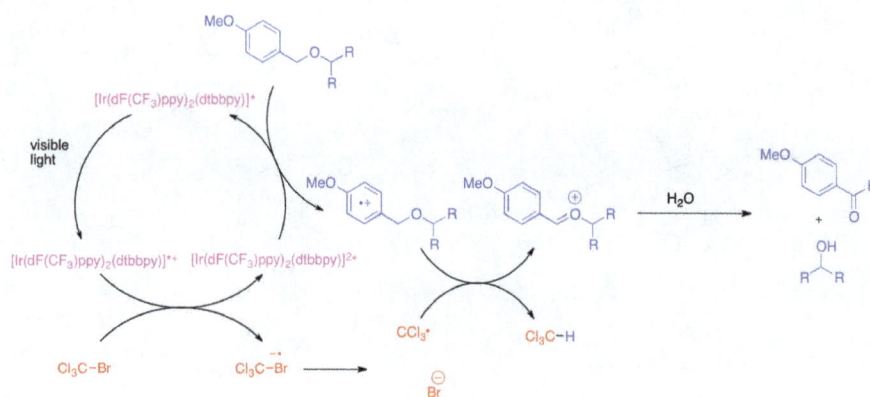

Cycloadditions

Cycloadditions and other pericyclic reactions are powerful transforms in organic synthesis because of their potential to rapidly generate complex molecular architectures and particularly because of their capacity to set multiple adjacent stereocenters in a highly controlled manner. However, only certain cycloadditions are allowed under thermal conditions according to the Woodward–Hoffmann rules of orbital symmetry, or other equivalent models such as frontier molecular orbital theory (FMO) or the Dewar-Zimmermann model. Cycloadditions that are not thermally allowed, such as the [2+2] cycloaddition, can be enabled by photochemical activation of the reaction. Under uncatalyzed conditions, this activation requires the use of high energy ultraviolet light capable of altering the orbital populations of the reactive compounds. Alternatively, metal catalysts such as cobalt and copper have been reported to catalyze thermally-forbidden [2+2] cycloadditions via single electron transfer.

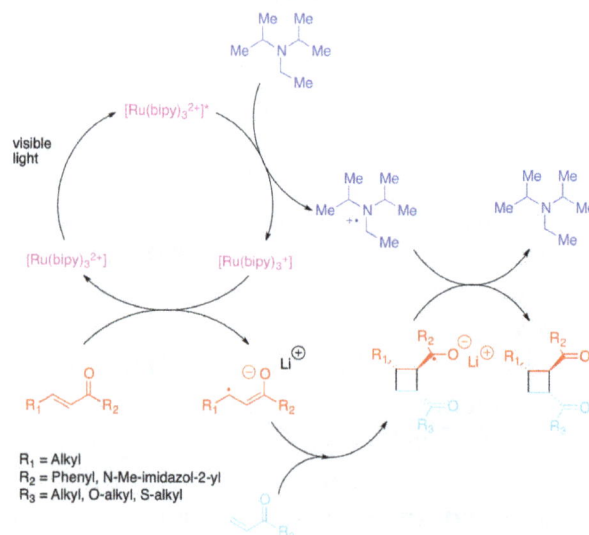

R_1 = Alkyl
R_2 = Phenyl, N-Me-imidazol-2-yl
R_3 = Alkyl, O-alkyl, S-alkyl

The required change in orbital populations can be achieved by electron transfer with a photocatalyst sensitive to lower energy visible light. Yoon demonstrated the efficient intra- and intermolecular [2+2] cycloadditions of activated olefins: particularly enones and styrenes. Enones, or electron-poor olefins, were discovered to react via a radical-anion pathway, utilizing diisopropylethylamine as a transient source of electrons. For this electron-transfer, Ru(bipy)$_3$$^{2+}$

was discovered to be an efficient photocatalyst. The anionic nature of the cyclization proved to be crucial: performing the reaction in acid rather than with a lithium counterion favored a non-cycloaddition pathway. Zhao et al. likewise discovered that a still different cyclization pathway is available to chalcones with a samarium counterion. Conversely, electron-rich styrenes were found to react via a radical-cation mechanism, utilizing methyl viologen or molecular oxygen as a transient electron sink. While Ru(bipy)$_3$$^{2+}$ proved to be a competent catalyst for intramolecular cyclizations using methyl viologen, it could not be used with molecular oxygen as an electron sink or for intermolecular cyclizations. For intermolecular cyclizations, Yoon et al. discovered that the more strongly oxidizing photocatalyst Ru(bpm)$_3$$^{2+}$ and molecular oxygen provided a catalytic system better suited to access the radical cation necessary for the cycloaddition to occur. Ru(bpz)$_3$$^{2+}$, a still more strongly oxidizing photocatalyst, proved to be problematic because although it could catalyze the desired [2+2] cycloaddition, it was also strong enough to oxidize the cycloadduct and catalyze the retro-[2+2] reaction. This comparison of photocatalysts highlights the importance of tuning the redox properties of a photocatalyst to the reaction system as well as demonstrating the value of polypyridyl compounds as ligands, due to the ease with which they can be modified to adjust the redox properties of their complexes.

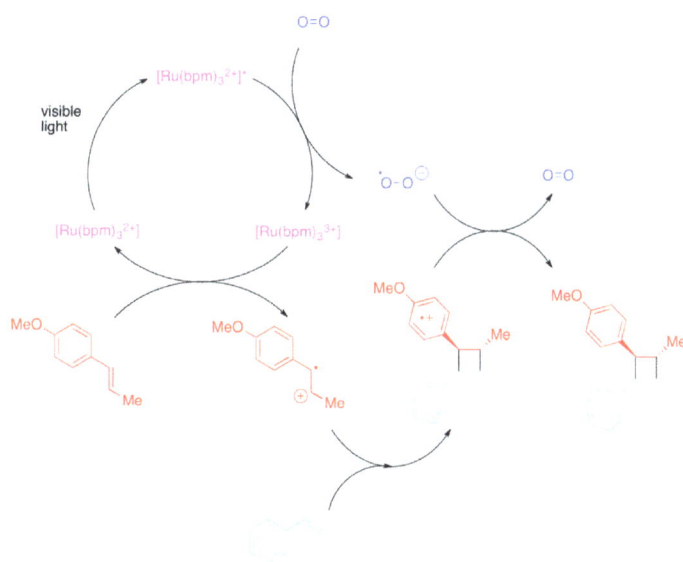

Photoredox-catalyzed [2+2] cycloadditions can also be effected with a triphenylpyrylium organic photoredox catalyst.

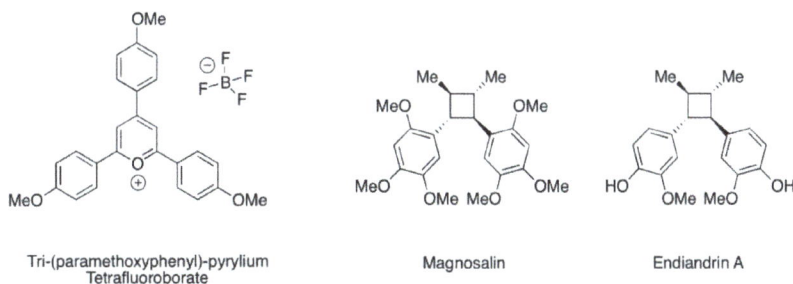

Tri-(paramethoxyphenyl)-pyrylium Tetrafluoroborate

Magnosalin

Endiandrin A

In addition to the thermally-forbidden [2+2] cycloaddition, photoredox catalysis can be applied to the [4+2] cyclization (Diels–Alder reaction). Bis-enones, similar to the substrates used for the photoredox [2+2] cyclization, but with a longer linker joining the two enone functional groups, undergo intramolecular radical-anion hetero-Diels–Alder reactions more rapidly than [2+2] cyc-loaddition.

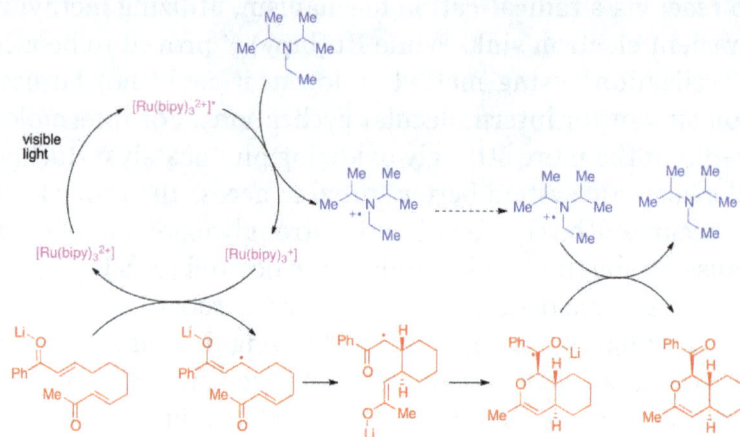

Similarly, electron-rich styrenes participate in intra- or intermolecular Diels–Alder cyclizations via a radical cation mechanism. Ru(bipy)$_3$$^{2+}$ was a competent catalyst for intermolecular, but not intramolecular, Diels–Alder cyclizations. This photoredox-catalyzed Diels–Alder reaction allows cycloaddition between two electronically mismatched substrates. The normal electronic demand for the Diels–Alder reaction calls for an electron-rich diene to react with an electron-poor olefin (or "dienophile"), while the inverse electron-demand Diels–Alder reaction takes place between the opposite case of an electron-poor diene and a very electron-rich dienophile. The photoredox case, since it takes place by a different mechanism than the thermal Diels–Alder reaction, allows cycloaddition between an electron-rich diene and an electron-rich dienophile, allowing access to new classes of Diels–Alder adducts.

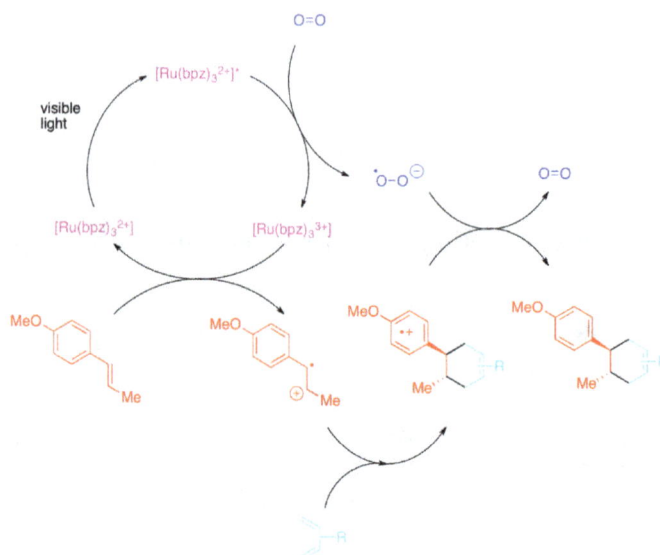

The synthetic value of Yoon's photoredox-catalyzed styrene Diels–Alder reaction was demonstrated via the total synthesis of the natural product Heitziamide A. This synthesis demonstrates that the thermal Diels–Alder reaction favors the undesired regioisomer, but the photoredox-catalyzed reaction gives the desired regioisomer in improved yield.

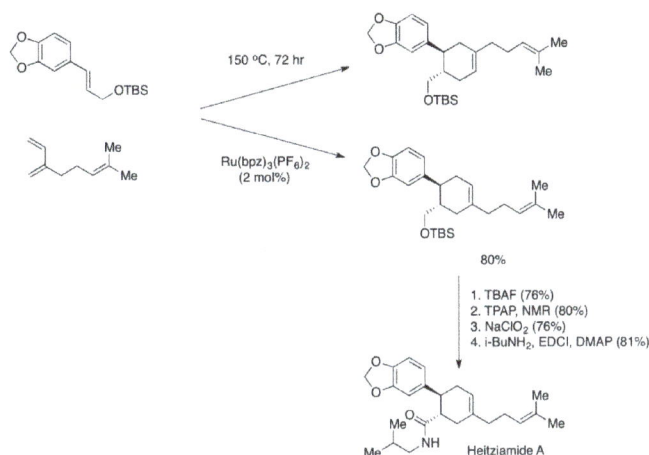

Photoredox Organocatalysis

Organocatalysis is a subfield of catalysis that explores the potential of organic small molecules as catalysts, particularly for the enantioselective creation of chiral molecules. One strategy in this subfield is the use of chiral secondary amines to activate carbonyl compounds. In this case, amine condensation with the carbonyl compound generates a nucleophilic enamine. The chiral amine is designed so that one face of the enamine is sterically shielded and so that only the unshielded face is free to react. Despite the power of this approach to catalyze the enantioselective functionalization of carbonyl compounds, certain valuable transformations, such as the catalytic enantioselective α-alkylation of aldehydes, remained elusive. The combination of organocatalysis and photoredox methods provides a catalytic solution to this problem. In this approach for the α-alkylation of aldehydes, $Ru(bipy)_3^{2+}$ reductively fragments an activated alkyl halide, such as bromomalonate or phenacyl bromide, which can then add to catalytically-generated enamine in an enantioselective manner. The oxidized photocatalyst then oxidatively quenches the resulting α-amino radical to form an iminium ion, which hydrolyzes to give the functionalized carbonyl compound. This photoredox transformation was shown to be mechanistically distinct from another organocatalytic radical process termed singly-occupied molecular orbital (SOMO) catalysis. SOMO catalysis employs superstoichiometric ceric ammonium nitrate (CAN) to oxidize the catalytically-generated enamine to the corresponding radical cation, which can then add to a suitable coupling partner such as allyl silane. This type of mechanism is excluded for the photocatalytic alkylation reaction because whereas enamine radical cation was observed to cyclize onto pendant olefins and open cyclopropane radical clocks in SOMO catalysis, these structures were unreactive in the photoredox reaction.

This transformation include alkylations with other classes of activated alkyl halides of synthetic interest. In particular, the use of the photocatalyst $Ir(dtbbpy)(ppy)_2^+$ allows the enantioselective α-trifluoromethylation of aldehydes while the use of $Ir(ppy)_3$ allowed the enantioselective coupling of aldehydes with electron-poor benzylic bromides. Zeitler et al. also investigated the productive merger of photoredox and organocatalytic methods to achieve enantioselective alkylation of aldehydes. The same chiral imidazolidinone organocatalyst was used to form enamine and introduce chirality. However, the organic photoredox catalyst Eosin Y was used rather than a ruthenium or iridium complex.

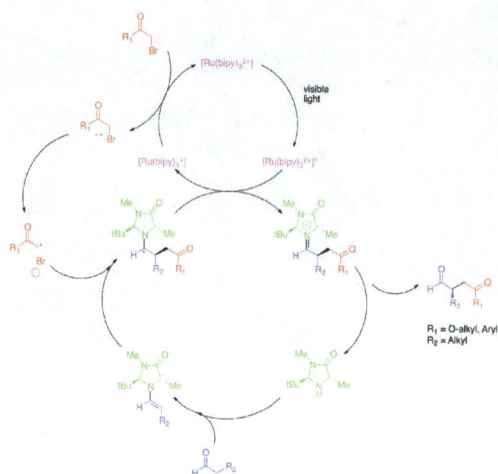

Direct β-arylation of saturated aldehydes and ketones can be effected through the combination of photoredox and organocatalytic methods. The previous method to accomplish direct β-functionalization of a saturated carbonyl consists of a one-pot consists of a two-step process, both catalyzed by a secondary amine organocatalyst: stoichiometric reduction of an aldehyde with IBX followed by addition of an activated alkyl nucleophile to the beta-position of the resulting enal. This transformation, which like other photoredox processes takes place by a radical mechanism, is limited to the addition of highly electrophilic arenes to the beta position. The severe limitations on the arene component scope in this reaction is due primarily to the need for an arene radical anion that is stable enough not to react directly with enamine or enamine radical cation. In the proposed mechanism, the activated photoredox catalyst is quenched oxidatively by an electron-deficient arene, such as 1,4-dicyanobenzene. The photocatalyst then oxidizes an enamine species, transiently generated by the condensation of an aldehyde with a secondary amine cocatalyst, such as the optimal isopropyl benzylamine. The resulting enamine radical cation usually reacts as a 3 π-electron system, but due to the stability of the radical coupling partners, deprotonation of the β-methylene position gives rise to a 5 π-electron system with strong radical character at the newly accessed β-carbon. Although this reaction relies on the use of a secondary amine organocatalyst to generate the enamine species which is oxidized in the proposed mechanism, no enantioselective variant of this reaction exists.

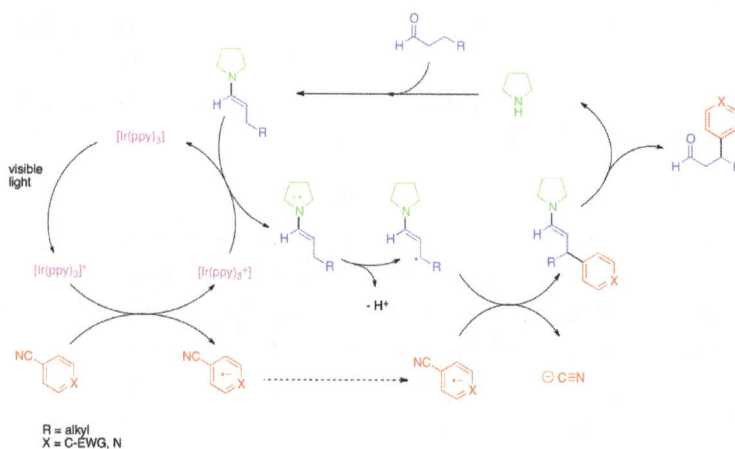

The development of this direct β-arylation of aldehydes led to related reactions for the β-functionalization of cyclic ketones. In particular, β-arylation of cyclic ketones has been achieved under similar reaction conditions, but using azepane as the secondary amine cocatalyst. A photocatalytic "homo-aldol" reaction works for cyclic ketones, allowing the coupling of the beta-position of the ketone to the ipso carbon of aryl ketones, such as benzophenone and acetophenone. In addition to the azepane cocatalyst, this reaction requires the use of the more strongly reducing photoredox catalyst Ir(ppy)$_3$ and the addition of lithium hexafluoroarsenide (LiAsF$_6$) to promote single-electron reduction of the aryl ketone.

Additions to Olefins

The use of photoredox catalysis to generate reactive heteroatom-centered radicals was first explored in the 1990s. Ru(bipy)$_3{}^{2+}$ was found to catalyze the fragmentation of tosylphenylselenide to phenylselenolate anion and tosyl radical and that a radical chain propagation mechanism allowed the addition of tosyl radical and phenylseleno- radical across the double bond of electron rich alkyl vinyl ethers. Since phenylselenolate anion is readily oxidized to diphenyldiselenide, the low quantities of diphenyldiselenide observed was taken as an indication that photoredox-catalyzed fragmentation of tosylphenylselenide was only important as an initiation step, and that most of the reactivity was due to a radical chain process.

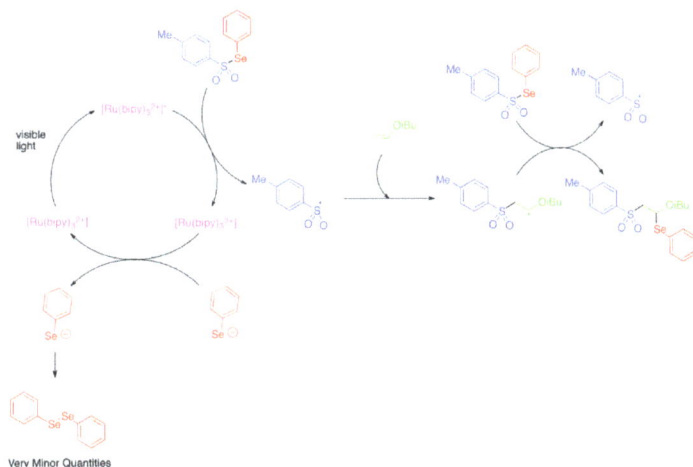

Heteroaromatic additions to olefins include multicomponent oxy- and aminotrifluoromethylation reactions. These reactions use Umemoto's reagent, a sulfonium salt that serves as an electrophilic source of the trifluoromethyl group and that is precedented to react via a single-electron transfer pathway. Thus, single-electron reduction of Umemoto's reagent releases trifluoromethyl radical, which adds to the reactive olefin. Subsequently, single-electron oxidation of the alkyl radical generated by this addition produces a cation which can be trapped by water, an alcohol, or a nitrile. In order to achieve high levels of regioselectivity, this reactivity has been explored mainly for styrenes, which are biased towards formation of the benzylic radical intermediate.

Hydrotrifluoromethylation of styrenes and aliphatic alkenes can be effected with a mesityl acridinium organic photoredox catalyst and Langlois' reagent as the source of CF$_3$ radical. In this reaction, it was found that trifluoroethanol and substoichiometric amounts of an aromatic thiol, such as methyl thiosalicylate, employed in tandem served as the best source of hydrogen radical to complete the catalytic cycle.

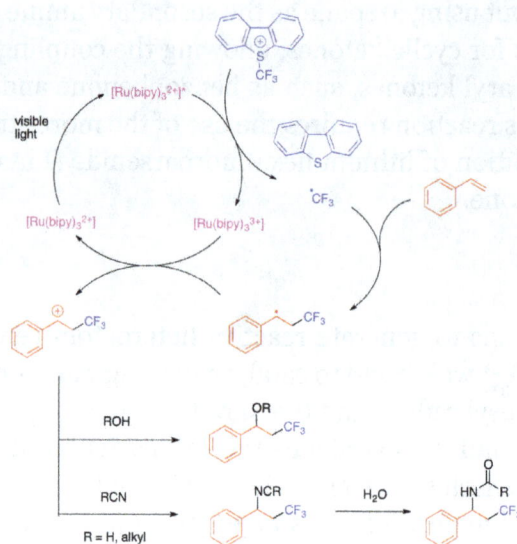

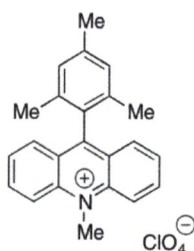

9-Mesityl-10-Methylacridinium Perchlorate

Langlois' Reagent

Intramolecular hydroetherifications and hydroaminations proceed with anti-Markovnikov selectivity. One mechanism invokes the single-electron oxidation of the olefin, trapping the radical cation by a pendant hydroxyl or amine functional group, and quenching the resulting alkyl radical by H-atom transfer from a highly labile donor species. Extensions of this reactivity to intermolecular systems have resulted in i) a new synthetic route to complex tetrahydrofurans by a "polar-radical-crossover cycloaddition" (PRCC reaction) of an allylic alcohol with an olefin, and ii) the anti-Markovnikov addition of carboxylic acids to olefins.

Photodissociation

Photodissociation, photolysis, or photodecomposition is a chemical reaction in which a chemical compound is broken down by photons. It is defined as the interaction of one or more photons with one target molecule. Photodissociation is not limited to visible light. Any photon with sufficient energy can affect the chemical bonds of a chemical compound. Since a photon's energy is inversely proportional to its wavelength, electromagnetic waves with the energy of visible light or higher, such as ultraviolet light, x-rays and gamma rays are usually involved in such reactions.

Photolysis in Photosynthesis

Photolysis is part of the light-dependent reactions of photosynthesis. The general reaction of photosynthetic photolysis can be given as

$$H_2A + 2 \text{ photons (light)} \rightarrow 2\ e^- + 2\ H^+ + A$$

The chemical nature of "A" depends on the type of organism. In purple sulfur bacteria, hydrogen sulfide (H_2S) is oxidized to sulfur (S). In oxygenic photosynthesis, water (H_2O) serves as a substrate for photolysis resulting in the generation of diatomic oxygen (O_2). This is the process which returns oxygen to Earth's atmosphere. Photolysis of water occurs in the thylakoids of cyanobacteria and the chloroplasts of green algae and plants.

Energy Transfer Models

The conventional, semi-classical, model describes the photosynthetic energy transfer process as one in which excitation energy hops from light-capturing pigment molecules to reaction center molecules step-by-step down the molecular energy ladder.

The effectiveness of photons of different wavelengths depends on the absorption spectra of the photosynthetic pigments in the organism. Chlorophylls absorb light in the violet-blue and red parts of the spectrum, while accessory pigments capture other wavelengths as well. The phycobilins of red algae absorb blue-green light which penetrates deeper into water than red light, enabling them to photosynthesize in deep waters. Each absorbed photon causes the formation of an exciton (an electron excited to a higher energy state) in the pigment molecule. The energy of the exciton is transferred to a chlorophyll molecule (P680, where P stands for pigment and 680 for its absorption maximum at 680 nm) in the reaction center of photosystem II via resonance energy transfer. P680 can also directly absorb a photon at a suitable wavelength.

Photolysis during photosynthesis occurs in a series of light-driven oxidation events. The energized electron (exciton) of P680 is captured by a primary electron acceptor of the photosynthetic electron transfer chain and thus exits photosystem II. In order to repeat the reaction, the electron in the reaction center needs to be replenished. This occurs by oxidation of water in the case of oxygenic photosynthesis. The electron-deficient reaction center of photosystem II (P680*) is the strongest biological oxidizing agent yet discovered, which allows it to break apart molecules as stable as water.

The water-splitting reaction is catalyzed by the oxygen evolving complex of photosystem II. This protein-bound inorganic complex contains four manganese ions, plus calcium and chloride ions as cofactors. Two water molecules are complexed by the manganese cluster, which then undergoes a series of four electron removals (oxidations) to replenish the reaction center of photosystem II. At the end of this cycle, free oxygen (O_2) is generated and the hydrogen of the water molecules has been converted to four protons released into the thylakoid lumen.

These protons, as well as additional protons pumped across the thylakoid membrane coupled with the electron transfer chain, form a proton gradient across the membrane that drives photophosphorylation and thus the generation of chemical energy in the form of adenosine triphosphate (ATP). The electrons reach the P700 reaction center of photosystem I where they are energized again by light. They are passed down another electron transfer chain and finally combine with the coenzyme NADP+ and protons outside the thylakoids to NADPH. Thus, the net oxidation reaction of water photolysis can be written as:

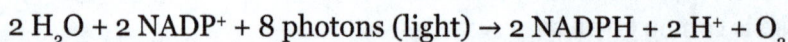

$$2\ H_2O + 2\ NADP^+ + 8\ photons\ (light) \rightarrow 2\ NADPH + 2\ H^+ + O_2$$

The free energy change (ΔG) for this reaction is 102 kilocalories per mole. Since the energy of light at 700 nm is about 40 kilocalories per mole of photons, approximately 320 kilocalories of light energy are available for the reaction. Therefore, approximately one-third of the available light energy is captured as NADPH during photolysis and electron transfer. An equal amount of ATP is generated by the resulting proton gradient. Oxygen as a byproduct is of no further use to the reaction and thus released into the atmosphere.

Quantum Models

In 2007 a quantum model was proposed by Graham Fleming and his co-workers which includes the possibility that photosynthetic energy transfer might involve quantum oscillations, explaining its unusually high efficiency.

According to Fleming there is direct evidence that remarkably long-lived wavelike electronic quantum coherence plays an important part in energy transfer processes during photosynthesis, which can explain the extreme efficiency of the energy transfer because it enables the system to sample all the potential energy pathways, with low loss, and choose the most efficient one. This claim has, however, since been proven wrong in several publications   .

This approach has been further investigated by Gregory Scholes and his team at the University of Toronto, which in early 2010 published research results that indicate that some marine algae make use of quantum-coherent electronic energy transfer (EET) to enhance the efficiency of their energy harnessing.

Photolysis in the Atmosphere

Photolysis occurs in the atmosphere as part of a series of reactions by which primary pollutants such as hydrocarbons and nitrogen oxides react to form secondary pollutants such as peroxyacyl nitrates.

The two most important photodissociaton reactions in the troposphere are firstly:

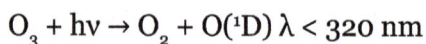

$$O_3 + hv \rightarrow O_2 + O(^1D)\ \lambda < 320\ nm$$

which generates an excited oxygen atom which can react with water to give the hydroxyl radical:

$$O(^1D) + H_2O \rightarrow 2\ {}^\cdot OH$$

The hydroxyl radical is central to atmospheric chemistry as it initiates the oxidation of hydrocarbons in the atmosphere and so acts as a detergent.

Secondly the reaction:

$$NO_2 + hv \rightarrow NO + O$$

is a key reaction in the formation of tropospheric ozone.

The formation of the ozone layer is also caused by photodissociation. Ozone in the Earth's strato-sphere is created by ultraviolet light striking oxygen molecules containing two oxygen atoms (O_2), splitting them into individual oxygen atoms (atomic oxygen). The atomic oxygen then combines with unbroken O_2 to create ozone, O_3. In addition, photolysis is the process by which CFCs are broken down in the upper atmosphere to form ozone-destroying chlorine free radicals.

Astrophysics

In astrophysics, photodissociation is one of the major processes through which molecules are broken down (but new molecules are being formed). Because of the vacuum of the interstellar medium, molecules and free radicals can exist for a long time. Photodissociation is the main path by which molecules are broken down. Photodissociation rates are important in the study of the composition of interstellar clouds in which stars are formed.

Examples of photodissociation in the interstellar medium are (hv is the energy of a single photon of frequency v):

$$H_2O + hv \rightarrow H + OH$$

$$CH_4 + hv \rightarrow CH_3 + H$$

Atmospheric Gamma-ray Bursts

Currently orbiting satellites detect an average of about one gamma-ray burst per day. Because gamma-ray bursts are visible to distances encompassing most of the observable universe, a volume encompassing many billions of galaxies, this suggests that gamma-ray bursts must be exceedingly rare events per galaxy.

Measuring the exact rate of gamma-ray bursts is difficult, but for a galaxy of approximately the same size as the Milky Way, the expected rate (for long GRBs) is about one burst every 100,000 to 1,000,000 years. Only a few percent of these would be beamed towards Earth. Estimates of rates of short GRBs are even more uncertain because of the unknown beaming fraction, but are probably comparable.

A gamma-ray burst in the Milky Way, if close enough to Earth and beamed towards it, could have significant effects on the biosphere. The absorption of radiation in the atmosphere would cause photodissociation of nitrogen, generating nitric oxide that would act as a catalyst to destroy ozone.

The atmospheric photodissociation

- N_2 -> $2N$

- O_2 -> $2O$

- CO_2 -> $C + 2O$

- H_2O -> $2H + O$

- $2NH_3$ -> $3H_2 + N_2$

would yield

- NO_2 (consumes up to 400 ozone molecules)
- CH_2 (nominal)
- CH_4 (nominal)
- CO_2

(incomplete)

According to a 2004 study, a GRB at a distance of about a kiloparsec could destroy up to half of Earth's ozone layer; the direct UV irradiation from the burst combined with additional solar UV radiation passing through the diminished ozone layer could then have potentially significant impacts on the food chain and potentially trigger a mass extinction. The authors estimate that one such burst is expected per billion years, and hypothesize that the Ordovician-Silurian extinction event could have been the result of such a burst.

There are strong indications that long gamma-ray bursts preferentially or exclusively occur in regions of low metallicity. Because the Milky Way has been metal-rich since before the Earth formed, this effect may diminish or even eliminate the possibility that a long gamma-ray burst has occurred within the Milky Way within the past billion years. No such metallicity biases are known for short gamma-ray bursts. Thus, depending on their local rate and beaming properties, the possibility for a nearby event to have had a large impact on Earth at some point in geological time may still be significant.

Multiple Photon Dissociation

Single photons in the infrared spectral range usually are not energetic enough for direct photodissociation of molecules. However, after absorption of multiple infrared photons a molecule may gain internal energy to overcome its barrier for dissociation. Multiple photon dissociation (MPD, IRMPD with infrared radiation) can be achieved by applying high power lasers, e.g. a carbon dioxide laser, or a free electron laser, or by long interaction times of the molecule with the radiation field without the possibility for rapid cooling, e.g. by collisions. The latter method allows even for MPD induced by black body radiation, a technique called blackbody infrared radiative dissociation (BIRD).

Photopolymer

A photopolymer is a polymer that changes its properties when exposed to light, often in the ultraviolet or visible region of the electromagnetic spectrum. These changes are often manifested structurally, for example hardening of the material occurs as a result of cross-linking when exposed to light. An example is shown below depicting a mixture of monomers, oligomers, and photoinitiators that conform into a hardened polymeric material through a process called curing. A wide variety of technologically useful applications rely on photopolymers, for example some enamels and varnishes depend on photopolymer formulation for proper hardening upon exposure to light.

In some instances, an enamel can cure in a fraction of a second when exposed to light, as opposed to thermally cured enamels which can require half an hour or longer. Curable materials are widely used for medical, printing, and photoresist technologies.

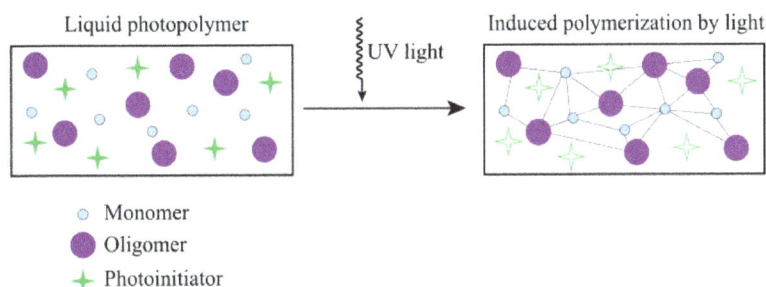

- ○ Monomer
- ● Oligomer
- + Photoinitiator

Changes in structural and chemical properties can be induced internally by chromophores that the polymer subunit already possesses, or externally by addition of photosensitive molecules. Typically a photopolymer consists of a mixture of multifunctional monomers and oligomers in order to achieve the desired physical properties, and therefore a wide variety of monomers and oligomers have been developed that can polymerize in the presence of light either through internal or external initiation. Photopolymers undergo a process called curing, where oligomers are cross-linked upon exposure to light, forming what is known as a network polymer. The result of photo curing is the formation of a thermoset network of polymers. One of the advantages of photo-curing is that it can be done selectively using high energy light sources, for example lasers, however, most systems are not readily activated by light, and in this case a photoinitiator is required. Photoinitiators are compounds that upon radiation of light decompose into reactive species that activate polymerization of specific functional groups on the oligomers. An example of a mixture that undergoes cross-linking when exposed to light is shown below. The mixture consists of monomeric styrene and oligomeric acrylates.

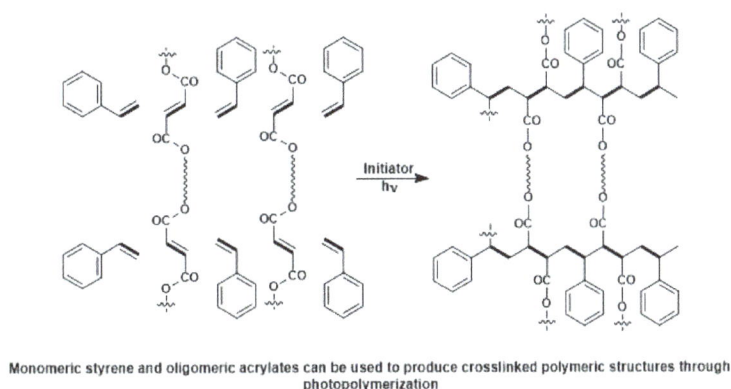

Monomeric styrene and oligomeric acrylates can be used to produce crosslinked polymeric structures through photopolymerization

Most commonly, photopolymerized systems are typically cured through UV radiation, since ultraviolet light is more energetic; however, the development of dye-based photoinitiator systems have allowed for the use of visible light, having potential advantages of processes that are more simple and safe to handle. UV curing in industrial processes has greatly expanded over the past several decades. Many traditional thermally cured and solvent-based technologies can be replaced by photopolymerization technologies. The advantages of photopolymerization over thermally cured polymerization include high rates of polymerization and environmental benefits from elimination of volatile organic solvents.

There are two general routes for photoinitiation: free radical and ionic. The general process involves doping a batch of neat polymer with small amounts of photoinitiator, followed by selective radiation of light, resulting a highly cross-linked product. Many of these reactions do not require solvent which eliminates termination path via reaction of initiators with solvent and impurities, in addition to decreasing the overall cost.

Mechanisms

Ionic Mechanism

In ionic curing process, an ionic photoinitiator is used to activate the functional group of the oligomers that are going to participate in cross-linking. Typically photopolymerization is a very selective process and it is crucial that the polymerization takes place only where it is desired to do so. In order to satisfy this liquid neat oligomer can be doped with either anionic or cationic photoinitiators that will initiate polymerization only when radiated with light. Monomers, or functional groups, employed in cationic photopolymerization include: styrenic compounds, vinyl ethers, N-vinyl carbazoles, lactones, lactams, cyclic ethers, cyclic acetals, and cyclic siloxanes. The majority of ionic photoinitiators fall under the cationic class, anionic photoinitiators are considerably less investigated. There are several classes of cationic initiators including: Onium salts, organometallic compounds and pyridinium salts. As mentioned earlier, one of the drawbacks of the photoinitiators used for photopolymerization is that they tend to absorb in the short UV region. Photosensitizers, or chromophores, that absorb in a much longer wavelength region can be employed to excite the photoinitiators through an energy transfer. Other modifications to these types of systems are free radical assisted cationic polymerization. In this case, a free radical is formed from another specie in solution that reacts with the photoinitiator in order to start polymerization. Although there are a diverse group of compounds activated by cationic photoinitiators, the compounds that find most industrial uses contain epoxides, oxetanes, and vinyl ethers. One of the advantages to using cationic photopolymerization is that once the polymerization has begun it is no longer sensitive to oxygen and does not require an inert atmosphere to perform well.

Photolysis

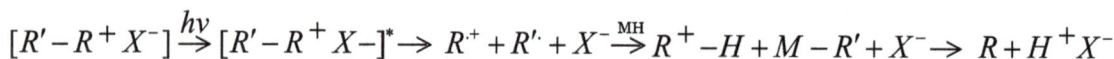

$$[R'-R^+X^-]\overset{h\nu}{\Rightarrow}[R'-R^+X-]^*\rightarrow R^{.+}+R^{'.}+X^-\overset{MH}{\Rightarrow}R^+-H+M-R'+X^-\rightarrow R+H^+X^-$$

M = Monomer

Cationic Photoinitiators

The proposed mechanism for cationic photopolymerization begins with the photoexcitation of the initiator. Once excited, both homolytic cleavage and dissociation of a counter anion takes place, generating cationic radical (R), an aryl radical(R') and unaltered counter anion (X). The abstraction of a lewis acid, in figure above a hydrogen, by the cationic radical produces a very weakly bound hydrogen and a free radical. The acid is further deprotonated by the anion(X) in solution generating a lewis acid with the starting anion (X) as a counter ion. It is thought that the acidic proton generated is what ultimately initiates the polymerization.

Onium Salts

Since their discovery in the 1970s aryl onium salts, more specifically iodonium and sulfonium salts, have received much attention and have found many industrial applications. Other, less common, onium salts not mentioned here include ammonium and phosphonium salts.

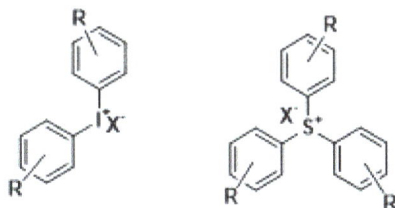

The typical onium compound used as a photoinitiator contains two or three arene groups for iodonium and sulfonium respectively. Onium salts generally absorb short wavelength light in the UV region spanning from 225–300 nm. One characteristic that is crucial to the performance of the onium photoinitiators is that the counter anion is non-nucleophilic. Since the Brønsted acid generated during the initiation step is considered the active initiator for polymerization, there is a termination route where the counter ion of the acid could act as the nucleophile instead of a functional groups on the oligomer. Common counter anions include: BF_4-, $PF-6$, $AsF-6$, $SbF-6$. There is an indirect relationship between the size of the counter ion and percent conversion.

Organometallic

Although less common, transition metal complexes can act as cationic photoinitiators as well. In general, the mechanism is more simplistic than the onium ions previously described. Most photoinitiators of this class consist of a metal salt with a non-nucleophilic counter anion. For example, ferrocinium salts have received much attention for commercial applications. The absorption band for ferrocinium salt derivatives are in a much longer, and sometimes visible, region. Upon radiation the metal center loses a ligand(s) and the ligand(s) is replaced by functional groups that begin the polymerization. One of the drawbacks of this method is a greater sensitivity to oxygen. There are also several organometallic anionic photoinitiators which react through a similar mechanism. For the anionic case, excitation of a metal center is followed by either heterolytic bond cleavage or electron transfer generating the active anionic initiator.

Pyridinium Salts

Generally pyridinium photoinitiators are N-substituted pyridine derivatives, with a positive charge placed on the nitrogen. The counter ion is in most cases a non-nucleophilic anion. Upon radiation, homolytic bond cleavage takes place generating a pyridinium cationic radical and a neutral free radical. In most cases, a hydrogen atom is abstracted from the oligomer by the pyridinium radical. The free radical generated from the hydrogen abstraction is then terminated by the free radical in solution. This results in a strong pyridinium acid that can initiate polymerization.

Free Radical Mechanism

Before the free radical nature of certain polymerizations was determined, certain monomers were observed to polymerize when exposed to light. The first to demonstrate the photoinduced free radical chain reaction of vinyl bromide was Ivan Ostromislensky, a Russian chemist who also studied the polymerization of synthetic rubber. Subsequently many compounds were found to become dissociated by light and found immediate use as photoinitiators in the polymerization industry. In the free radical mechanism of radiation curable systems light absorbed by a photoinitiator generates free-radicals which induce cross-linking reactions of a mixture of functionalized oligomers and monomers to generate the cured film Photocurable materials that form through the free-radical mechanism undergo chain-growth polymerization, which includes three basic steps: initiation, chain propagation, and chain termination. The three steps are depicted in the scheme below, where R• represents the radical that forms upon interaction with radiation during initiation, and M is a monomer. The active monomer that is formed is then propagated to create growing polymeric chain radicals. In photocurable materials the propagation step involves reactions of the chain radicals with reactive double bonds of the prepolymers or oligomers. The termination reaction usually proceeds through combination, in which two chain radicals are joined together, or through disproportionation, which occurs when an atom (typically hydrogen) is transferred from one radical chain to another resulting in two polymeric chains.

Initiation

$$\{Initiatior + h_v -> R{\cdot}R^{\cdot} + M -> RM^{\cdot}$$

Propagation

$$RM^{\cdot} + M_n -> RM^{\cdot}_{n+1}$$

Termination

    combination

$$RM^{\cdot}_n + {\cdot}M_m R -> RM_n M_m R$$

    dispropotionation

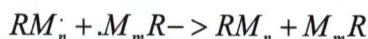

$$RM^{\cdot}_n + {\cdot}M_m R -> RM_n + M_m R$$

Most composites that cure through radical chain growth contain a diverse mixture of oligomers and monomers with functionality that can range from 2-8 and molecular weights from 500-3000. In general, monomers with higher functionality result is a tighter crosslinking density of the finished material. Typically these oligomers and monomers alone do not absorb sufficient energy for the commercial light sources used, therefore photoinitiators are included.,

Photoinitiators

There are two types of free-radical photoinitiators: A two component system where the radical is generated through abstraction of a hydrogen atom from a donor compound (also called co-initiator), and a one component system where two radicals are generated by cleavage. Examples of each type of free-radical photoinitiator is shown below.

Abstraction

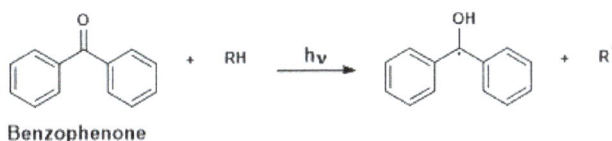

Benzophenone

Cleavage

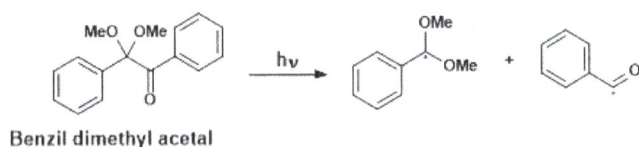

Benzil dimethyl acetal

Benzophenone, Xanthones, and Quinones are examples of abstraction type photoinitiators, with common donor compounds being aliphatic amines. The resulting R• species from the donor compound becomes the initiator for the free radical polymerization process, while the radical resulting from the starting photoinitiator (benzophenone in the example shown above) is typically unreactive.

Benzoin ethers, Acetophenones, Benzoyl Oximes, and Acylphosphines are some examples of cleavage-type photoinitiators. Cleavage readily occurs for the species to give two radicals upon absorption of light, and both radicals generated can typically initiate polymerization. Cleavage type photoinitiators do not require a co-initiator, such as aliphatic amines. This can be beneficial since amines are also effective chain transfer species. Chain-transfer processes reduce the chain length and ultimately the crosslink density of the resulting film.

Oligomers and Monomers

The properties of a photocured material, such as flexibility, adhesion, and chemical resistance are provided by the functionalized oligomers present in the photocurable composite. Oligomers are typically epoxides, urethanes, polyethers, or polyesters, each of which provide specific properties to the resulting material. Each of these oligomers are typically functionallized by an acrylate. An example shown below is an epoxy oligomer that has been functionalized by acrylic acid. Acrylated epoxies are useful as coatings on metallic substrates, and result in glossy hard coatings. Acrylated urethane oligomers are typically abrasion resistant, tough, and flexible making ideal coatings for floors, paper, printing plates, and packaging materials. Acrylated polyethers and polyesters result in very hard solvent resistant films, however, polyethers are prone to UV degradation and therefore are rarely used in UV curable material. Often formulations are composed of several types of oligomers to achieve the desirable properties for a material.

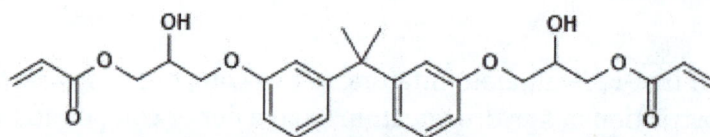

acrylated epoxy oligomer

The monomers used in radiation curable systems help control the speed of cure, crosslink density, final surface properties of the film, and viscosity of the resin. Examples of monomers include styrene, N-Vinylpyrrolidone, and acrylates. Styrene is a low cost monomer and provides a fast cure, N-vinylpyrrolidone results in a material that is highly flexible when cured, has low toxicity, and acrylates are highly reactive, allowing for rapid cure rates, and are highly versatile with monomer functionality ranging from monofunctional to tetrafunctional. Like oligomers, several types of monomers can be employed to achieve the desirable properties of the final material.

Applications

Photopolymerization is a widely used technology, used in applications ranging from imaging to biomedical uses. Below is a description of just some photopolymerization applications.

Medical Uses

Dentistry is one market where free radical photopolymers have found wide usage as adhesives, sealant composites, and protective coatings. These dental composites are based on a camphorquinone photoinitiator and a matrix containing methacrylate oligomers with inorganic fillers such as silicon dioxide. Photocurable adhesives are also used in the production of catheters, hearing aids, surgical masks, medical filters, and blood analysis sensors. Photopolymers have also been explored for uses in drug delivery, tissue engineering and cell encapsulation systems. Photopolymerization processes for these applications are being developed to be carried out *in vivo* or *ex vivo*. *In vivo* photopolymerization would provide the advantages of production and implantation with minimal invasive surgery.*Ex vivo* photopolymerization would allow for fabrication of complex matrices, and versatility of formulation. Although photopolymers show promise for a wide range of new biomedical applications, biocompatibility with photopolymeric materials must still be addressed and developed.

3D-Imaging

Stereolithography, digital imaging, and 3D inkjet printing are just a few 3D imaging technologies that make use of photopolymers. 3D imaging usually proceeds with CAD-CAM software, which creates a 3D image to be translated into a 3D plastic object. The image is cut in slices, where each slice is reconstructed through radiation curing of the liquid polymer,converting the image into a solid object. Photopolymers used in 3D imaging processes must be designed to have a low volume shrinkage upon polymerization in order to avoid distortion of the solid object. Common monomers utilized for 3D imaging include multifunctional acrylates and methacrylates combined with a non-polymeric component in order to reduce volume shrinkage. A competing composite mixture of epoxide resins with cationic photoinitiators is becoming increasingly used since their volume shrinkage upon ring-opening polymerization is significantly below those of acrylates and methacrylates. Free-radical and cationic polymerizations composed of both epoxide and acrylate monomers have also been employed, gaining the high rate of polymerization from the acrylic monomer, and better mechanical properties from the epoxy matrix.

Photoresists

Photoresists are coatings, or oligomers, that are deposited on a surface and are designed to change properties upon irradiation of light. These changes either polymerize the liquid oligomers into insoluble cross-linked network polymers or decompose the already solid polymers into liquid products. Polymers that form networks during photopolymerization are referred to as negative resist. Conversely, polymers that decompose during photopolymerization are referred to as positive resists. Both positive and negative resists have found many applications including the design and production of micro fabicated chips. The ability to pattern the resist using a focused light source has driven the field of photolithography.

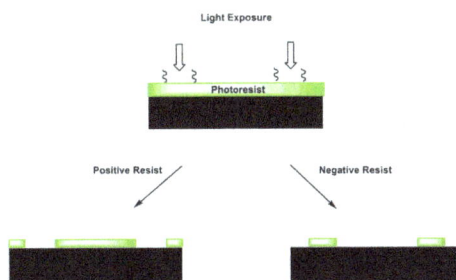

Negative Resists

As mentioned, negative resists are photopolymers that become insoluble upon exposure to radiation. They have found a variety of commercial applications. Especially in the area of designing and printing small chips for electronics. A characteristic found in most negative tone resists is the presence of multifunctional branches on the polymers used. Radiation of the polymers in the presence of an intiator results in the formation of chemically resistant network polymer. A common functional group used in negative resist is epoxy functional groups. An example of a widely used polymer of this class is SU-8. SU-8 was one of the first polymers used in this field, and found applications in wire board printing. In the presence of a cationic photoinitiator photopolymer SU-8 forms networks with other polymers in solution. Basic scheme shown below.

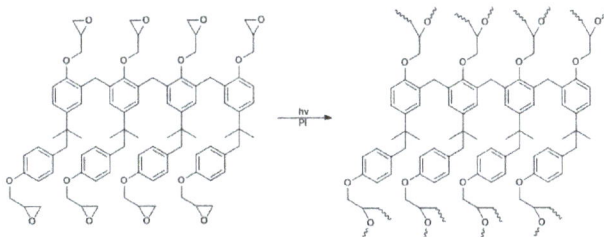

SU-8 is an example of an intramolecular photopolymerization forming a matrix of cross-linked material. Negative resists can also be made using co-polymerization. In the event that you have two different monomers, or oligomers, in solution with multiple functionalities it is possible for the two to polymerize and form a less soluble polymer.

Positive Resists

As mentioned, positive resist exposure to radiation changes the chemical structure such that it becomes a liquid or more soluble. These changes in chemical structure are often rooted in the cleavage of specific linkers in the polymer. Once irradiated the "decomposed" polymers can be washed away using a developer solvent leaving behind the polymer that was not exposed to light. This type of technology allows the production of very fine stencils for applications such as microelectronics. In order to have these types of qualities, positive resist utilize polymers with labile linkers in their back bone that can be cleaved upon irradiation or using a photo-generated acid to hydrolyze bonds in the polymer. A polymer that decomposes upon irradiation to a liquid, or more soluble product is referred to as a positive tone resist. Common functional groups that can be hydrolyzed by photo-generated acid catalyst include polycarbonates and polyesters.

Fine Printing

A printing plate of a city map created in photopolymer.

Photopolymer can be used to generate printing plates, which are then pressed onto paper like metal type. This is often used in modern fine printing to achieve the effect of embossing (or the more subtly three-dimensional effect of letterpress printing) from designs created on a computer without needing to engrave designs into metal or cast metal type. It is often used for business cards.

Scintillator

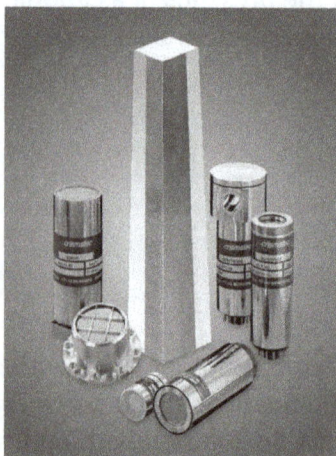

Scintillation crystal surrounded by various scintillation detector assemblies.

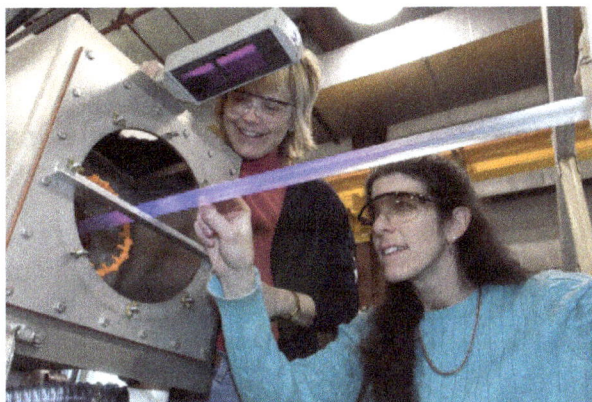

Extruded plastic scintillator material fluorescing under a UV inspection lamp at Fermilab for the MINERvA project

A scintillator is a material that exhibits scintillation — the property of luminescence, when excited by ionizing radiation. Luminescent materials, when struck by an incoming particle, absorb its energy and scintillate, (i.e. re-emit the absorbed energy in the form of light).[a] Sometimes, the excited state is metastable, so the relaxation back down from the excited state to lower states is delayed (necessitating anywhere from a few nanoseconds to hours depending on the material): the process then corresponds to either one of two phenomena, depending on the type of transition and hence the wavelength of the emitted optical photon: delayed fluorescence or phosphorescence, also called after-glow.

Principle of Operation

A scintillation detector or scintillation counter is obtained when a scintillator is coupled to an electronic light sensor such as a photomultiplier tube (PMT), photodiode, or silicon photomultiplier. PMTs absorb the light emitted by the scintillator and re-emit it in the form of electrons via the photoelectric effect. The subsequent multiplication of those electrons (sometimes called photo-electrons) results in an electrical pulse which can then be analyzed and yield meaningful information about the particle that originally struck the scintillator. Vacuum photodiodes are similar but do not amplify the signal while silicon photodiodes, on the other hand, detect incoming photons by the excitation of charge carriers directly in the silicon. Silicon photomultipliers consist of an array of photodiodes which are reverse-biased with sufficient voltage to operate in avalanche mode, enabling each pixel of the array to be sensitive to single photons.

History

The first device which used a scintillator was built in 1903 by Sir William Crookes and used a ZnS screen. The scintillations produced by the screen were visible to the naked eye if viewed by a microscope in a darkened room; the device was known as a spinthariscope. The technique led to a number of important discoveries but was obviously tedious. Scintillators gained additional attention in 1944, when Curran and Baker replaced the naked eye measurement with the newly developed PMT. This was the birth of the modern scintillation detector.

Applications for Scintillators

Alpha scintillation probe for detecting surface contamination under calibration

Scintillators are used by the American government as Homeland Security radiation detectors. Scintillators can also be used in neutron and high energy particle physics experiments, new energy resource exploration, X-ray security, nuclear cameras, computed tomography and gas exploration. Other applications of scintillators include CT scanners and gamma cameras in medical diagnostics, and screens in older style CRT computer monitors and television sets.

The use of a scintillator in conjunction with a photomultiplier tube finds wide use in hand-held survey meters used for detecting and measuring radioactive contamination and monitoring nuclear material. Scintillators generate light in fluorescent tubes, to convert the ultra-violet of the discharge into visible light. Scintillation detectors are also used in the petroleum industry as detectors for Gamma Ray logs.

Properties of Scintillators

There are many desired properties of scintillators, such as high density, fast operation speed, low cost, radiation hardness, production capability and durability of operational parameters. High density reduces the material size of showers for high-energy γ-quanta and electrons. The range of Compton scattered photons for lower energy γ-rays is also decreased via high density materials. This results in high segmentation of the detector and leads to better spatial resolution. Usually high density materials have heavy ions in the lattice (e.g., lead, cadmium), significantly increasing the photo-fraction ($\sim Z^4$). The increased photo-fraction is important for some applications such as positron emission tomography. High stopping power for electromagnetic component of the ionizing radiation needs greater photo-fraction; this allows for a compact detector. High operating speed is needed for good resolution of spectra. Precision of time measurement with a scintillation detector is proportional to $\sqrt{\tau_{sc}}$. Short decay times are important for the measurement of time intervals and for the operation in fast coincidence circuits. High density and fast response time can allow detection of rare events in particle physics. Particle energy deposited in the material of a scintillator is proportional to the scintillator's response. Charged particles, γ-quanta and ions have different slopes when their response is measured. Thus, scintillators could be used to identify various types of γ-quanta and particles in fluxes of mixed radiation. Another consideration of scintillators is the cost of producing them. Most crystal scintillators require high-purity chemicals and sometimes rare-earth metals that are fairly expensive. Not only are the materials an expenditure, but many crystals require expensive furnaces and almost six months of growth and analyzing time. Currently, other scintillators are being researched for reduced production cost.

Several other properties are also desirable in a good detector scintillator: a low gamma output (i.e., a high efficiency for converting the energy of incident radiation into scintillation photons), transparency to its own scintillation light (for good light collection), efficient detection of the radiation being studied, a high stopping power, good linearity over a wide range of energy, a short rise time for fast timing applications (e.g., coincidence measurements), a short decay time to reduce detector dead-time and accommodate high event rates, emission in a spectral range matching the spectral sensitivity of existing PMTs (although wavelength shifters can sometimes be used), an index of refraction near that of glass (≈ 1.5) to allow optimum coupling to the PMT window. Ruggedness and good behavior under high temperature may be desirable where resistance to vibration and high temperature is necessary (e.g., oil exploration). The practical choice of a scintillator material is usually a compromise among those properties to best fit a given application.

Among the properties listed above, the light output is the most important, as it affects both the efficiency and the resolution of the detector (the efficiency is the ratio of detected particles to the total number of particles impinging upon the detector; the energy resolution is the ratio of the full width at half maximum of a given energy peak to the peak position, usually expressed in %). The light output is a strong function of the type of incident particle or photon and of its energy, which therefore strongly influences the type of scintillation material to be used for a particular application. The presence of quenching effects results in reduced light output (i.e., reduced scintillation efficiency). Quenching refers to all radiationless deexcitation processes in which the excitation is degraded mainly to heat. The overall signal production efficiency of the detector, however, also depends on the quantum efficiency of the PMT (typically ~30% at peak), and on the efficiency of light transmission and collection (which depends on the type of reflector material covering the scintillator and light guides, the length/shape of the light guides, any light absorption, etc.). The light output is often quantified as a number of scintillation photons produced per keV of deposited energy. Typical numbers are (when the incident particle is an electron): ≈40 photons/keV for NaI(Tl), ~10 photons/keV for plastic scintillators, and ~8 photons/keV for bismuth germanate (BGO).

Scintillation detectors are generally assumed to be linear. This assumption is based on two requirements: (1) that the light output of the scintillator is proportional to the energy of the incident radiation; (2) that the electrical pulse produced by the photomultiplier tube is proportional to the emitted scintillation light. The linearity assumption is usually a good rough approximation, although deviations can occur (especially pronounced for particles heavier than the proton at low energies).

Resistance and good behavior under high-temperature, high-vibration environments is especially important for applications such as oil exploration (wireline logging, measurement while drilling). For most scintillators, light output and scintillation decay time depends on the temperature. This dependence can largely be ignored for room-temperature applications since it is usually weak. The dependence on the temperature is also weaker for organic scintillators than it is for inorganic crystals, such as NaI-Tl or BGO. The coupled PMTs also exhibit temperature sensitivity, and can be damaged if submitted to mechanical shock. Hence, high temperature rugged PMTs should be used for high-temperature, high-vibration applications.

The time evolution of the number of emitted scintillation photons N in a single scintillation event can often be described by linear superposition of one or two exponential decays. For two decays, we have the form:

$$N = A\exp\left(-\frac{t}{\tau_f}\right) + B\exp\left(-\frac{t}{\tau_s}\right)$$

where τ_f and τ_s are the fast (or prompt) and the slow (or delayed) decay constants. Many scintillators are characterized by 2 time components: one fast (or prompt), the other slow (or delayed). While the fast component usually dominates, the relative amplitude A and B of the two components depend on the scintillating material. Both of these components can also be a function of the energy loss dE/dx. In cases where this energy loss dependence is strong, the overall decay time constant varies with the type of incident particle. Such scintillators enable pulse shape discrimination, i.e., particle identification based on the decay characteristics of the PMT electric pulse. For instance, when BaF_2 is used, γ rays typically excite the fast component, while α particles excite the slow component: it is thus possible to identify them based on the decay time of the PMT signal.

Types of Scintillators

Organic Crystals

Organic scintillators are aromatic hydrocarbon compounds which contain benzene ring structures interlinked in various ways. Their luminescence typically decays within a few nanoseconds.

Some organic scintillators are pure crystals. The most common types are anthracene (C 14H 10, decay time ≈30 ns), stilbene (C14 H12, 4.5 ns decay time), and naphthalene (C 10H 8, few ns decay time). They are very durable, but their response is anisotropic (which spoils energy resolution when the source is not collimated), and they cannot be easily machined, nor can they be grown in large sizes; hence they are not very often used. Anthracene has the highest light output of all organic scintillators and is therefore chosen as a reference: the light outputs of other scintillators are sometimes expressed as a percent of anthracene light.

Organic Liquids

These are liquid solutions of one or more organic scintillators in an organic solvent. The typical solutes are fluors such as p-terphenyl (C 18H 14), PBD (C 20H 14N 2O), butyl PBD (C 24H 22N 2O), PPO (C 15H 11NO), and wavelength shifter such as POPOP (C 24H 16N 2O). The most widely used solvents are toluene, xylene, benzene, phenylcyclohexane, triethylbenzene, and decalin. Liquid scintillators are easily loaded with other additives such as wavelength shifters to match the spectral sensitivity range of a particular PMT, or [10]B to increase the neutron detection efficiency of the scintillation counter itself (since [10]B has a high interaction cross section with thermal neutrons). For many liquids, dissolved oxygen can act as a quenching agent and lead to reduced light output, hence the necessity to seal the solution in an oxygen-free, airtight enclosure.

Plastic Scintillators

The term "plastic scintillator" typically refers to a scintillating material in which the primary fluorescent emitter, called a fluor, is suspended in the base, a solid polymer matrix. While this combination is typically accomplished through the dissolution of the fluor prior to bulk polymerization, the fluor is sometimes associated with the polymer directly, either covalently or through coordination, as is the case with many Li6 plastic scintillators. Polyethylene naphthalate has been found to exhibit scintillation by itself without any additives and is expected to replace existing plastic scintillators due to higher performance and lower price. The advantages of plastic scintillators include fairly high light output and a relatively quick signal, with a decay time of 2–4 nanoseconds, but perhaps the biggest advantage of plastic scintillators is their ability to be shaped, through the use of molds or other means, into almost any desired form with what is often a high degree of durability. Plastic scintillators are known to show light output saturation when the energy density is large (Birks' Law).

Bases

The most common bases used in plastic scintillators are the aromatic plastics, polymers with aromatic rings as pendant groups along the polymer backbone, amongst which polyvinyltoluene (PVT) and polystyrene (PS) are the most prominent. While the base does fluoresce in the presence of ionizing radiation, its low yield and negligible transparency to its own emission make the use of fluors necessary in the construction of a practical scintillator. Aside from the aromatic plastics, the most common base is polymethylmethacrylate (PMMA), which carries two advantages over many other bases: high ultraviolet and visible light transparency and mechanical properties and higher durability with respect to brittleness. The lack of fluorescence associated with PMMA is often compensated through the addition of an aromatic co-solvent, usually naphthalene. A plastic scintillator based on PMMA in this way boasts transparency to its own radiation, helping to ensure uniform collection of light.

Other common bases include polyvinyl xylene (PVX) polymethyl, 2,4-dimethyl, 2,4,5-trimethyl styrenes, polyvinyl diphenyl, polyvinyl naphthalene, polyvinyl tetrahydronaphthalene, and copolymers of these and other bases.

Fluors

Also known as luminophors, these compounds absorb the scintillation of the base and then emit at larger wavelength, effectively converting the ultraviolet radiation of the base into the more easily transferred visible light. Further increasing the attenuation length can be accomplished through the addition of a second fluor, referred to as a spectrum shifter or converter, often resulting in the emission of blue or green light.

Common fluors include polyphenyl hydrocarbons, oxazole and oxadiazole aryls, especially, n-terphenyl (PPP), 2,5-diphenyloxazole (PPO), 1,4-di-(5-phenyl-2-oxazolyl)-benzene (POPOP), 2-phenyl-5-(4-biphenylyl)-1,3,4-oxadiazole (PBD), and 2-(4'-tert-butylphenyl)-5-(4"-biphenylyl)-1,3,4-oxadiazole (B-PBD).

Inorganic Crystals

Inorganic scintillators are usually crystals grown in high temperature furnaces, for example, alkali metal halides, often with a small amount of activator impurity. The most widely used is NaI(Tl) (sodium iodide doped with thallium). Other inorganic alkali halide crystals are: CsI(Tl), CsI(Na), CsI(pure), CsF, KI(Tl), LiI(Eu). Some non-alkali crystals include: BaF_2, CaF_2(Eu), ZnS(Ag), $CaWO_4$, $CdWO_4$, YAG(Ce) ($Y_3Al_5O_{12}$(Ce)), GSO, LSO.

Newly developed products include $LaCl_3$(Ce), lanthanum chloride doped with cerium, as well as a cerium-doped lanthanum bromide, $LaBr_3$(Ce). They are both very hygroscopic (i.e., damaged when exposed to moisture in the air) but offer excellent light output and energy resolution (63 photons/keV γ for $LaBr_3$(Ce) versus 38 photons/keV γ for NaI(Tl)), a fast response (16 ns for $LaBr_3$(Ce) versus 230 ns for NaI(Tl)), excellent linearity, and a very stable light output over a wide range of temperatures. In addition $LaBr_3$(Ce) offers a higher stopping power for γ rays (density of 5.08 g/cm³ versus 3.67 g/cm³ for NaI(Tl)). LYSO ($Lu_{1.8}Y_{0.2}SiO_5$(Ce)) has an even higher density (7.1 g/cm³, comparable to BGO), is non-hygroscopic, and has a higher light output than BGO (32 photons/keV γ), in addition to being rather fast (41 ns decay time versus 300 ns for BGO).

A disadvantage of some inorganic crystals, e.g., NaI, is their hygroscopicity, a property which requires them to be housed in an airtight enclosure to protect them from moisture. CsI(Tl) and BaF_2 are only slightly hygroscopic and do not usually need protection. CsF, NaI(Tl), LaCl 3(Ce), LaBr 3(Ce) are hygroscopic, while BGO, CaF 2(Eu), LYSO, and YAG(Ce) are not.

Inorganic crystals can be cut to small sizes and arranged in an array configuration so as to provide position sensitivity. Such arrays are often used in medical physics or security applications to detect X-rays or γ rays: high-Z, high density materials (e.g. LYSO, BGO) are typically preferred for this type of applications.

Scintillation in inorganic crystals is typically slower than in organic ones, ranging typically from 1.48 ns for ZnO(Ga) to 9000 ns for CaWO 4. Exceptions are CsF} (~5 ns), fast BaF 2 (0.7 ns; the slow component is at 630 ns), as well as the newer products (LaCl 3(Ce), 28 ns; LaBr 3(Ce), 16 ns; LYSO, 41 ns).

Gaseous Scintillators

Gaseous scintillators consist of nitrogen and the noble gases helium, argon, krypton, and xenon, with helium and xenon receiving the most attention. The scintillation process is due to the de-excitation of single atoms excited by the passage of an incoming particle. This de-excitation is very rapid (~1 ns), so the detector response is quite fast. Coating the walls of the container with a wavelength shifter is generally necessary as those gases typically emit in the ultraviolet and PMTs respond better to the visible blue-green region. In nuclear physics, gaseous detectors have been used to detect fission fragments or heavy charged particles.

Glasses

The most common glass scintillators are cerium-activated lithium or boron silicates. Since both lithium and boron have large neutron cross-sections, glass detectors are particularly well suited to the detection of thermal (slow) neutrons. Lithium is more widely used than boron since it has a greater energy release on capturing a neutron and therefore greater light output. Glass scintillators are however sensitive to electrons and γ rays as well (pulse height discrimination can be used for particle identification). Being very robust, they are also well-suited to harsh environmental conditions. Their response time is ≈10 ns, their light output is however low, typically ≈30% of that of anthracene.

Physics of Scintillation

Organic Scintillators

Transitions made by the free valence electrons of the molecules are responsible for the production of scintillation light in organic crystals. These electrons are associated with the whole molecule rather than any particular atom and occupy the so-called -molecular orbitals. The ground state S_0 is a singlet state above which are the excited singlet states (S^*, S^{**}, ...), the lowest triplet state (T_0), and its excited levels (T^*, T^{**}, ...). A fine structure corresponding to molecular vibrational modes is associated with each of those electron levels. The energy spacing between electron levels is ≈1 eV; the spacing between the vibrational levels is about 1/10 of that for electron levels.

An incoming particle can excite either an electron level or a vibrational level. The singlet excitations immediately decay (< 10 ps) to the S^* state without the emission of radiation (internal degradation). The S^* state then decays to the ground state S_0 (typically to one of the vibrational levels above S_0) by emitting a scintillation photon. This is the prompt component or fluorescence. The transparency of the scintillator to the emitted photon is due to the fact that the energy of the photon is less than that required for a $S^* \to S_0$ transition (the transition is usually being to a vibrational level above S_0).

When one of the triplet states gets excited, it immediately decays to the T_0 state with no emission of radiation (internal degradation). Since the $T_0 \to S_0$ transition is very improbable, the T_0 state instead decays by interacting with another T_0 molecule:

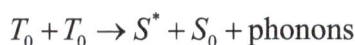

$$T_0 + T_0 \to S^* + S_0 + \text{phonons}$$

and leaves one of the molecules in the S^* state, which then decays to S_0 with the release of a scintillation photon. Since the T_0-T_0 interaction takes time, the scintillation light is delayed: this is the slow or delayed component (corresponding to delayed fluorescence). Sometimes, a direct $T_0 \to S_0$ transition occurs (also delayed), and corresponds to the phenomenon of phosphorescence. Note that the observational difference between delayed-fluorescence and phosphorescence is the difference in the wavelengths of the emitted optical photon in a $S^* \to S_0$ transition versus a $T_0 \to S_0$ transition.

Organic scintillators can be dissolved in an organic solvent to form either a liquid or plastic scintillator. The scintillation process is the same as described for organic crystals (above); what differs is the mechanism of energy absorption: energy is first absorbed by the solvent, then passed onto the scintillation solute (the details of the transfer are not clearly understood).

Inorganic Scintillators

The scintillation process in inorganic materials is due to the electronic band structure found in crystals and is not molecular in nature as is the case with organic scintillators. An incoming particle can excite an electron from the valence band to either the conduction band or the exciton band (located just below the conduction band and separated from the valence band by an energy gap;). This leaves an associated hole behind, in the valence band. Impurities create electronic levels in the forbidden gap. The excitons are loosely bound electron-hole pairs which wander through the crystal lattice until they are captured as a whole by impurity centers. The latter then rapidly de-excite by emitting scintillation light (fast component). The activator impurities are typically chosen so that the emitted light is in the visible range or near-UV where photomultipliers are effective. The holes associated with electrons in the conduction band are independent from the latter. Those holes and electrons are captured successively by impurity centers exciting certain metastable states not accessible to the excitons. The delayed de-excitation of those metastable impurity states again results in scintillation light (slow component).

BGO is a pure inorganic scintillator without any activator impurity. There, the scintillation process is due to an optical transition of the Bi^{3+} ion, a major constituent of the crystal. A similar process exists in $CdWO_4$.

Gases

In gases, the scintillation process is due to the de-excitation of single atoms excited by the passage of an incoming particle (a very rapid process: $\approx$1 ns).

Response to Various Radiations

Heavy Ions

Scintillation counters are usually not ideal for the detection of heavy ions for three reasons:

1. the very high ionizing power of heavy ions induces quenching effects which result in a reduced light output (e.g. for equal energies, a proton will produce 1/4 to 1/2 the light of an electron, while alphas will produce only about 1/10 the light;

2. the high stopping power of the particles also results in a reduction of the fast component relative to the slow component, increasing detector dead-time;

3. strong non-linearities are observed in the detector response especially at lower energies.

The reduction in light output is stronger for organics than for inorganic crystals. Therefore, where needed, inorganic crystals, e.g. CsI(Tl), ZnS(Ag) (typically used in thin sheets as α-particle monitors), CaF 2(Eu), should be preferred to organic materials. Typical applications are α-survey instruments, dosimetry instruments, and heavy ion dE/dx detectors. Gaseous scintillators have also been used in nuclear physics experiments.

Electrons

The detection efficiency for electrons is essentially 100% for most scintillators. But because electrons can make large angle scatterings (sometimes backscatterings), they can exit the detector without depositing their full energy in it. The back-scattering is a rapidly increasing function of the atomic number Z of the scintillator material. Organic scintillators, having a lower Z than inorganic crystals, are therefore best suited for the detection of low-energy ($<$ 10 MeV) beta particles. The situation is different for high energy electrons: since they mostly lose their energy by bremsstrahlung at the higher energies, a higher-Z material is better suited for the detection of the bremsstrahlung photon and the production of the electromagnetic shower which it can induce.

Gamma Rays

High-Z materials, e.g. inorganic crystals, are best suited for the detection of gamma rays. The three basic ways that a gamma ray interacts with matter are: the photoelectric effect, Compton scattering, and pair production. The photon is completely absorbed in photoelectric effect and pair production, while only partial energy is deposited in any given Compton scattering. The cross section for the photoelectric process is proportional to Z^5, that for pair production proportional to Z^2, whereas Compton scattering goes roughly as Z. A high-Z material therefore favors the former two processes, enabling the detection of the full energy of the gamma ray. If the gamma rays are at higher energies ($>$5 MeV), pair production dominates.

Neutrons

Since the neutron is not charged it does not interact via the Coulomb force and therefore does not ionize the scintillation material. It must first transfer some or all of its energy via the strong force to a charged atomic nucleus. The positively charged nucleus then produces ionization. Fast neutrons (generally >0.5 MeV) primarily rely on the recoil proton in (n,p) reactions; materials rich in hydrogen, e.g. plastic scintillators, are therefore best suited for their detection. Slow neutrons rely on nuclear reactions such as the (n,γ) or (n,α) reactions, to produce ionization. Their mean free path is therefore quite large unless the scintillator material contains nuclides having a high cross section for these nuclear reactions such as ^{6}Li or ^{10}B. Materials such as LiI(Eu) or glass silicates are therefore particularly well-suited for the detection of slow (thermal) neutrons.

List of Inorganic Scintillators

The following is a list of commonly used inorganic crystals:

- BaF 2 or barium fluoride: BaF 2 contains a very fast and a slow component. The fast scintillation light is emitted in the UV band (220 nm) and has a 0.7 ns decay time (smallest decay time for any scintillator), while the slow scintillation light is emitted at longer wavelengths (310 nm) and has a 630 ns decay time. It is used for fast timing applications, as well as applications for which pulse shape discrimination is needed. The light yield of BaF 2 is about 12 photons/keV. BaF 2 is not hygroscopic.

- BGO or bismuth germanate: bismuth germanate has a higher stopping power, but a lower optical yield than NaI(Tl). It is often used in coincidence detectors for detecting back-to-back gamma rays emitted upon positron annihilation in positron emission tomography machines.

- CdWO 4 or cadmium tungstate: a high density, high atomic number scintillator with a very long decay time (14 μs), and relatively high light output (about 1/3 of that of NaI(Tl)). CdWO 4 is routinely used for X-ray detection (CT scanners). Having very little ^{228}Th and ^{226}Ra contamination, it is also suitable for low activity counting applications.

- CaF 2(Eu) or calcium fluoride doped with europium: The material is not hygroscopic, has a 940 ns decay time, and is relatively low-Z. The latter property makes it ideal for detection of low energy β particles because of low backscattering, but not very suitable for γ detection. Thin layers of CaF 2(Eu) have also been used with a thicker slab of NaI(Tl) to make phoswiches capable of discriminating between α, β, and γ particles.

- CaWO 4 or calcium tungstate: exhibits long decay time 9 μs and short wavelength emission with maximum at 420 nm matching sensitivity curve of bialkali PMT. The light yield and energy resolution of the scintillator (6.6% for ^{137}Cs) is comparable with that of CdWO 4.

- CsI: undoped cesium iodide emits predominantly at 315 nm, is only slightly hygroscopic, and has a very short decay time (16 ns), making it suitable for fast timing applications. The light output is quite low, however, and very sensitive to variations in temperature.

- CsI(Na) or cesium iodide doped with sodium: the crystal is less bright than CsI(Tl), but comparable in light output to NaI(Tl). The wavelength of maximum emission is at 420 nm, well matched to the photocathode sensitivity of bialkali PMTs. It has a slightly shorter decay time than CsI(Tl) (630 ns versus 1000 ns for CsI(Tl)). CsI(Na) is hygroscopic and needs an airtight enclosure for protection against moisture.

- CsI(Tl) or cesium iodide doped with thallium: these crystals are one of the brightest scintillators. Its maximum wavelength of light emission is in the green region at 550 nm. CsI(Tl) is only slightly hygroscopic and does not usually require an airtight enclosure.

- Gd 2O 2S or gadolinium oxysulfide has a high stopping power due to its relatively high density ($7.32 \, g/cm^3$) and the high atomic number of gadolinium. The light output is also good, making it useful as a scintillator for x-ray imaging applications.

- LaBr 3(Ce) (or lanthanum bromide doped with cerium): a better (novel) alternative to NaI(Tl); denser, more efficient, much faster (having a decay time about ~20ns), offers superior energy resolution due to its very high light output. Moreover, the light output is very stable and quite high over a very wide range of temperatures, making it particularly attractive for high temperature applications. Depending on the application, the intrinsic activity of [138]La can be a disadvantage. LaBr 3(Ce) is very hygroscopic.

- LaCl 3(Ce) (or lanthanum chloride doped with cerium): very fast, high light output. LaCl 3(Ce) is a cheaper alternative to LaBr 3(Ce). It is also quite hygroscopic.

- PbWO 4 or lead tungstate: due to its high-Z, PbWO 4 is suitable for applications where a high stopping power is required (e.g. γ ray detection).

- LuI 3 or lutetium iodide

- LSO or lutetium oxyorthosilicate (Lu 2SiO 5): used in positron emission tomography because it exhibits properties similar to bismuth germanate (BGO), but with a higher light yield. Its only disadvantage is the intrinsic background from the beta decay of natural [176]Lu.

- LYSO (Lu 1.8Y 0.2SiO 5(Ce)): comparable in density to BGO, but much faster and with much higher light output; excellent for medical imaging applications. LYSO is non-hygroscopic.

- NaI(Tl) or sodium iodide doped with thallium: NaI(Tl) is by far the most widely used scintillator material. It is available in single crystal form or the more rugged polycrystalline form (used in high vibration environments, e.g. wireline logging in the oil industry). Other applications include nuclear medicine, basic research, environmental monitoring, and aerial surveys. NaI(Tl) is very hygroscopic and needs to be housed in an airtight enclosure.

- YAG(Ce) or yttrium aluminum garnet: YAG(Ce) is non-hygroscopic. The wavelength of maximum emission is at 550 nm, well-matched to red-resistive PMTs or photo-diodes. It is relatively fast (70 ns decay time). Its light output is about 1/3 of that of NaI(Tl). The material exhibits some properties that make it particularly attractive for electron microscopy applications (e.g. high electron conversion efficiency, good resolution, mechanical ruggedness and long lifetime).

- ZnS(Ag) or zinc sulfide: ZnS(Ag) is one of the older inorganic scintillators (the first experiment making use of a scintillator by Sir William Crookes (1903) involved a ZnS screen). It is only available as a polycrystalline powder, however. Its use is therefore limited to thin screens used primarily for α particle detection.

- ZnWO 4 or zinc tungstate is similar to CdWO 4 scintillator exhibiting long decay constant 25 µs and slightly lower light yield.

Radiolysis

Radiolysis is the dissociation of molecules by nuclear radiation. It is the cleavage of one or several chemical bonds resulting from exposure to high-energy flux. The radiation in this context is associated with ionizing radiation; radiolysis is therefore distinguished from, for example, photolysis of the Cl_2 molecule into two Cl-radicals, where (ultraviolet or visible) light is used.

For example, water dissociates under alpha radiation into a hydrogen radical and a hydroxyl radical, unlike ionization of water which produces a hydrogen ion and a hydroxide ion. The chemistry of concentrated solutions under ionizing radiation is extremely complex. Radiolysis can locally modify redox conditions, and therefore the speciation and the solubility of the compounds.

Water Decomposition

Of all the radiation-chemical reactions that have been studied, the most important is the decomposition of water. When exposed to radiation, water undergoes a breakdown sequence into hydrogen peroxide, hydrogen radicals, and assorted oxygen compounds, such as ozone, which when converted back into oxygen releases great amounts of energy. Some of these are explosive. This decomposition is produced mainly by the alpha particles, which can be entirely absorbed by very thin layers of water.

Applications

Corrosion Prediction and Prevention in Nuclear Power Plants

The possibility that enhanced concentration of hydroxyl present in irradiated water in the inner coolant loops of a light-water reactor must be taken into account when designing nuclear power plants, to prevent coolant loss resulting from corrosion.

Hydrogen Production

The current interest in nontraditional methods for the generation of hydrogen has prompted a revisit of radiolytic splitting of water, where the interaction of various types of ionizing radiation (α, β, and γ) with water produces molecular hydrogen. This reevaluation was further prompted by the current availability of large amounts of radiation sources contained in the fuel discharged from nuclear reactors. This spent fuel is usually stored in water pools, awaiting permanent disposal or reprocessing. The yield of hydrogen resulting from the irradiation of water with β and γ radiation is low (G-values = <1 molecule per 100 electronvolts of absorbed energy) but this is largely due to the rapid reassociation of the species arising during the initial radiolysis. If impurities are present or if physical conditions are created that prevent the establishment of a chemical equilibrium, the net production of hydrogen can be greatly enhanced.

Another approach uses radioactive waste as an energy source for regeneration of spent fuel by converting sodium borate into sodium borohydride. By applying the proper combination of controls, stable borohydride compounds may be produced and used as hydrogen fuel storage medium.

Spent Nuclear Fuel

Gas generation by radiolytic decomposition of hydrogen-containing materials, has been an area of concern for the transport and storage of radioactive materials and waste for a number of years. Potentially combustible and corrosive gases can be generated while at the same time, chemical reactions can remove hydrogen, and these reactions can be enhanced by the presence of radiation. The balance between these competing reactions is not well known at this time.

Earth's History

A suggestion has been made that in the early stages of the Earth's development when its radio-activity was almost two orders of magnitude higher than at present, radiolysis could have been the principal source of atmospheric oxygen, which ensured the conditions for the origin and development of life. Molecular hydrogen and oxidants produced by the radiolysis of water may also provide a continuous source of energy to subsurface microbial communities (Pedersen, 1999). Such speculation is supported by a discovery in the Mponeng Gold Mine in South Africa, where the researchers found a community dominated by a new phylotype of *Desulfotomaculum*, feeding on primarily radiolytically produced H_2.

Methods

Pulse Radiolysis

Pulse radiolysis is a recent method of initiating fast reactions to study reactions occurring on a timescale faster than approximately one hundred microseconds, when simple mixing of reagents is too slow and other methods of initiating reactions have to be used.

The technique involves exposing a sample of material to a beam of highly accelerated electrons, where the beam is generated by a linac. It has many applications. It was developed in the late 1950s and early 1960s by John Keene in Manchester and Jack W. Boag in London.

Flash Photolysis

Flash photolysis is an alternative to pulse radiolysis that uses high-power light pulses (e.g. from an excimer laser) rather than beams of electrons to initiate chemical reactions. Typically ultraviolet light is used which requires less radiation shielding than required for the X-rays emitted in pulse radiolysis.

References

- Campbell, Neil A.; Reece, Jane B. (2005). Biology (7th ed.). San Francisco: Pearson - Benjamin Cummings. pp. 186–191. ISBN 0-8053-7171-0.

- Raven, Peter H.; Ray F. Evert; Susan E. Eichhorn (2005). Biology of Plants (7th ed.). New York: W.H. Freeman and Company Publishers. pp. 115–127. ISBN 0-7167-1007-2.

- Ravve, A. (2006). Light-Associated Reactions of Synthetic Polymers. Spring Street, New York, NY 10013, USA:

Springer Science+Business Media, LLC. ISBN 0-387-31803-8.

- E. Thyrhaug; K. Zidek; J. Dostal; D. Bina; D. Zigmantas (2016). "Exciton Structure and Energy Transfer in the Fenna–Matthews– Olson Complex". J. Phys. Chem. Lett. 7: 1653–1660. doi:10.1021/acs.jpclett.6b00534.

- Reichmanis, Elsa; Crivello, James (2014). "Photopolymer Materials and Processes for Advanced Technologies". Chem. Mater. 26: 533–548.

- Crivello, J.; E. Reichmanis (2014). "Photopolymer Materials and Processes for Advanced Technologies". Chemistry of Materials. 26: 533–548. doi:10.1021/cm402262g.

- Manuel Nuño, Richard J. Ball and Chris R. Bowen. "Study of solid/gas phase photocatalytic reactions by electron ionization mass spectrometry" J Mass Spec, 2014, 49 (8), p. 716-726

- Hari, Durga Prasad; König, Burkhard (5 August 2011). "Eosin Y Catalyzed Visible Light Oxidative C–C and C–P bond Formation". Organic Letters. 13 (15): 3852–3855. doi:10.1021/ol201376v. PMID 21744842.

- Riener, Michelle; Nicewicz, David A. (2013). "Synthesis of cyclobutane lignans via an organic single electron oxidant–electron relay system". Chemical Science. 4 (6): 2625. doi:10.1039/c3sc50643f.

Photoelectrochemical Process: An Overview

The processes that are used in photoelectrochemistry are referred to as photoelectrochemical processes. It involves the transformation of light to other forms. Fluorescence, photoelectric effect, photoionization mode and absorption have been briefly explained in the following text.

Photoelectrochemical Process

Photoelectrochemical processes are processes in photoelectrochemistry; they usually involve transforming light into other forms of energy. These processes apply to photochemistry, optically pumped lasers, sensitized solar cells, luminescence, and photochromism.

Electron Excitation

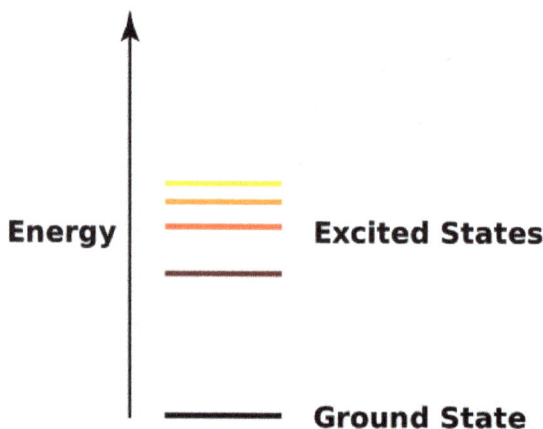

After absorbing energy, an electron may jump from the ground state to a higher energy excited state.

Electron excitation is the movement of an electron to a higher energy state. This can either be done by photoexcitation (PE), where the original electron absorbs the photon and gains all the photon's energy or by electrical excitation (EE), where the original electron absorbs the energy of another, energetic electron. Within a semiconductor crystal lattice, thermal excitation is a process where lattice vibrations provide enough energy to move electrons to a higher energy band. When an excited electron falls back to a lower energy state again, it is called electron relaxation. This can be done by radiation of a photon or giving the energy to a third spectator particle as well.

In physics there is a specific technical definition for energy level which is often associated with an atom being excited to an excited state. The excited state, in general, is in relation to the ground state, where the excited state is at a higher energy level than the ground state.

Photoexcitation

Photoexcitation is the mechanism of electron excitation by photon absorption, when the energy of the photon is too low to cause photoionization. The absorption of the photon takes place in accordance with Planck's quantum theory.

Photoexcitation plays role in photoisomerization. Photoexcitation is exploited in dye-sensitized solar cells, photochemistry, luminescence, optically pumped lasers, and in some photochromic applications.

Photoisomerization

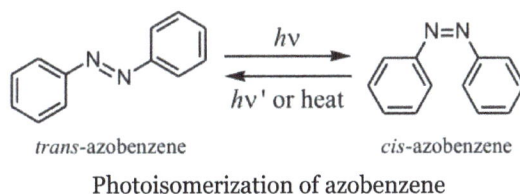

trans-azobenzene   cis-azobenzene

Photoisomerization of azobenzene

In chemistry, photoisomerization is molecular behavior in which structural change between isomers is caused by photoexcitation. Both reversible and irreversible photoisomerization reactions exist. However, the word "photoisomerization" usually indicates a reversible process. Photoisomerizable molecules are already put to practical use, for instance, in pigments for rewritable CDs, DVDs, and 3D optical data storage solutions. In addition, recent interest in photoisomerizable molecules has been aimed at molecular devices, such as molecular switches, molecular motors, and molecular electronics.

Photoisomerization behavior can be roughly categorized into several classes. Two major classes are trans-cis (or 'E-'Z) conversion, and open-closed ring transition. Examples of the former include stilbene and azobenzene. This type of compounds has a double bond, and rotation or inversion around the double bond affords isomerization between the two states. Examples of the latter include fulgide and diarylethene. This type of compounds undergoes bond cleavage and bond creation upon irradiation with particular wavelengths of light. Still another class is the Di-pi-methane rearrangement.

Photoionization

Photoionization is the physical process in which an incident photon ejects one or more electrons from an atom, ion or molecule. This is essentially the same process that occurs with the photoelectric effect with metals. In the case of a gas or single atoms, the term photoionization is more common.

The ejected electrons, known as photoelectrons, carry information about their pre-ionized states. For example, a single electron can have a kinetic energy equal to the energy of the incident photon minus the electron binding energy of the state it left. Photons with energies less than the electron binding energy may be absorbed or scattered but will not photoionize the atom or ion.

For example, to ionize hydrogen, photons need an energy greater than 13.6 electronvolts, which corresponds to a wavelength of 91.2 nm. For photons with greater energy than this, the energy of the emitted photoelectron is given by:

$$\frac{mv^2}{2} = hv - 13.6eV$$

where h is Planck's constant and v is the frequency of the photon.

This formula defines the photoelectric effect.

Not every photon which encounters an atom or ion will photoionize it. The probability of photoionization is related to the photoionization cross-section, which depends on the energy of the photon and the target being considered. For photon energies below the ionization threshold, the photoionization cross-section is near zero. But with the development of pulsed lasers it has become possible to create extremely intense, coherent light where multi-photon ionization may occur. At even higher intensities (around 10^{15} - 10^{16} W/cm² of infrared or visible light), non-perturbative phenomena such as *barrier suppression ionization* and *rescattering ionization* are observed.

Multi-photon Ionization

Several photons of energy below the ionization threshold may actually combine their energies to ionize an atom. This probability decreases rapidly with the number of photons required, but the development of very intense, pulsed lasers still makes it possible. In the perturbative regime (below about 10^{14} W/cm² at optical frequencies), the probability of absorbing N photons depends on the laser-light intensity I as I^N.

Above threshold ionization (ATI) is an extension of multi-photon ionization where even more photons are absorbed than actually would be necessary to ionize the atom. The excess energy gives the released electron higher kinetic energy than the usual case of just-above threshold ionization. More precisely, The system will have multiple peaks in its photoelectron spectrum which are separated by the photon energies, this indicates that the emitted electron has more kinetic energy than in the normal (lowest possible number of photons) ionization case. The electrons released from the target will have approximately an integer number of photon-energies more kinetic energy. In intensity regions between 10^{14} W/cm² and 10^{18} W/cm², each of MPI, ATI, and barrier suppression ionization can occur simultaneously, each contributing to the overall ionization of the atoms involved.

Photo-dember

In semiconductor physics the Photo-Dember effect (named after its discoverer H. Dember) consists in the formation of a charge dipole in the vicinity of a semiconductor surface after ultra-fast photo-generation of charge carriers. The dipole forms owing to the difference of mobilities (or diffusion constants) for holes and electrons which combined with the break of symmetry provided by the surface lead to an effective charge separation in the direction perpendicular to the surface.

Grotthuss–Draper Law

The Grotthuss–Draper law (also called the Principle of Photochemical Activation) states that only that light which is absorbed by a system can bring about a photochemical change. Materials such as dyes and phosphors must be able to absorb "light" at optical frequencies. This law provides a basis for fluorescence and phosphorescence. The law was first proposed in 1817 by Theodor Grotthuss and in 1842, independently, by John William Draper.

This is considered to be one of the two basic laws of photochemistry. The second law is the Stark–Einstein law, which says that primary chemical or physical reactions occur with each photon absorbed.

Stark–einstein Law

The Stark–Einstein law is named after German-born physicists Johannes Stark and Albert Einstein, who independently formulated the law between 1908 and 1913. It is also known as the photochemical equivalence law or photoequivalence law. In essence it says that every photon that is absorbed will cause a (primary) chemical or physical reaction.

The photon is a quantum of radiation, or one unit of radiation. Therefore, this is a single unit of EM radiation that is equal to Planck's constant (h) times the frequency of light. This quantity is symbolized by γ, $h\nu$, or $\hbar\omega$.

The photochemical equivalence law is also restated as follows: for every mole of a substance that reacts, an equivalent mole of quanta of light are absorbed. The formula is:

$$\Delta E_{mol} = N_A h\nu$$

where N_A is Avogadro's number.

The photochemical equivalence law applies to the part of a light-induced reaction that is referred to as the primary process (i.e. absorption or fluorescence).

In most photochemical reactions the primary process is usually followed by so-called secondary photochemical processes that are normal interactions between reactants not requiring absorption of light. As a result such reactions do not appear to obey the one quantum–one molecule reactant relationship.

The law is further restricted to conventional photochemical processes using light sources with moderate intensities; high-intensity light sources such as those used in flash photolysis and in laser experiments are known to cause so-called biphotonic processes; i.e., the absorption by a molecule of a substance of two photons of light.

Absorption

In physics, absorption of electromagnetic radiation is the way by which the energy of a photon is taken up by matter, typically the electrons of an atom. Thus, the electromagnetic energy is transformed to other forms of energy, for example, to heat. The absorption of light during wave propagation is often called attenuation. Usually, the absorption of waves does not depend on their intensity (linear absorption), although in certain conditions (usually, in optics), the medium changes its transparency dependently on the intensity of waves going through, and the Saturable absorption (or nonlinear absorption) occurs.

Photosensitization

Photosensitization is a process of transferring the energy of absorbed light. After absorption, the energy is transferred to the (chosen) reactants. This is part of the work of photochemistry in general. In particular this process is commonly employed where reactions require light sources of certain wavelengths that are not readily available.

For example, mercury absorbs radiation at 1849 and 2537 angstroms, and the source is often high-intensity mercury lamps. It is a commonly used sensitizer. When mercury vapor is mixed with ethylene, and the compound is irradiated with a mercury lamp, this results in the photo-decomposition of ethylene to acetylene. This occurs on absorption of light to yield excited state mercury atoms, which are able to transfer this energy to the ethylene molecules, and are in turn deactivated to their initial energy state.

Cadmium; some of the noble gases, for example xenon; zinc; benzophenone; and a large number of organic dyes, are also used as sensitizers.

Photosensitisers are a key component of photodynamic therapy used to treat cancers.

Sensitizer

A sensitizer in chemiluminescence is a chemical compound, capable of light emission after it has received energy from a molecule, which became excited previously in the chemical reaction. A good example is this:

When an alkaline solution of sodium hypochlorite and a concentrated solution of hydrogen peroxide are mixed, a reaction occurs:

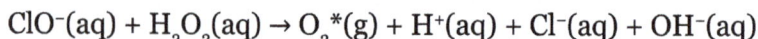

$$ClO^-(aq) + H_2O_2(aq) \rightarrow O_2^*(g) + H^+(aq) + Cl^-(aq) + OH^-(aq)$$

O_2^* is excited oxygen – meaning, one or more electrons in the O_2 molecule have been promoted to higher-energy molecular orbitals. Hence, oxygen produced by this chemical reaction somehow 'absorbed' the energy released by the reaction and became excited. This energy state is unstable, therefore it will return to the ground state by lowering its energy. It can do that in more than one way:

- it can react further, without any light emission

- it can lose energy without emission, for example, giving off heat to the surroundings or transferring energy to another molecule

- it can emit light

The intensity, duration and color of emitted light depend on quantum and kinetical factors. However, excited molecules are frequently less capable of light emission in terms of brightness and duration when compared to sensitizers. This is because sensitizers can store energy (that is, be excited) for longer periods of time than other excited molecules. The energy is stored through means of quantum vibration, so sensitizers are usually compounds which either include systems of aromatic rings or many conjugated double and triple bonds in their structure. Hence, if an excited molecule transfers its energy to a sensitizer thus exciting it, longer and easier to quantify light emission is often observed.

The color (that is, the wavelength), brightness and duration of emission depend upon the sensitizer used. Usually, for a certain chemical reaction, many different sensitizers can be used.

List of Some Common Sensitizers

- Violanthrone
- Isoviolanthrone
- Fluorescein
- Rubrene
- 9,10-Diphenylanthracene
- Tetracene
- 13,13'-Dibenzantronile
- Levulinic Acid

Fluorescence Spectroscopy

Fluorescence spectroscopy aka fluorometry or spectrofluorometry, is a type of electromagnetic spectroscopy which analyzes fluorescence from a sample. It involves using a beam of light, usually ultraviolet light, that excites the electrons in molecules of certain compounds and causes them to emit light of a lower energy, typically, but not necessarily, visible light. A complementary technique is absorption spectroscopy.

Devices that measure fluorescence are called fluorometers or fluorimeters.

Absorption Spectroscopy

Absorption spectroscopy refers to spectroscopic techniques that measure the absorption of radiation, as a function of frequency or wavelength, due to its interaction with a sample. The sample absorbs energy, i.e., photons, from the radiating field. The intensity of the absorption varies as a function of frequency, and this variation is the absorption spectrum. Absorption spectroscopy is performed across the electromagnetic spectrum.

Fluorescence

Fluorescence is the emission of light by a substance that has absorbed light or other electromagnetic radiation. It is a form of luminescence. In most cases, the emitted light has a longer wavelength, and therefore lower energy, than the absorbed radiation. The most striking example of fluorescence occurs when the absorbed radiation is in the ultraviolet region of the spectrum, and thus invisible to the human eye, while the emitted light is in the visible region, which gives the fluorescent substance a distinct color that can only be seen when exposed to UV light. Fluorescent materials cease to glow immediately when the radiation source stops, unlike phosphorescence, where it continues to emit light for some time after.

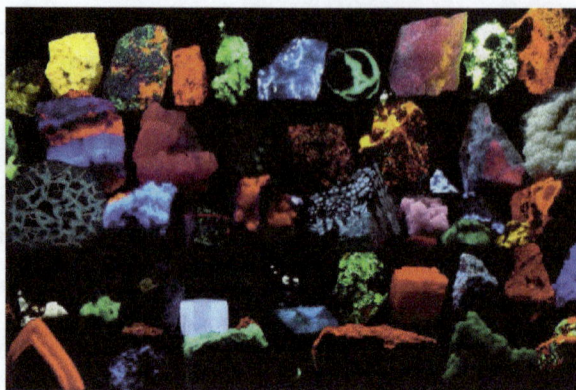

Fluorescent minerals emit visible light when exposed to ultraviolet light
Biofluorescent marine organisms.

Fluorescence has many practical applications, including mineralogy, gemology, medicine, chemical sensors (fluorescence spectroscopy), fluorescent labelling, dyes, biological detectors, cosmic-ray detection, and, most commonly, fluorescent lamps. Fluorescence also occurs frequently in nature in some minerals and in various biological states in many branches of the animal kingdom.

History

Lignum nephriticum cup made from the wood of the narra tree (*Pterocarpus indicus*), and a flask containing its fluorescent solution

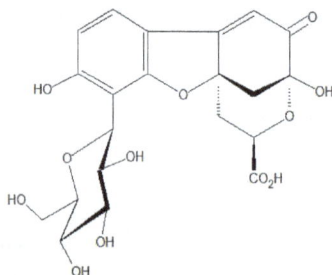

Matlaline, the fluorescent substance in the wood of the tree *Eysenhardtia polystachya*

An early observation of fluorescence was described in 1560 by Bernardino de Sahagún and in 1565 by Nicolás Monardes in the infusion known as *lignum nephriticum*. It was derived from the wood of two tree species, *Pterocarpus indicus* and *Eysenhardtia polystachya*. The chemical compound responsible for this fluorescence is matlaline, which is the oxidation product of one of the flavonoids found in this wood.

In 1819, Edward D. Clarke and in 1822 René Just Haüy described fluorescence in fluorites, Sir David Brewster described the phenomenon for chlorophyll in 1833 and Sir John Herschel did the same for quinine in 1845.

In his 1852 paper on the "Refrangibility" (wavelength change) of light, George Gabriel Stokes described the ability of fluorspar and uranium glass to change invisible light beyond the violet end of the visible spectrum into blue light. He named this phenomenon *fluorescence* : «I am almost inclined to coin a word, and call the appearance *fluorescence*, from fluor-spar [i.e., fluorite], as the analogous term *opalescence* is derived from the name of a mineral." The name was derived from the mineral fluorite (calcium difluoride), some examples of which contain traces of divalent europium, which serves as the fluorescent activator to emit blue light. In a key experiment he used a prism to isolate ultraviolet radiation from sunlight and observed blue light emitted by an ethanol solution of quinine exposed by it.

Physical Principles

Photochemistry

Fluorescence occurs when an orbital electron of a molecule, atom, or nanostructure, relaxes to its ground state by emitting a photon from an excited singlet state:

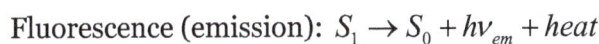

Excitation: $S_0 + h\nu_{ex} \rightarrow S_1$

Fluorescence (emission): $S_1 \rightarrow S_0 + h\nu_{em} + heat$

Here $h\nu$ is a generic term for photon energy with h = Planck's constant and ν = frequency of light. The specific frequencies of exciting and emitted light are dependent on the particular system.

S_0 is called the ground state of the fluorophore (fluorescent molecule), and S_1 is its first (electronically) excited singlet state.

A molecule in S_1 can relax by various competing pathways. It can undergo *non-radiative* relaxation in which the excitation energy is dissipated as heat (vibrations) to the solvent. Excited organic molecules can also relax via conversion to a triplet state, which may subsequently relax via phosphorescence, or by a secondary non-radiative relaxation step.

Relaxation from S_1 can also occur through interaction with a second molecule through fluorescence quenching. Molecular oxygen (O_2) is an extremely efficient quencher of fluorescence just because of its unusual triplet ground state.

In most cases, the emitted light has a longer wavelength, and therefore lower energy, than the absorbed radiation; this phenomenon is known as the Stokes shift. However, when the absorbed electromagnetic radiation is intense, it is possible for one electron to absorb two photons; this two-photon absorption can lead to emission of radiation having a shorter wavelength than the absorbed radiation. The emitted radiation may also be of the same wavelength as the absorbed radiation, termed "resonance fluorescence".

Molecules that are excited through light absorption or via a different process (e.g. as the product of a reaction) can transfer energy to a second 'sensitized' molecule, which is converted to its excited state and can then fluoresce.

Quantum Yield

The fluorescence quantum yield gives the efficiency of the fluorescence process. It is defined as the ratio of the number of photons emitted to the number of photons absorbed.

$$\Phi = \frac{\text{Number of photons emitted}}{\text{Number of photons absorbed}}$$

The maximum fluorescence quantum yield is 1.0 (100%); each photon absorbed results in a photon emitted. Compounds with quantum yields of 0.10 are still considered quite fluorescent. Another way to define the quantum yield of fluorescence, is by the rate of excited state decay:

$$\Phi = \frac{k_f}{\sum_i k_i}$$

where k_f is the rate constant of spontaneous emission of radiation and

$$\sum_i k_i$$

is the sum of all rates of excited state decay. Other rates of excited state decay are caused by mechanisms other than photon emission and are, therefore, often called "non-radiative rates", which can include: dynamic collisional quenching, near-field dipole-dipole interaction (or resonance energy transfer), internal conversion, and intersystem crossing. Thus, if the rate of any pathway changes, both the excited state lifetime and the fluorescence quantum yield will be affected.

Fluorescence quantum yields are measured by comparison to a standard. The quinine salt *quinine sulfate* in a sulfuric acid solution is a common fluorescence standard.

Lifetime

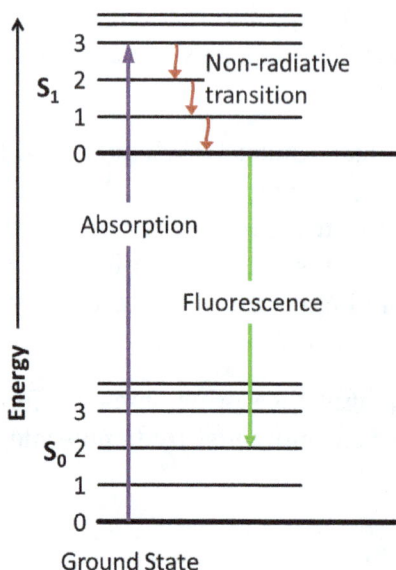

Jablonski diagram. After an electron absorbs a high energy photon the system is excited electronically and vibrationally. The system relaxes vibrationally, and eventually fluoresces at a longer wavelength.

The fluorescence lifetime refers to the average time the molecule stays in its excited state before emitting a photon. Fluorescence typically follows first-order kinetics:

$$[S1] = [S1]_0 \, e^{-\Gamma t}$$

where $[S1]$ is the concentration of excited state molecules at time t, $[S1]_0$ is the initial concentration and Γ is the decay rate or the inverse of the fluorescence lifetime. This is an instance of exponential decay. Various radiative and non-radiative processes can de-populate the excited state. In such case the total decay rate is the sum over all rates:

$$\Gamma_{tot} = \Gamma_{rad} + \Gamma_{nrad}$$

where Γ_{tot} is the total decay rate, Γ_{rad} the radiative decay rate and Γ_{nrad} the non-radiative decay rate. It is similar to a first-order chemical reaction in which the first-order rate constant is the sum of all of the rates (a parallel kinetic model). If the rate of spontaneous emission, or any of the other rates are fast, the lifetime is short. For commonly used fluorescent compounds, typical excited state decay times for photon emissions with energies from the UV to near infrared are within the range of 0.5 to 20 nanoseconds. The fluorescence lifetime is an important parameter for practical applications of fluorescence such as fluorescence resonance energy transfer and Fluorescence-lifetime imaging microscopy.

Fluorescence Anisotropy

Fluorophores are more likely to be excited by photons if the transition moment of the fluorophore is parallel to the electric vector of the photon. The polarization of the emitted light will also depend on the transition moment. The transition moment is dependent on the physical orientation of the fluorophore molecule. For fluorophores in solution this means that the intensity and polarization of the emitted light is dependent on rotational diffusion. Therefore, anisotropy measurements can be used to investigate how freely a fluorescent molecule moves in a particular environment.

Fluorescence anisotropy can be defined quantitatively as

$$r = \frac{I_\| - I_\perp}{I_\| + 2I_\perp}$$

where $I_\|$ is the emitted intensity parallel to polarization of the excitation light and $I_\perp$ is the emitted intensity perpendicular to the polarization of the excitation light.

Fluorence

Strongly fluorescent pigments often have an unusual appearance which is often described colloquially as a "neon color." This phenomenon was termed "Farbenglut" by Hermann von Helmholtz and "fluorence" by Ralph M. Evans. It is generally thought to be related to the high brightness of the color relative to what it would be as a component of white. Fluorescence shifts energy in the incident illumination from shorter wavelengths to longer (such as blue to yellow) and thus can make the fluorescent color appear brighter (more saturated) than it could possibly be by reflection alone.

Rules

There are several general rules that deal with fluorescence. Each of the following rules has exceptions but they are useful guidelines for understanding fluorescence (these rules do not necessarily apply to two-photon absorption).

Kasha–Vavilov Rule

The Kasha–Vavilov rule dictates that the quantum yield of luminescence is independent of the wavelength of exciting radiation. This occurs because excited molecules usually decay to the lowest vibrational level of the excited state before fluorescence emission takes place. The Kasha–Vavilov rule does not always apply and is violated severely in many simple molecules. A somewhat more reliable statement, although still with exceptions, would be that the fluorescence spectrum shows very little dependence on the wavelength of exciting radiation.

Mirror image Rule

For many fluorophores the absorption spectrum is a mirror image of the emission spectrum. This is known as the mirror image rule and is related to the Franck–Condon principle which states that electronic transitions are vertical, that is energy changes without distance changing as can be represented with a vertical line in Jablonski diagram. This means the nucleus does not move and the vibration levels of the excited state resemble the vibration levels of the ground state.

Stokes Shift

In general, emitted fluorescent light has a longer wavelength and lower energy than the absorbed light. This phenomenon, known as Stokes shift, is due to energy loss between the time a photon is absorbed and when it is emitted. The causes and magnitude of Stokes shift can be complex and are dependent on the fluorophore and its environment. However, there are some common causes. It is frequently due to non-radiative decay to the lowest vibrational energy level of the excited state. Another factor is that the emission of fluorescence frequently leaves a fluorophore in a higher vibrational level of the ground state.

Fluorescence in Nature

There are many natural compounds that exhibit fluorescence, and they have a number of applications. Some deep-sea animals, such as the greeneye, use fluorescence.

Biofluorescence vs. Bioluminescence vs. Biophosphorescence

Biofluorescence

Biofluorescence is the absorption of electromagnetic wavelengths from the visible light spectrum by fluorescent proteins in a living organism, and the reemission of that light at a lower energy level. This causes the light that is re-emitted to be a different color than the light that is absorbed. Stimulating light excites an electron, raising energy to an unstable level. This instability is unfavorable, so the energized electron is returned to a stable state almost as immediately as it becomes unstable. This return to stability corresponds with the release of excess energy in the form of fluorescent light. This emission of light is only observable when the stimulant light is still providing light to the organism/object and is typically yellow, pink, orange, red, green, or purple. Biofluorescence is often confused with the following forms of biotic light, bioluminescence and biophosphorescence.

Bioluminescence

Bioluminescence differs from biofluorescence in that it is the natural production of light by chemical reactions within an organism, whereas biofluorescence is the absorption and reemission of light from the environment.

Biophosphorescence

Biophosphorescence is similar to biofluorescence in its requirement of light wavelengths as a provider of excitation energy. The difference here lies in the relative stability of the energized electron. Unlike with biofluorescence, here the electron retains stability, emitting light that continues to "glow-in-the-dark" even long after the stimulating light source has been removed.

Mechanisms of Biofluorescence

Epidermal Chromatophores

Pigment cells that exhibit fluorescence are called fluorescent chromatophores, and function somatically similar to regular chromatophores. These cells are dendritic, and contain pigments called fluorosomes. These pigments contain fluorescent proteins are activated by K+ (potassium) ions, and it is their movement, aggregation, and dispersion within the fluorescent chromatophore that cause directed fluorescence patterning. Fluorescent cells are innervated the same as other chromatphores, like melanophores, pigment cells that contain melanin. Short term fluorescent patterning and signaling is controlled by the nervous system. Fluorescent chromatophores can be found in the skin (e.g. in fish) just below the epidermis, amongst other chromatophores.

Epidermal fluorescent cells in fish also respond to hormonal stimuli by the α–MSH and MCH hormones much the same as melanophores. This suggests that fluorescent cells may be have color changes throughout the day that coincide with their circadian rhythm. Fish may also be sensitive to cortisol induced stress responses to environmental stimuli, such as interaction with a predator or engaging in a mating ritual.

Phylogenetics

Evolutionary Origins

It is suspected by some scientists that GFPs and GFP like proteins began as electron donors activated by light. These electrons were then used for reactions requiring light energy. Functions of fluorescent proteins, such as protection from the sun, conversion of light into different wavelengths, or for signaling are thought to have evolved secondarily.

Fluorescence has multiple origins in the tree of life. This diagram displays the origins within actinopterygians (ray finned fish).

The incidence of fluorescence across the tree of life is widespread, and has been studied most extensively in a phylogenetic sense in fish. The phenomenon appears to have evolved multiple times in multiple taxa such as in the anguilliformes (eels), gobioidei (gobies and cardinalfishes), and tetradontiformes (triggerfishes), along with the other taxa discussed later in the article. Fluorescence is highly genotypically and phenotypically variable even within ecosystems, in regards to the wavelengths emitted, the patterns displayed, and the intensity of the fluorescence. Generally, the species relying upon camouflage exhibit the greatest diversity in fluorescence, likely because camouflage is one of the most common uses of fluorescence.

Adaptive Functions

Currently, relatively little is known about the functional significance of fluorescence and fluorescent proteins. However, it is suspected that biofluorescence may serve important functions in signaling and communication, mating, lures, camouflage, UV protection and antioxidation, photoacclimation, dinoflagellate regulation, and in coral health.

Aquatic Biofluorescence

Water absorbs light of long wavelengths, so less light from these wavelengths reflects back to reach the eye. Therefore, warm colors from the visual light spectrum appear less vibrant at increasing depths. Water scatters light of shorter wavelengths, meaning cooler colors dominate the visual field in the photic zone. Light intensity decreases 10 fold with every 75 m of depth, so at depths of 75 m, light is 10% as intense as it is on the surface, and is only 1% as intense at 150 m as it is on the surface. Because the water filters out the wavelengths and intensity of water reaching certain depths, different proteins, because of the wavelengths and intensities of light they are capable of absorbing, are better suited to different depths. Theoretically, some fish eyes can detect light as deep as 1000 m. At these depths of the aphotic zone, the only sources of light are organisms themselves, giving off light through chemical reactions in a process called bioluminescence.

Fluorescence is simply defined as the absorption of electromagnetic radiation at one wavelength and its reemission at another, lower energy wavelength. Thus any type of fluorescence depends on the presence of external sources of light. Biologically functional fluorescence is found in the photic zone, where there is not only enough light to cause biofluorescence, but enough light for other organisms to detect it. The visual field in the photic zone is naturally blue, so colors of fluorescence can be detected as bright reds, oranges, yellows, and greens. Green is the most commonly found color in the biofluorescent spectrum, yellow the second most, orange the third, and red is the rarest. Fluorescence can occur in organisms in the aphotic zone as a byproduct of that same organism's bioluminescence. Some biofluorescence in the aphotic zone is merely a byproduct of the organism's tissue biochemistry and does not have a functional purpose. However, some cases of functional and adaptive significance of biofluorescence in the aphotic zone of the deep ocean is an active area of research.

Photic Zone

Fish

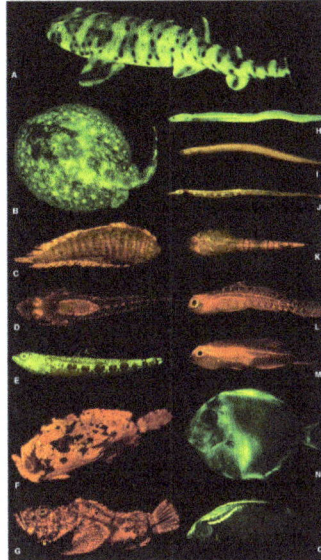

Fluorescent marine fish

Bony fishes living in shallow water, due to living in a colorful environment, generally have good color vision. Thus, in shallow-water fishes, red, orange, and green fluorescence most likely serves as a means of communication with conspecifics, especially given the great phenotypic variance of the phenomenon.

Many fish that exhibit biofluorescence, such as sharks, lizardfish, scorpionfish, wrasses, and flatfishes, also possess yellow intraocular filters. Yellow intraocular filters in the lenses and cornea of certain fishes function as long-pass filters, thus enabling the species that possess them to visualize and potentially exploit fluorescence to enhance visual contrast and patterns that are unseen to other fishes and predators that lack this visual specialization. Fishes that possess the necessary yellow intraocular filters for visualizing biofluorescence potentially exploit a light signal from members of it or a similar functional role. Biofluorescent patterning was especially prominent in cryptically patterned fishes possessing complex camouflage, and that many of these lineages also possess yellow long-pass intraocular filters that could enable visualization of such patterns.

Another adaptive use of fluorescence is to generate red light from the ambient blue light of the photic zone to aid vision. Red light can only be seen across short distances due to attenuation of red light wavelengths by water. Many fish species that fluoresce are small, group-living, or benthic/aphotic, and have conspicuous patterning. This patterning is caused by fluorescent tissue and is visible to other members of the species, however the patterning is invisible at other visual spectra. These intraspecific fluorescent patterns also coincide with intra-species signaling. The patterns present in ocular rings to indicate directionality of an individual's gaze, and along fins to indicate directionality of an individual's movement. Current research suspects that this red fluorescence is used for private communication between members of the same species. Due to the prominence of blue light at ocean depths, red light and light of longer wavelengths are muddled, and many predatory reef fish have little to no sensitivity for light at these wavelengths. Fish such as the fairy wrasse that have developed visual sensitivity to longer wavelengths are able to display red fluorescent signals that give a high contrast to the blue environment and are conspicuous to conspecifics in short ranges, yet are relatively invisible to other common fish that have reduced sensitivities to long wavelengths. Thus, fluorescence can be used as adaptive signaling and intra-species communication in reef fish.

Additionally, it is suggested that fluorescent tissues that surround an organism's eyes are used to convert blue light from the photic zone or green bioluminescence in the aphotic zone into red light to aid vision.

Coral

Fluorescence serves a wide variety of functions in coral. Fluorescent proteins in corals may contribute to photosynthesis by converting otherwise unusable wavelengths of light into ones for which the coral's symbiotic algae are able to conduct photosynthesis. Also, the proteins may fluctuate in number as more or less light becomes available as a means of photoacclimation. Similarly, these fluorescent proteins may possess antioxidant capacities to eliminate oxygen radicals produced by photosynthesis. Finally, through modulating photosynthesis, the fluorescent proteins may also serve as a means of regulating the activity of the coral's photosynthetic algal symbionts.

Cephalopods

Alloteuthis subulata and *Loligo vulgaris*, two types of nearly transparent squid, have fluorescent spots above their eyes. These spots reflect incident light, which may serve as a means of camouflage, but also for signaling to other squids for schooling purposes.

Jellyfish

Another, well-studied example of biofluorescence in the ocean is the hydrozoan Aequorea victoria. This jellyfish lives in the photic zone off the west coast of North America and was identified as a carrier of green fluorescent protein (GFP) by Osamu Shimomura. The gene for these green fluorescent proteins has been isolated and is scientifically significant because it is widely used in genetic studies to indicate the expression of other genes.

Mantis Shrimp

Several species of mantis shrimp, which are stomatopod crustaceans, including *Lysiosquillina glabriuscula*, have yellow fluorescent markings along their antennal scales and carapace (shell) that males present during threat displays to predators and other males. The display involves raising the head and thorax, spreading the striking appendages and other maxillipeds, and extending the prominent, oval antennal scales laterally, which makes the animal appear larger and accentuates its yellow fluorescent markings. Furthermore, as depth increases, mantis shrimp fluorescence accounts for a greater part of the visible light available. During mating rituals, mantis shrimp actively fluoresce, and the wavelength of this fluorescence matches the wavelengths detected by their eye pigments.

Aphotic Zone

Siphonophores

Siphonophorae is an order of marine animals from the phylum Hydrozoa that consist of a specialized medusoid and polyp zooid. Some siphonophores, including the genus Erenna that live in the aphotic zone between depths of 1600 m and 2300 m, exhibit yellow to red fluorescence in the photophores of their tentacle-like tentilla. This fluorescence occurs as a by-product of bioluminescence from these same photophores. The siphonophores exhibit the fluorescence in a flicking pattern that is used as a lure to attract prey.

Dragonfish

The predatory deep-sea dragonfish *Malacosteus niger*, the closely related *Aristostomias* genus and the species *Pachystomias microdon* are capable of harnessing the blue light emitted from their own bioluminescence to generate red biofluorescence from suborbital photophores. This red fluorescence is invisible to other animals, which allows these dragonfish extra light at dark ocean depths without attracting or signaling predators.

Terrestrial Biofluorescence

Butterflies

Swallowtail (*Papilio*) butterflies have complex systems for emitting fluorescent light. Their wings contain pigment-infused crystals that provide directed fluorescent light. These crystals function to produce fluorescent light best when they absorb radiance from sky-blue light (wavelength about 420 nm). The wavelengths of light that the butterflies see the best correspond to the absorbance of the crystals in the butterfly's wings. This likely functions to enhance the capacity for signaling.

Parrots

Parrots have fluorescent plumage that may be used in mate signaling. A study using mate-choice experiments on budgerigars (*Melopsittacus undulates*) found compelling support for fluorescent sexual signaling, with both males and females significantly preferring birds with the fluorescent experimental stimulus. This study suggests that the fluorescent plumage of parrots is not simply a by-product of pigmentation, but instead an adapted sexual signal. Considering the intricacies of the pathways that produce fluorescent pigments, there may be significant costs involved. Therefore, individuals exhibiting strong fluorescence may be honest indicators of high individual quality, since they can deal with the associated costs.

Arachnids

Spiders fluoresce under UV light and possess a huge diversity of fluorophores. Remarkably, spiders are the only known group in which fluorescence is "taxonomically widespread, variably expressed, evolutionarily labile, and probably under selection and potentially of ecological importance for intraspecific and interspecific signaling." A study by Andrews et al. (2007) reveals that fluorescence has evolved multiple times across spider taxa, with novel fluorophores evolving during spider diversification. In some spiders, ultraviolet cues are important for predator-prey interactions, intraspecific communication, and camouflaging with matching fluorescent flowers. Differing ecological contexts could favor inhibition or enhancement of fluorescence expression, depending upon whether fluorescence helps spiders be cryptic or makes them more conspicuous to predators. Therefore, natural selection could be acting on expression of fluorescence across spider species.

Scorpions also fluoresce.

Fluorescing scorpion

Plants

The *Mirabilis jalapa* flower contains violet, fluorescent betacyanins and yellow, fluorescent betaxanthins. Under white light, parts of the flower containing only betaxanthins appear yellow, but in areas where both betaxanthins and betacyanins are present, the visible fluorescence of the flower is faded due to internal light-filtering mechanisms. Fluorescence was previously suggested to play a role in pollinator attraction, however, it was later found that the visual signal by fluorescence is negligible compared to the visual signal of light reflected by the flower.

Chlorophyll fluoresces a weak red under ultraviolet light.

Abiotic Fluorescence

Gemology, Mineralogy and Geology

Fluorescence of Aragonite

Gemstones, minerals, may have a distinctive fluorescence or may fluoresce differently under short-wave ultraviolet, long-wave ultraviolet, visible light, or X-rays.

Many types of calcite and amber will fluoresce under shortwave UV, longwave UV and visible light.

Rubies, emeralds, and diamonds exhibit red fluorescence under long-wave UV, blue and sometimes green light; diamonds also emit light under X-ray radiation.

Fluorescence in minerals is caused by a wide range of activators. In some cases, the concentration of the activator must be restricted to below a certain level, to prevent quenching of the fluorescent emission. Furthermore, the mineral must be free of impurities such as iron or copper, to prevent quenching of possible fluorescence. Divalent manganese, in concentrations of up to several percent, is responsible for the red or orange fluorescence of calcite, the green fluorescence of willemite, the yellow fluorescence of esperite, and the orange fluorescence of wollastonite and clinohedrite. Hexavalent uranium, in the form of the uranyl cation, fluoresces at all concentrations in a yellow green, and is the cause of fluorescence of minerals such as autunite or andersonite, and, at low concentration, is the cause of the fluorescence of such materials as some samples of hyalite opal. Trivalent chromium at low concentration is the source of the red fluorescence of ruby. Divalent europium is the source of the blue fluorescence, when seen in the mineral fluorite. Trivalent lanthanides such as terbium and dysprosium are the principal activators of the creamy yellow fluorescence exhibited by the yttrofluorite variety of the mineral fluorite, and contribute to the orange fluorescence of zircon. Powellite (calcium molybdate) and scheelite (calcium tungstate) fluoresce intrinsically in yellow and blue, respectively. When present together in solid solution, energy is transferred from the higher-energy tungsten to the lower-energy molybdenum, such that fairly low levels of molybdenum are sufficient to cause a yellow emission for scheelite, instead of blue. Low-iron sphalerite (zinc sulfide), fluoresces and phosphoresces in a range of colors, influenced by the presence of various trace impurities.

Crude oil (petroleum) fluoresces in a range of colors, from dull-brown for heavy oils and tars through to bright-yellowish and bluish-white for very light oils and condensates. This phenomenon is used in oil exploration drilling to identify very small amounts of oil in drill cuttings and core samples.

Organic Liquids

Organic solutions such anthracene or stilbene, dissolved in benzene or toluene, fluoresce with ultraviolet or gamma ray irradiation. The decay times of this fluorescence are of the order of nanoseconds, since the duration of the light depends on the lifetime of the excited states of the fluorescent material, in this case anthracene or stilbene.

Scintillation is defined a flash of light produced in a transparent material by the passage of a particle (an electron, an alpha particle, an ion, or a high-energy photon). Stilbene and derivatives are used in scintillation counters to detect such particles. Stilbene is also one of the gain mediums used in dye lasers.

Atmosphere

Fluorescence is observed in the atmosphere when the air is under energetic electron bombardment. In cases such as the natural aurora, high-altitude nuclear explosions, and rocket-borne electron gun experiments, the molecules and ions formed have a fluorescent response to light.

Common Materials that Fluoresce

- Vitamin B2 fluoresces yellow.

- Tonic water fluoresces blue due to the presence of quinine.

- Highlighter ink is often fluorescent due to the presence of pyranine.

- Banknotes, postage stamps and credit cards often have fluorescent security features.

Applications of Fluorescence

Lighting

Fluorescent paint and plastic lit by UV tubes. Paintings by Beo Beyond

The common fluorescent lamp relies on fluorescence. Inside the glass tube is a partial vacuum and a small amount of mercury. An electric discharge in the tube causes the mercury atoms to emit mostly ultraviolet light. The tube is lined with a coating of a fluorescent material, called the *phosphor*, which absorbs the ultraviolet and re-emits visible light. Fluorescent lighting is more energy-efficient than incandescent lighting elements. However, the uneven spectrum of traditional fluorescent lamps may cause certain colors to appear different than when illuminated by incandescent light or daylight. The mercury vapor emission spectrum is dominated by a short-wave UV line at 254 nm (which provides most of the energy to the phosphors), accompanied by visible light emission at 436 nm (blue), 546 nm (green) and 579 nm (yellow-orange). These three lines can be observed superimposed on the white continuum using a hand spectroscope, for light emitted by the usual white fluorescent tubes. These same visible lines, accompanied by the emission lines of trivalent europium and trivalent terbium, and further accompanied by the emission continuum of divalent europium in the blue region, comprise the more discontinuous light emission of the modern trichromatic phosphor systems used in many compact fluorescent lamp and traditional lamps where better color rendition is a goal.

Fluorescent lights were first available to the public at the 1939 New York World's Fair. Improvements since then have largely been better phosphors, longer life, and more consistent internal discharge, and easier-to-use shapes (such as compact fluorescent lamps). Some high-intensity discharge (HID) lamps couple their even-greater electrical efficiency with phosphor enhancement for better color rendition.

White light-emitting diodes (LEDs) became available in the mid-1990s as LED lamps, in which blue light emitted from the semiconductor strikes phosphors deposited on the tiny chip. The combination of the blue light that continues through the phosphor and the green to red fluorescence from the phosphors produces a net emission of white light.

Glow sticks sometimes utilize fluorescent materials to absorb light from the chemiluminescent reaction and emit light of a different color.

Analytical Chemistry

Many analytical procedures involve the use of a fluorometer, usually with a single exciting wavelength and single detection wavelength. Because of the sensitivity that the method affords, fluorescent molecule concentrations as low as 1 part per trillion can be measured.

Fluorescence in several wavelengths can be detected by an array detector, to detect compounds from HPLC flow. Also, TLC plates can be visualized if the compounds or a coloring reagent is fluorescent. Fluorescence is most effective when there is a larger ratio of atoms at lower energy levels in a Boltzmann distribution. There is, then, a higher probability of excitement and release of photons by lower-energy atoms, making analysis more efficient.

Spectroscopy

Usually the setup of a fluorescence assay involves a light source, which may emit many different wavelengths of light. In general, a single wavelength is required for proper analysis, so, in order to selectively filter the light, it is passed through an excitation monochromator, and then that chosen wavelength is passed through the sample cell. After absorption and re-emission of the energy, many wavelengths may emerge due to Stokes shift and various electron transitions. To separate and analyze them, the fluorescent radiation is passed through an emission monochromator, and observed selectively by a detector.

Biochemistry and Medicine

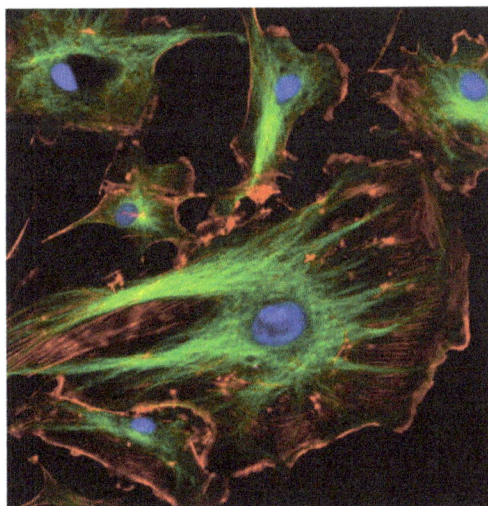

Endothelial cells under the microscope with three separate channels marking specific cellular components

Fluorescence in the life sciences is used generally as a non-destructive way of tracking or analysis of biological molecules by means of the fluorescent emission at a specific frequency where there is no background from the excitation light, as relatively few cellular components are naturally fluorescent (called intrinsic or autofluorescence). In fact, a protein or other component can be "labelled" with an extrinsic fluorophore, a fluorescent dye that can be a small molecule, protein, or quantum dot, finding a large use in many biological applications.

The quantification of a dye is done with a spectrofluorometer and finds additional applications in:

Microscopy

- When scanning the fluorescent intensity across a plane one has fluorescence microscopy of tissues, cells, or subcellular structures, which is accomplished by labeling an antibody with a fluorophore and allowing the antibody to find its target antigen within the sample. Labelling multiple antibodies with different fluorophores allows visualization of multiple targets within a single image (multiple channels). DNA microarrays are a variant of this.

- Immunology: An antibody is first prepared by having a fluorescent chemical group attached, and the sites (e.g., on a microscopic specimen) where the antibody has bound can be seen, and even quantified, by the fluorescence.

- FLIM (Fluorescence Lifetime Imaging Microscopy) can be used to detect certain bio-molecular interactions that manifest themselves by influencing fluorescence lifetimes.

- Cell and molecular biology: detection of colocalization using fluorescence-labelled antibodies for selective detection of the antigens of interest using specialized software, such as CoLocalizer Pro.

Other Techniques

- FRET (fluorescence resonance energy transfer or Förster resonance energy transfer) is used to study protein interactions, detect specific nucleic acid sequences and used as biosensors, while fluorescence lifetime (FLIM) can give an additional layer of information.

- Biotechnology: biosensors using fluorescence are being studied as possible Fluorescent glucose biosensors.

- Automated sequencing of DNA by the chain termination method; each of four different chain terminating bases has its own specific fluorescent tag. As the labelled DNA molecules are separated, the fluorescent label is excited by a UV source, and the identity of the base terminating the molecule is identified by the wavelength of the emitted light.

- FACS (fluorescence-activated cell sorting). One of several important cell sorting techniques used in the separation of different cell lines (especially those isolated from animal tissues).

- DNA detection: the compound ethidium bromide, in aqueous solution, has very little fluorescence, as it is quenched by water. Ethidium bromide's fluorescence is greatly enhanced after it binds to DNA, so this compound is very useful in visualising the location of DNA fragments in agarose gel electrophoresis. Intercalated ethidium is in a hydrophobic environment when it is between the base pairs of the DNA, protected from quenching by water which is excluded from the local environment of the intercalated ethidium. Ethidium bromide may be carcinogenic – an arguably safer alternative is the dye SYBR Green.

- FIGS (Fluorescence image-guided surgery) is a medical imaging technique that uses fluorescence to detect properly labeled structures during surgery.

- Intravascular fluorescence is a catheter-based medical imaging technique that uses fluorescence to detect high-risk features features of atherosclerosis and unhealed vascular stent devices. Plaque autofluorescence has been used in a first-in-man study in coronary arteries in combination with optical coherence tomography. Molecular agents has been also used to detect specific features, such as stent fibrin accumulation and enzymatic activity related to artery inflammation.

- SAFI (species altered fluorescence imaging) an imaging technique in electrokinetics and microfluidics. It uses non-electromigrating dyes whose fluorescence is easily quenched by migrating chemical species of interest. The dye(s) are usually seeded everywhere in the flow and differential quenching of their fluorescence by analytes is directly observed.

- Fluorescence-based assays for screening toxic chemicals. The optical assays consist of a mixture of environmental-sensitive fluorescent dyes and human skin cells that generate fluorescence spectra patterns. This approach can reduce the need for laboratory animals in biomedical research and pharmaceutical industry.

Forensics

Fingerprints can be visualized with fluorescent compounds such as ninhydrin. Blood and other substances are sometimes detected by fluorescent reagents, like fluorescein. Fibers, and other materials that may be encountered in forensics or with a relationship to various collectibles, are sometimes fluorescent.

Mechanical Engineering

Fluorescent penetrant inspection is used to find cracks and other defects on the surface of a part. Dye tracing, using fluorescent dyes, is used to find leaks in liquid and gas plumbing systems.

Signage

Fluorescent colors are frequently used in signage, particularly road signs. Fluorescent colors are generally recognizable at longer ranges than their non-fluorescent counterparts, with fluorescent orange being particularly noticeable. This property has led to its frequent use in safety signs and labels.

Optical Brighteners

Fluorescent compounds are often used to enhance the appearance of fabric and paper, causing a "whitening" effect. A white surface treated with an optical brightener can emit more visible light than that which shines on it, making it appear brighter. The blue light emitted by the brightener compensates for the diminishing blue of the treated material and changes the hue away from yellow or brown and toward white. Optical brighteners are used in laundry detergents, high brightness paper, cosmetics, high-visibility clothing and more.

Photoelectric Effect

The photoelectric effect or *photoemission* (given by *Albert Einstein*) is the production of electrons or other free carriers when light is shone onto a material. Electrons emitted in this manner can be called *photoelectrons*. The phenomenon is commonly studied in electronic physics, as well as in fields of chemistry, such as quantum chemistry or electrochemistry.

According to classical electromagnetic theory, this effect can be attributed to the transfer of energy from the light to an electron. From this perspective, an alteration in the intensity of light would induce changes in the rate of emission of electrons from the metal. Furthermore, according to this theory, a sufficiently dim light would be expected to show a time lag between the initial shining of its light and the subsequent emission of an electron. However, the experimental results did not correlate with either of the two predictions made by classical theory.

Instead, electrons are dislodged only by the impingement of photons when those photons reach or exceed a threshold frequency (energy). Below that threshold, no electrons are emitted from the metal regardless of the light intensity or the length of time of exposure to the light. To make sense of the fact that light can eject electrons even if its intensity is low, Albert Einstein proposed that a beam of light is not a wave propagating through space, but rather a collection of discrete wave packets (photons), each with energy hf. This shed light on Max Planck's previous discovery of the Planck relation ($E = hf$) linking energy (E) and frequency (f) as arising from quantization of energy. The factor h is known as the Planck constant.

In 1887, Heinrich Hertz discovered that electrodes illuminated with ultraviolet light create electric sparks more easily. In 1905 Albert Einstein published a paper that explained experimental data from the photoelectric effect as the result of light energy being carried in discrete quantized packets. This discovery led to the quantum revolution. In 1914, Robert Millikan's experiment confirmed Einstein's law on photoelectric effect. Einstein was awarded the Nobel Prize in 1921 for "his discovery of the law of the photoelectric effect", and Millikan was awarded the Nobel Prize in 1923 for "his work on the elementary charge of electricity and on the photoelectric effect".

The photoelectric effect requires photons with energies approaching zero (in the case of negative electron affinity) to over 1 MeV for core electrons in elements with a high atomic number. Emission of conduction electrons from typical metals usually requires a few electron-volts, corresponding to short-wavelength visible or ultraviolet light. Study of the photoelectric effect led to important steps in understanding the quantum nature of light and electrons and influenced the formation of the concept of wave–particle duality. Other phenomena where light affects the movement of electric charges include the photoconductive effect (also known as photoconductivity or photoresistivity), the photovoltaic effect, and the photoelectrochemical effect.

Photoemission can occur from any material, but it is most easily observable from metals or other conductors because the process produces a charge imbalance, and if this charge imbalance is not neutralized by current flow (enabled by conductivity), the potential barrier to emission increases until the emission current ceases. It is also usual to have the emitting surface in a vacuum, since gases impede the flow of photoelectrons and make them difficult to observe. Additionally, the energy barrier to photoemission is usually increased by thin oxide layers on metal surfaces if the metal has been exposed to oxygen, so most practical experiments and devices based on the photoelectric effect use clean metal surfaces in a vacuum.

When the photoelectron is emitted into a solid rather than into a vacuum, the term *internal photoemission* is often used, and emission into a vacuum distinguished as *external photoemission*.

Emission Mechanism

The photons of a light beam have a characteristic energy proportional to the frequency of the light. In the photoemission process, if an electron within some material absorbs the energy of one photon and acquires more energy than the work function (the electron binding energy) of the material, it is ejected. If the photon energy is too low, the electron is unable to escape the material. Since an increase in the intensity of low-frequency light will only increase the number of low-energy photons sent over a given interval of time, this change in intensity will not create any single photon with enough energy to dislodge an electron. Thus, the energy of the emitted electrons does not depend on the intensity of the incoming light, but only on the energy (equivalently frequency) of the individual photons. It is an interaction between the incident photon and the outermost electrons.

Electrons can absorb energy from photons when irradiated, but they usually follow an "all or nothing" principle. All of the energy from one photon must be absorbed and used to liberate one electron from atomic binding, or else the energy is re-emitted. If the photon energy is absorbed, some of the energy liberates the electron from the atom, and the rest contributes to the electron's kinetic energy as a free particle.

Experimental Observations of Photoelectric Emission

The theory of the photoelectric effect must explain the experimental observations of the emission of electrons from an illuminated metal surface.

For a given metal, there exists a certain minimum frequency of incident radiation below which no photoelectrons are emitted. This frequency is called the *threshold frequency*. Increasing the frequency of the incident beam, keeping the number of incident photons fixed (this would result in a proportionate increase in energy) increases the maximum kinetic energy of the photoelectrons emitted. Thus the stopping voltage increases. The number of electrons also changes because the probability that each photon results in an emitted electron is a function of photon energy. If the intensity of the incident radiation of a given frequency is increased, there is no effect on the kinetic energy of each photoelectron.

Above the threshold frequency, the maximum kinetic energy of the emitted photoelectron depends on the frequency of the incident light, but is independent of the intensity of the incident light so long as the latter is not too high.

For a given metal and frequency of incident radiation, the rate at which photoelectrons are ejected is directly proportional to the intensity of the incident light. An increase in the intensity of the incident beam (keeping the frequency fixed) increases the magnitude of the photoelectric current, although the stopping voltage remains the same.

The time lag between the incidence of radiation and the emission of a photoelectron is very small, less than 10^{-9} second.

The direction of distribution of emitted electrons peaks in the direction of polarization (the direction of the electric field) of the incident light, if it is linearly polarized.

Mathematical Description

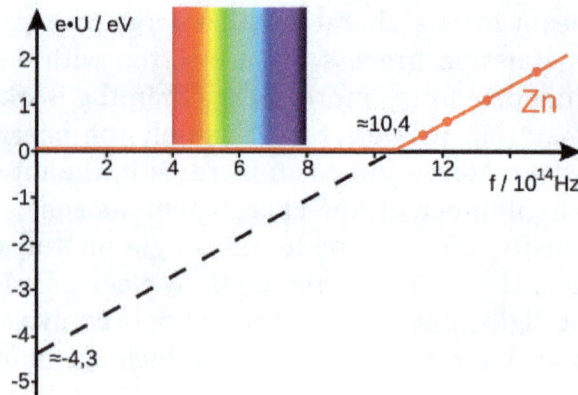

Diagram of the maximum kinetic energy as a function of the frequency of light on zinc

The maximum kinetic energy K_{max} of an ejected electron is given by

$$K_{max} = hf - \varphi,$$

where h is the Planck constant and f is the frequency of the incident photon. The term φ is the work function (sometimes denoted W, or ϕ), which gives the minimum energy required to remove a delocalised electron from the surface of the metal. The work function satisfies

$$\varphi = hf_0,$$

where f_0 is the threshold frequency for the metal. The maximum kinetic energy of an ejected electron is then

$$K_{max} = h(f - f_0).$$

Kinetic energy is positive, so we must have $f > f_0$ for the photoelectric effect to occur.

Stopping Potential

The relation between current and applied voltage illustrates the nature of the photoelectric effect. For discussion, a light source illuminates a plate P, and another plate electrode Q collects any emitted electrons. We vary the potential between P and Q and measure the current flowing in the external circuit between the two plates.

$$E = \frac{hc}{\lambda}$$

Work function = E.

Cut-off wavelength = λ

Work function and cut off frequency

If the frequency and the intensity of the incident radiation are fixed, the photoelectric current increases gradually with an increase in the positive potential on the collector electrode until all the photoelectrons emitted are collected. The photoelectric current attains a saturation value and does not increase further for any increase in the positive potential. The saturation current increases with the increase of the light intensity. It also increases with greater frequencies due to a greater probability of electron emission when collisions happen with higher energy photons.

If we apply a negative potential to the collector plate Q with respect to the plate P and gradually increase it, the photoelectric current decreases, becoming zero at a certain negative potential. The negative potential on the collector at which the photoelectric current becomes zero is called the *stopping potential* or *cut off* potential

i. For a given frequency of incident radiation, the stopping potential is independent of its intensity.

ii. For a given frequency of incident radiation, the stopping potential is determined by the maximum kinetic energy K_{max} of the photoelectrons that are emitted. If q_e is the charge on the electron and V_0 is the stopping potential, then the work done by the retarding potential in stopping the electron is $q_e V_0$, so we have

$$q_e V_0 = K_{max}.$$

Recalling

$$K_{max} = h(f - f_0),$$

we see that the stopping voltage varies linearly with frequency of light, but depends on the type of material. For any particular material, there is a threshold frequency that must be exceeded, independent of light intensity, to observe any electron emission.

Three-step Model

In the X-ray regime, the photoelectric effect in crystalline material is often decomposed into three steps:

1. Inner photoelectric effect (see photodiode below). The hole left behind can give rise to Auger effect, which is visible even when the electron does not leave the material. In molecular solids phonons are excited in this step and may be visible as lines in the final electron energy. The inner photoeffect has to be dipole allowed. The transition rules for atoms translate via the tight-binding model onto the crystal. They are similar in geometry to plasma oscillations in that they have to be transversal.

2. Ballistic transport of half of the electrons to the surface. Some electrons are scattered.

3. Electrons escape from the material at the surface.

In the three-step model, an electron can take multiple paths through these three steps. All paths can interfere in the sense of the path integral formulation. For surface states and molecules the three-step model does still make some sense as even most atoms have multiple electrons which can scatter the one electron leaving.

History

When a surface is exposed to electromagnetic radiation above a certain threshold frequency (typically visible light for alkali metals, near ultraviolet for other metals, and extreme ultraviolet for non-metals), the radiation is absorbed and electrons are emitted. Light, and especially ultra-violet light, discharges negatively electrified bodies with the production of rays of the same nature as cathode rays. Under certain circumstances it can directly ionize gases. The first of these phenomena was discovered by Hertz and Hallwachs in 1887. The second was announced first by Philipp Lenard in 1900.

The ultra-violet light to produce these effects may be obtained from an arc lamp, or by burning magnesium, or by sparking with an induction coil between zinc or cadmium terminals, the light from which is very rich in ultra-violet rays. Sunlight is not rich in ultra-violet rays, as these have been absorbed by the atmosphere, and it does not produce nearly so large an effect as the arc-light. Many substances besides metals discharge negative electricity under the action of ultraviolet light: lists of these substances will be found in papers by G. C. Schmidt and O. Knoblauch.

19th Century

In 1839, Alexandre Edmond Becquerel discovered the photovoltaic effect while studying the effect of light on electrolytic cells. Though not equivalent to the photoelectric effect, his work on photovoltaics was instrumental in showing a strong relationship between light and electronic properties of materials. In 1873, Willoughby Smith discovered photoconductivity in selenium while testing the metal for its high resistance properties in conjunction with his work involving submarine telegraph cables.

Johann Elster (1854–1920) and Hans Geitel (1855–1923), students in Heidelberg, developed the first practical photoelectric cells that could be used to measure the intensity of light. Elster and Geitel had investigated with great success the effects produced by light on electrified bodies.

Heinrich Rudolf Hertz

In 1887, Heinrich Hertz observed the photoelectric effect and the production and reception of electromagnetic waves. He published these observations in the journal Annalen der Physik. His receiver consisted of a coil with a spark gap, where a spark would be seen upon detection of electromagnetic waves. He placed the apparatus in a darkened box to see the spark better. However, he noticed that the maximum spark length was reduced when in the box. A glass panel placed between the source of electromagnetic waves and the receiver absorbed ultraviolet radiation that assisted the electrons in jumping across the gap. When removed, the spark length would increase. He observed no decrease in spark length when he replaced glass with quartz, as quartz does not absorb UV radiation. Hertz concluded his months of investigation and reported the results obtained. He did not further pursue investigation of this effect.

The discovery by Hertz in 1887 that the incidence of ultra-violet light on a spark gap facilitated the passage of the spark, led immediately to a series of investigations by Hallwachs, Hoor, Righi and Stoletow on the effect of light, and especially of ultra-violet light, on charged bodies. It was proved by these investigations that a newly cleaned surface of zinc, if charged with negative electricity, rapidly loses this charge however small it may be when ultra-violet light falls upon the surface; while if the surface is uncharged to begin with, it acquires a positive charge when exposed to the light, the negative electrification going out into the gas by which the metal is surrounded; this positive electrification can be much increased by directing a strong airblast against the surface. If however the zinc surface is positively electrified it suffers no loss of charge when exposed to the light: this result has been questioned, but a very careful examination of the phenomenon by Elster and Geitel has shown that the loss observed under certain circumstances is due to the discharge by the light reflected from the zinc surface of negative electrification on neighbouring conductors induced by the positive charge, the negative electricity under the influence of the electric field moving up to the positively electrified surface.

With regard to the *Hertz effect*, the researches from the start showed a great complexity of the phenomenon of photoelectric fatigue — that is, the progressive diminution of the effect observed upon fresh metallic surfaces. According to an important research by Wilhelm Hallwachs, ozone played an important part in the phenomenon. However, other elements enter such as oxidation, the humidity, the mode of polish of the surface, etc. It was at the time not even sure that the fatigue is absent in a vacuum.

In the period from February 1888 and until 1891, a detailed analysis of photoeffect was performed by Aleksandr Stoletov with results published in 6 works; four of them in *Comptes Rendus*, one review in *Physikalische Revue* (translated from Russian), and the last work in *Journal de Physique*. First, in these works Stoletov invented a new experimental setup which was more suitable for a quantitative analysis of photoeffect. Using this setup, he discovered the direct proportionality between the intensity of light and the induced photo electric current (the first law of photoeffect or Stoletov's law). One of his other findings resulted from measurements of the dependence of the intensity of the electric photo current on the gas pressure, where he found the existence of an optimal gas pressure P_m corresponding to a maximum photocurrent; this property was used for a creation of solar cells.

In 1899, J. J. Thomson investigated ultraviolet light in Crookes tubes. Thomson deduced that the ejected particles were the same as those previously found in the cathode ray, later called electrons, which he called "corpuscles". In the research, Thomson enclosed a metal plate (a cathode) in a vacuum tube, and exposed it to high frequency radiation. It was thought that the oscillating electromagnetic fields caused the atoms' field to resonate and, after reaching a certain amplitude, caused a subatomic "corpuscle" to be emitted, and current to be detected. The amount of this current varied with the intensity and colour of the radiation. Larger radiation intensity or frequency would produce more current.

20th Century

The discovery of the ionization of gases by ultra-violet light was made by Philipp Lenard in 1900. As the effect was produced across several centimeters of air and made very great positive and small negative ions, it was natural to interpret the phenomenon, as did J. J. Thomson, as a *Hertz effect* upon the solid or liquid particles present in the gas.

German physicist Philipp Lenard

In 1902, Lenard observed that the energy of individual emitted electrons increased with the frequency (which is related to the color) of the light.

This appeared to be at odds with Maxwell's wave theory of light, which predicted that the electron energy would be proportional to the intensity of the radiation.

Lenard observed the variation in electron energy with light frequency using a powerful electric arc lamp which enabled him to investigate large changes in intensity, and that had sufficient power to enable him to investigate the variation of potential with light frequency. His experiment directly measured potentials, not electron kinetic energy: he found the electron energy by relating it to the maximum stopping potential (voltage) in a phototube. He found that the calculated maximum electron kinetic energy is determined by the frequency of the light. For example, an increase in frequency results in an increase in the maximum kinetic energy calculated for an electron upon liberation – ultraviolet radiation would require a higher applied stopping potential to stop current in a phototube than blue light. However Lenard's results were qualitative rather than quantitative because of the difficulty in performing the experiments: the experiments needed to be done on freshly cut metal so that the pure metal was observed, but it oxidised in a matter of minutes even in the partial vacuums he used. The current emitted by the surface was determined by the light's intensity, or brightness: doubling the intensity of the light doubled the number of electrons emitted from the surface.

The researches of Langevin and those of Eugene Bloch have shown that the greater part of the Lenard effect is certainly due to this 'Hertz effect'. The Lenard effect upon the gas itself nevertheless does exist. Refound by J. J. Thomson and then more decisively by Frederic Palmer, Jr., it was studied and showed very different characteristics than those at first attributed to it by Lenard.

Einstein, in 1905, when he wrote the *Annus Mirabilis* papers

In 1905, Albert Einstein solved this apparent paradox by describing light as composed of discrete quanta, now called photons, rather than continuous waves. Based upon Max Planck's theory of black-body radiation, Einstein theorized that the energy in each quantum of light was equal to the frequency multiplied by a constant, later called Planck's constant. A photon above a threshold frequency has the required energy to eject a single electron, creating the observed effect. This discovery led to the quantum revolution in physics and earned Einstein the Nobel Prize in Physics in 1921. By wave-particle duality the effect can be analyzed purely in terms of waves though not as conveniently.

Albert Einstein's mathematical description of how the photoelectric effect was caused by absorption of quanta of light was in one of his 1905 papers, named "*On a Heuristic Viewpoint Concerning the Production and Transformation of Light*". This paper proposed the simple description of "light quanta", or photons, and showed how they explained such phenomena as the photoelectric effect. His simple explanation in terms of absorption of discrete quanta of light explained the features of the phenomenon and the characteristic frequency.

The idea of light quanta began with Max Planck's published law of black-body radiation ("*On the Law of Distribution of Energy in the Normal Spectrum*") by assuming that Hertzian oscillators could only exist at energies E proportional to the frequency f of the oscillator by $E = hf$, where h is Planck's constant. By assuming that light actually consisted of discrete energy packets, Einstein wrote an equation for the photoelectric effect that agreed with experimental results. It explained why the energy of photoelectrons was dependent only on the *frequency* of the incident light and not on its *intensity*: a low-intensity, high-frequency source could supply a few high energy photons, whereas a high-intensity, low-frequency source would supply no photons of sufficient individual energy to dislodge any electrons. This was an enormous theoretical leap, but the concept was strongly resisted at first because it contradicted the wave theory of light that followed naturally from James Clerk Maxwell's equations for electromagnetic behavior, and more generally, the assumption of infinite divisibility of energy in physical systems. Even after experiments showed that Einstein's equations for the photoelectric effect were accurate, resistance to the idea of photons continued, since it appeared to contradict Maxwell's equations, which were well-understood and verified.

Robert Millikan (picture around 1923), who first experimentally
showed Einstein's prediction about the photoelectric effect was correct.

Einstein's work predicted that the energy of individual ejected electrons increases linearly with the frequency of the light. Perhaps surprisingly, the precise relationship had not at that time been tested. By 1905 it was known that the energy of photoelectrons increases with increasing *frequency* of incident light and is independent of the *intensity* of the light. However, the manner of the increase was not experimentally determined until 1914 when Robert Andrews Millikan showed that Einstein's prediction was correct.

The photoelectric effect helped to propel the then-emerging concept of wave–particle duality in the nature of light. Light simultaneously possesses the characteristics of both waves and particles, each being manifested according to the circumstances. The effect was impossible to understand in terms of the classical wave description of light, as the energy of the emitted electrons did not depend on the intensity of the incident radiation. Classical theory predicted that the electrons would 'gather up' energy over a period of time, and then be emitted.

Uses and Effects

Photomultipliers

These are extremely light-sensitive vacuum tubes with a photocathode coated onto part (an end or side) of the inside of the envelope. The photocathode contains combinations of materials such as caesium, rubidium and antimony specially selected to provide a low work function, so when illuminated even by very low levels of light, the photocathode readily releases electrons. By means of a series of electrodes (dynodes) at ever-higher potentials, these electrons are accelerated and substantially increased in number through secondary emission to provide a readily detectable output current. Photomultipliers are still commonly used wherever low levels of light must be detected.

Image Sensors

Video camera tubes in the early days of television used the photoelectric effect, for example, Philo Farnsworth's "Image dissector" used a screen charged by the photoelectric effect to transform an optical image into a scanned electronic signal.

Gold-leaf Electroscope

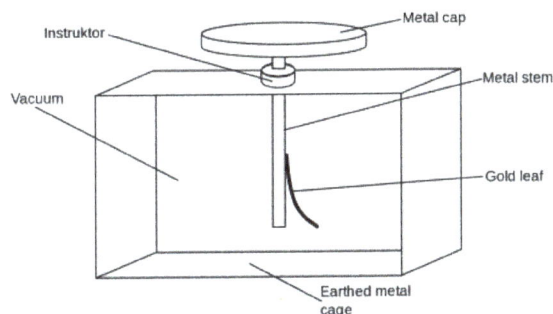

The gold leaf electroscope

Gold-leaf electroscopes are designed to detect static electricity. Charge placed on the metal cap spreads to the stem and the gold leaf of the electroscope. Because they then have the same charge, the stem and leaf repel each other. This will cause the leaf to bend away from the stem.

The electroscope is an important tool in illustrating the photoelectric effect. For example, if the electroscope is negatively charged throughout, there is an excess of electrons and the leaf is separated from the stem. If high-frequency light shines on the cap, the electroscope discharges and the leaf will fall limp. This is because the frequency of the light shining on the cap is above the cap's threshold frequency. The photons in the light have enough energy to liberate electrons from the cap, reducing its negative charge. This will discharge a negatively charged electroscope and further charge a positive electroscope. However, if the electromagnetic radiation hitting the metal cap does not have a high enough frequency (its frequency is below the threshold value for the cap), then the leaf will never discharge, no matter how long one shines the low-frequency light at the cap.

Photoelectron Spectroscopy

Since the energy of the photoelectrons emitted is exactly the energy of the incident photon minus the material's work function or binding energy, the work function of a sample can be determined by bombarding it with a monochromatic X-ray source or UV source, and measuring the kinetic energy distribution of the electrons emitted.

Photoelectron spectroscopy is usually done in a high-vacuum environment, since the electrons would be scattered by gas molecules if they were present. However, some companies are now selling products that allow photoemission in air. The light source can be a laser, a discharge tube, or a synchrotron radiation source.

The concentric hemispherical analyser (CHA) is a typical electron energy analyzer, and uses an electric field to change the directions of incident electrons, depending on their kinetic energies. For every element and core (atomic orbital) there will be a different binding energy. The many electrons created from each of these combinations will show up as spikes in the analyzer output, and these can be used to determine the elemental composition of the sample.

Spacecraft

The photoelectric effect will cause spacecraft exposed to sunlight to develop a positive charge. This can be a major problem, as other parts of the spacecraft in shadow develop a negative charge from nearby plasma, and the imbalance can discharge through delicate electrical components. The static charge created by the photoelectric effect is self-limiting, though, because a more highly charged object gives up its electrons less easily.

Moon dust

Light from the sun hitting lunar dust causes it to become charged through the photoelectric effect. The charged dust then repels itself and lifts off the surface of the Moon by electrostatic levitation. This manifests itself almost like an "atmosphere of dust", visible as a thin haze and blurring of distant features, and visible as a dim glow after the sun has set. This was first photographed by the Surveyor program probes in the 1960s. It is thought that the smallest particles are repelled up to kilometers high, and that the particles move in "fountains" as they charge and discharge.

Night Vision Devices

Photons hitting a thin film of alkali metal or semiconductor material such as gallium arsenide in an image intensifier tube cause the ejection of photoelectrons due to the photoelectric effect. These are accelerated by an electrostatic field where they strike a phosphor coated screen, converting the electrons back into photons. Intensification of the signal is achieved either through acceleration of the electrons or by increasing the number of electrons through secondary emissions, such as with a micro-channel plate. Sometimes a combination of both methods is used. Additional kinetic energy is required to move an electron out of the conduction band and into the vacuum level. This is known as the electron affinity of the photocathode and is another barrier to photoemission other than the forbidden band, explained by the band gap model. Some materials such as Gallium Arsenide have an effective electron affinity that is below the level of the conduction band. In these materials, electrons that move to the conduction band are all of sufficient energy to be emitted from the material and as such, the film that absorbs photons can be quite thick. These materials are known as negative electron affinity materials.

Cross Section

The photoelectric effect is one interaction mechanism between photons and atoms. It is one of 12 theoretically possible interactions.

At the high photon energies comparable to the electron rest energy of 511 keV, Compton scattering, another process, may take place. Above twice this (1.022 MeV) pair production may take place. Compton scattering and pair production are examples of two other competing mechanisms.

Indeed, even if the photoelectric effect is the favoured reaction for a particular single-photon bound-electron interaction, the result is also subject to statistical processes and is not guaranteed, albeit the photon has certainly disappeared and a bound electron has been excited (usually K or L shell electrons at gamma ray energies). The probability of the photoelectric effect occurring is measured by the cross section of interaction, σ. This has been found to be a function of the atomic number of the target atom and photon energy. A crude approximation, for photon energies above the highest atomic binding energy, is given by:

$$\sigma = \text{constant} \cdot \frac{Z^n}{E^3}$$

Here Z is atomic number and n is a number which varies between 4 and 5. (At lower photon energies a characteristic structure with edges appears, K edge, L edges, M edges, etc.) The obvious interpretation follows that the photoelectric effect rapidly decreases in significance, in the gamma ray region of the spectrum, with increasing photon energy, and that photoelectric effect increases steeply with atomic number. The corollary is that high-Z materials make good gamma-ray shields, which is the principal reason that lead ($Z = 82$) is a preferred and ubiquitous gamma radiation shield.

Photoionization Mode

A photoionization mode is a mode of interaction between a laser beam and matter involving photoionization.

General Considerations

Laser light affects materials of all types through fundamental processes such as excitation, ionization, and dissociation of atoms and molecules. These processes depend on the properties of the light, as well as on the properties of the material. Using lasers for material processing requires understanding and being able to control these fundamental effects. A better understanding can be achieved by defining distinct interaction regimes, hence the definition of four photoionization modes.

This new way of looking at the laser interaction with matter was first proposed by Tiberius Brastaviceanu in 2006, after his description of the "filamentary ionization mode" (Sherbrooke University, 2005). In his Master's work he provided the empirical proof of the formation of filamentary distributions of solvated electrons in water, induced by high-power fs (femtosecond, one trillionth of a second) laser pulses in the self-focusing propagation regime, and described the theoretical context in which this phenomenon can be explained and controlled. Refer to main article on filament propagation.

Single-photon Photoionization Mode

The SP mode is obtained at small wavelengths (UV, X-ray), or high energy per photon, and at low intensity levels. The only photoionization process involved in this case is the single-photon ionization.

Optical Breakdown Photoionization Mode

The OB mode is observed when a material is subjected to powerful laser pulses. It manifests a power threshold in the range of MW for the majority of dielectric materials, which depends on the duration and on the wavelength of the laser pulse. Optical breakdown is related to the dielectric breakdown phenomenon which was studied and modeled successfully towards the end of the 1950s. One describes the effect as a strong local ionization of the medium, where the plasma reaches densities beyond the critical value (between 10^{20} and 10^{22} electrons/cm³). Once the plasma critical density is achieved, energy is very efficiently absorbed from the light pulse, and the local plasma temperature increases dramatically. An explosive Coulombian expansion follows, and forms a very powerful and damaging shockwave through the material that develops on ns timescale. In liquids one talks about cavitation bubbles. If the rate of plasma formation is relatively slow, in the nanosecond time regime (for nanosecond excitation laser pulses), energy is transferred from the plasma to the lattice, and thermal damages can occur. In the femtosecond time regime (for femtosecond excitation laser pulses) the plasma expansion happens on a timescale smaller than the rate of energy transfer to the lattice, and thermal damages are reduced or eliminated. This is the basis of cold laser machining using high-power sub-ps laser sources.

The optical breakdown is a very "violent" phenomenon and changes drastically the structure of the surrounding medium. To the naked eye, optical breakdown looks like a spark and if the event happens in air or some other fluid, it is even possible to hear a short noise (burst) caused by the explosive plasma expansion.

There are several photoionization processes involved in optical breakdown, which depend on the wavelength, local intensity, and pulse duration, as well as on the electronic structure of the material. First, we should mention that optical breakdown is only observed at very high intensities. For pulse durations greater than a few tens of fs avalanche ionization plays a role. The longer pulse duration, the greater the avalanche ionization's contribution. Multi-photon ionization processes are important in the fs time regime, and their role increases as the pulse duration decreases. The type of multi-photon ionization processes involved is also wavelength dependent.

The theory needed to understand the most important features of optical breakdown are:

- the physics of strong-(laser)field interaction with matter, to account for the plasma formation;

- the physics of strong-(laser)field interaction with plasma, to account for plasma expansion, and for thermal and mechanical effects;

- the geometrical/linear optical theory, to account at the first approximation for the spatial intensity distribution. Non-linear propagation theory is usually invoked to account for self-focusing that occurs in experiments conducted at low numerical aperture, and to account for detail features of the plasma density spatial distribution.

Below Optical Breakdown Threshold Photoionization Mode

B/OB mode is an intermediary between the optical breakdown mode (OB mode) and the filamentary mode (F mode). The plasma density generated in this mode can go from 0 to the critical value i.e. optical breakdown threshold. Intensities reached inside the B/OB zone can range from multi-photon ionization threshold to the optical breakdown threshold. In the visible-IR domain, B/OB mode is obtained under very tight external focusing (high numerical aperture), to avoid self-focusing, and for intensities below optical breakdown threshold. In the UV regime, where optical breakdown intensity threshold is below self-focusing intensity threshold, tight focusing is not necessary. The shape of the ionization area is similar to that of the focal area of the beam, and can be very small (only a few micrometres). B/OB mode is possible only at short pulse durations, where AI's contribution to the total free electron population is very small. As the pulse duration becomes even shorter, the intensity domain where B/OB is possible becomes even wider.

The principles governing this mode of ionization are very simple. Localized plasma must be generated in predictable fashion, under the optical breakdown threshold. Optical breakdown intensity threshold is strongly correlated to the input intensity only at short pulse durations. Therefore, one important requirement, in order to systematically avoid the optical breakdown, is to operate at short pulse durations. In order for the ionization to take place, multi-photon ionization (MPI) intensity threshold must be reached. The idea is to adjust the duration of the laser pulse so that multi-photon ionization, and perhaps to a lesser extent avalanche ionization, have no time to raise the plasma's density above the critical value. In the UV, the distinction between single-photon mode (SP) and B/OB is that for the latter multi-photon ionization, single-photon ionization, and perhaps to a lesser extent avalanche ionization, are operating, whereas for the former, only single-photon ionization is operating.

B/OB relies mostly on MPI processes. Therefore, it is more selective than OB in terms of which type of atom or molecule is ionized or dissociated. The theory needed to understand the most important features of B/OB are:

- The physics of strong-(laser)field interaction with matter, to account for the plasma formation. As opposed to the OB mode, in this case the role of avalanche ionization is greatly reduced, and the effects are dominated by multi-photon ionization processes.

- The geometrical/linear optical theory, to account at the first approximation for the spatial intensity distribution. Non-linear propagation theory is usually invoked to account for self-focusing that occurs in experiments conducted at low numerical aperture, and to account for detailed features of plasma spatial distribution.

The B/OB mode was described by A. Vogel et al. [ref 2].

Filamentary Photoionization Mode

In the F mode, filamentary or linear ionization patterns are formed. The plasma density within these filaments is below the critical value.

The self-focusing effect is responsible for the most important characteristics of the dose distribution. The diameter of these filamentary ionization traces is the same within 20% (in the order of a few micrometres). Their length, their number, and their relative position are controllable parameters. The plasma density and the yield of photolytic species are believed to be homogeneously distributed along these filaments. The local intensity reached by the laser light during propagation is also practically constant along their length. The power range of operation of the F mode is above self-focusing threshold and below optical breakdown threshold. Consequently, a necessary condition for it to exist is that the self-focusing threshold must be smaller than the optical breakdown threshold.

The F mode exhibits very important characteristics, which in combination with the other three photoionization modes makes possible the generation of a wide range of dose distributions, expanding the application range of lasers in the domain of material processing. The F mode is the only mode capable of generating linear ionization traces.

The theory needed to understand the most important features of the F mode are:

- The physics of high-(laser)field interaction with matter, to account for the plasma formation

- The theory of non-linear propagation, to account for the spatial redistribution of the laser light, intensity clamping, and the formation of filaments, as well as for frequency conversion processes.

The first concrete connection between non-linear optical effects, such as the supercontinuum generation, and photoionization was established by A. Brodeur and S.L. Chin [ref 4] in 1999, based on optical experimental data and modeling. In 2002 T. Brastaviceanu published the first direct measurement of the spatial distribution of photoionization induced in the self-focusing regime, in water.

Superposition of Photoionization Modes

It is possible to control the spatial distribution of the dose induced by laser pulses, and the relative yields of primary photolytic species, by controlling the properties of the laser beam. The dose distribution can be conveniently shaped by inducing a superposition of the four modes of photoionization. The mixed ionization modes are: SP-OB, SP-B/OB, and F-OB.

Photosensitizer

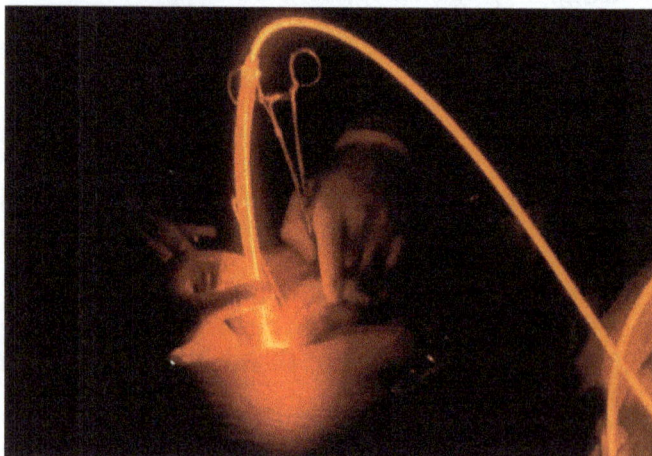

A photosensitizer being used in photodynamic therapy.

A photosensitizer is a molecule that produces a chemical change in another molecule in a photo-chemical process. Photosensitizers are commonly used in polymer chemistry in reactions such as photopolymerization, photocrosslinking, and photodegradation. Photosensitizers generally act by absorbing ultraviolet or visible region of electromagnetic radiation and transferring it to adjacent molecules. Photosensitizers usually have large de-located π systems, which lower the energy of HOMO orbitals and its absorption of light might be able to ionize the molecule.

Chlorophyll act as a photosensitizer in photosynthesis of carbohydrates in plants:

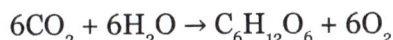

$$6CO_2 + 6H_2O \rightarrow C_6H_{12}O_6 + 6O_2$$

Applications

Medical

Photosensitisers are a key part of Photodynamic therapy (PDT) which is used to treat some cancers. They help to produce singlet oxygen to damage tumours. They can be divided into porphyrins, chlorophylls and dyes.

Absorption (Electromagnetic Radiation)

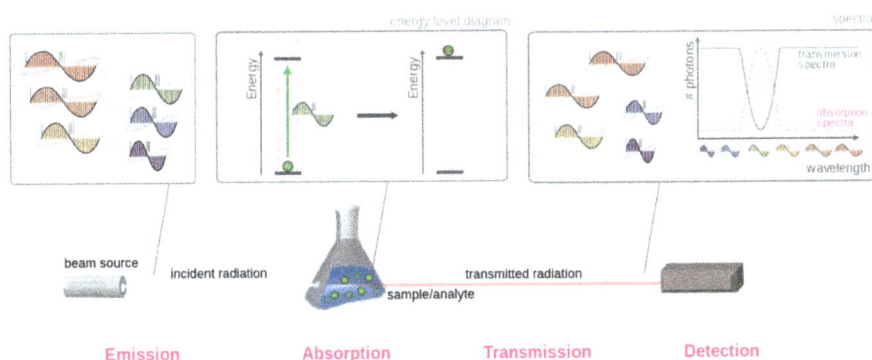

Emission   Absorption   Transmission   Detection

An overview of electromagnetic radiation absorption. This example discusses the general principle using visible light as specific example. A white light source — emitting light of multiple wavelengths — is focused on a sample (the pairs of complementary colors are indicated by the yellow dotted lines). Upon striking the sample, photons that match the energy gap of the molecules present (green light in this example) are *absorbed*, exciting the molecules. Other photons are transmitted unaffected and, if the radiation is in the visible region (400–700 nm), the transmitted light appears as the complementary color (here red). By recording the attenuation of light for various wavelengths, an absorption spectrum can be obtained.

In physics, absorption of electromagnetic radiation is the way in which the energy of a photon is taken up by matter, typically the electrons of an atom. Thus, the electromagnetic energy is transformed into internal energy of the absorber, for example thermal energy. The reduction in intensity of a light wave propagating through a medium by absorption of a part of its photons is often called attenuation. Usually, the absorption of waves does not depend on their intensity (linear absorption), although in certain conditions (usually, in optics), the medium changes its transparency dependently on the intensity of waves going through, and saturable absorption (or nonlinear absorption) occurs.

Quantifying Absorption

There are a number of ways to quantify how quickly and effectively radiation is absorbed in a certain medium, for example:

- The absorption coefficient, and some closely related derived quantities:

  o The attenuation coefficient, which is sometimes but not always synonymous with the absorption coefficient

  o Molar absorptivity, also called "molar extinction coefficient", which is the absorption coefficient divided by molarity.

  o The mass attenuation coefficient, also called "mass extinction coefficient", which is the absorption coefficient divided by density.

  o The absorption cross section and scattering cross-section are closely related to the absorption and attenuation coefficients, respectively.

  o "Extinction" in astronomy is equivalent to the attenuation coefficient.

- Penetration depth and skin effect,

- Propagation constant, attenuation constant, phase constant, and complex wavenumber,

- Complex refractive index and extinction coefficient,

- Complex dielectric constant,

- Electrical resistivity and conductivity.

- Absorbance (also called "optical density") and optical depth (also called "optical thickness") are two related measures

All these quantities measure, at least to some extent, how well a medium absorbs radiation. However, practitioners of different fields and techniques tend to conventionally use different quantities drawn from the list above.

Measuring Absorption

The absorbance of an object quantifies how much of the incident light is absorbed by it (instead of being reflected or refracted). This may be related to other properties of the object through the Beer–Lambert law.

Precise measurements of the absorbance at many wavelengths allow the identification of a substance via absorption spectroscopy, where a sample is illuminated from one side, and the intensity of the light that exits from the sample in every direction is measured. A few examples of absorption are ultraviolet–visible spectroscopy, infrared spectroscopy, and X-ray absorption spectroscopy.

Applications

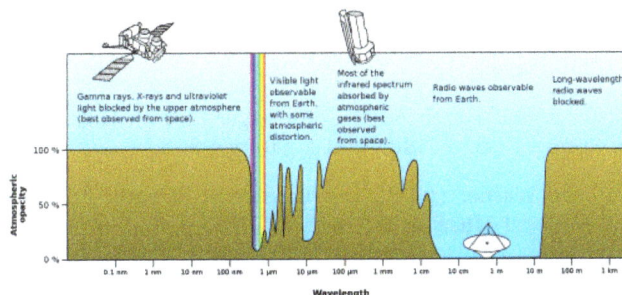

Rough plot of Earth's atmospheric transmittance (or opacity) to
various wavelengths of electromagnetic radiation, including visible light.

Understanding and measuring the absorption of electromagnetic radiation has a variety of applications. Here are a few examples:

- In meteorology and climatology, global and local temperatures depend in part on the absorption of radiation by atmospheric gases (such as in the greenhouse effect) and land and ocean surfaces.

- In medicine, X-rays are absorbed to different extents by different tissues (bone in particular), which is the basis for X-ray imaging. For example, see computation of radiowave attenuation in the atmosphere used in satellite link design.

- In chemistry and materials science, because different materials and molecules will absorb radiation to different extents at different frequencies, which allows for material identification.

- In optics, sunglasses, colored filters, dyes, and other such materials are designed specifically with respect to which visible wavelengths they absorb, and in what proportions.

- In biology, photosynthetic organisms require that light of the appropriate wavelengths be absorbed within the active area of chloroplasts, so that the light energy can be converted into chemical energy within sugars and other molecules.

References

- Gerischer, Heinz (1985). "Semiconductor electrodes and their interaction with light". In Schiavello, Mario. Photoelectrochemistry, Photocatalysis and Photoreactors Fundamentals and Developments. Springer. p. 39. ISBN 978-90-277-1946-1.

- Harris, D. C.; Bertolucci, M. D. (1978). Symmetry and Spectroscopy: An introduction to vibrational and electronic spectroscopy (Reprint ed.). Dover Publications. ISBN 0-486-66144-X.

- Holler, F. James; Skoog, Douglas A. and Crouch, Stanley R. (2006) Principles Of Instrumental Analysis. Cengage Learning. ISBN 0495012017

- Valeur, Bernard, Berberan-Santos, Mario (2012). Molecular Fluorescence: Principles and Applications. Wiley-VCH. ISBN 978-3-527-32837-6. p. 64

- Sears, F. W.; Zemansky, M. W.; Young, H. D. (1983). University Physics (6th ed.). Addison-Wesley. pp. 843–844. ISBN 0-201-07195-9.

- Mee, C.; Crundell, M.; Arnold, B.; Brown, W. (2011). International A/AS Level Physics. Hodder Education. p. 241. ISBN 978-0-340-94564-3.

- Fromhold, A. T. (1991). Quantum Mechanics for Applied Physics and Engineering. Courier Dover Publications. pp. 5–6. ISBN 978-0-486-66741-6.

- Vesselinka Petrova-Koch; Rudolf Hezel; Adolf Goetzberger (2009). High-Efficient Low-Cost Photovoltaics: Recent Developments. Springer. pp. 1–. ISBN 978-3-540-79358-8.

- Robert Bud; Deborah Jean Warner (1998). Instruments of Science: An Historical Encyclopedia. Science Museum, London, and National Museum of American History, Smithsonian Institution. ISBN 978-0-8153-1561-2.

- Thomson, J. J. (2005). Conduction of Electricity Through Gases. Watchmaker Publishing. ISBN 978-1-929148-49-3. Retrieved 9 July 2011.

- Buchwald, Jed; Warwick, Andrew, eds. (2004). Histories of the Electron: The Birth of Microphysics (PDF) (illustrated, reprint ed.). MIT Press. pp. 21–23. ISBN 978-0-262-52424-7.

5

Photosynthesis: An Integrated Study

The process used by plants to convert light into energy as a mechanism of feeding is known as photosynthesis. The major cause of the production of oxygen is photosynthesis. This section helps the readers in understanding all the topics related to photosynthesis.

Photosynthesis

Photosynthesis is a process used by plants and other organisms to convert light energy into chemical energy that can later be released to fuel the organisms' activities (energy transformation). This chemical energy is stored in carbohydrate molecules, such as sugars, which are synthesized from carbon dioxide and water – hence the name *photosynthesis*, from the "light", and *synthesis*, "putting together". In most cases, oxygen is also released as a waste product. Most plants, most algae, and cyanobacteria perform photosynthesis; such organisms are called photoautotrophs. Photosynthesis is largely responsible for producing and maintaining the oxygen content of the Earth's atmosphere, and supplies all of the organic compounds and most of the energy necessary for life on Earth.

Schematic of photosynthesis in plants. The carbohydrates produced are stored in or used by the plant.

Although photosynthesis is performed differently by different species, the process always begins when energy from light is absorbed by proteins called reaction centres that contain green chlorophyll pigments. In plants, these proteins are held inside organelles called chloroplasts, which are most abundant in leaf cells, while in bacteria they are embedded in the plasma membrane. In these light-dependent reactions, some energy is used to strip electrons from suitable substances, such as water, producing oxygen gas. The hydrogen freed by the splitting of water is used in the creation of two further compounds that act as an immediate energy storage means: reduced nicotinamide adenine dinucleotide phosphate (NADPH) and adenosine triphosphate (ATP), the "energy currency" of cells.

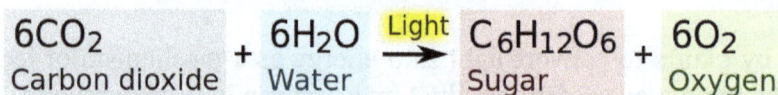

$$6CO_2 + 6H_2O \xrightarrow{\text{Light}} C_6H_{12}O_6 + 6O_2$$

Carbon dioxide   Water   Sugar   Oxygen

Overall equation for the type of photosynthesis that occurs in plants

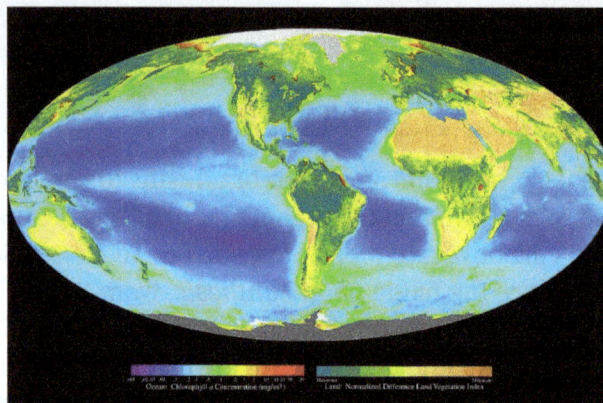

Composite image showing the global distribution of photosynthesis, including both oceanic phytoplankton and terrestrial vegetation. Dark red and blue-green indicate regions of high photosynthetic activity in the ocean and on land, respectively.

In plants, algae and cyanobacteria, long-term energy storage in the form of sugars is produced by a subsequent sequence of light-independent reactions called the Calvin cycle; some bacteria use different mechanisms, such as the reverse Krebs cycle, to achieve the same end. In the Calvin cycle, atmospheric carbon dioxide is incorporated into already existing organic carbon compounds, such as ribulose bisphosphate (RuBP). Using the ATP and NADPH produced by the light-dependent reactions, the resulting compounds are then reduced and removed to form further carbohydrates, such as glucose.

The first photosynthetic organisms probably evolved early in the evolutionary history of life and most likely used reducing agents such as hydrogen or hydrogen sulfide, rather than water, as sources of electrons. Cyanobacteria appeared later; the excess oxygen they produced contributed directly to the oxygenation of the Earth, which rendered the evolution of complex life possible. Today, the average rate of energy capture by photosynthesis globally is approximately 130 terawatts, which is about three times the current power consumption of human civilization. Photosynthetic organisms also convert around 100–115 thousand million metric tonnes of carbon into biomass per year.

Overview

Photosynthetic organisms are photoautotrophs, which means that they are able to synthesize food directly from carbon dioxide and water using energy from light. However, not all organisms that use light as a source of energy carry out photosynthesis; photoheterotrophs use organic compounds, rather than carbon dioxide, as a source of carbon. In plants, algae and cyanobacteria, photosynthesis releases oxygen. This is called *oxygenic photosynthesis* and is by far the most common type of photosynthesis used by living organisms. Although there are some differences between oxygenic photosynthesis in plants, algae, and cyanobacteria, the overall process is quite similar in these organisms. There are also many varieties of anoxygenic photosynthesis, used mostly by certain types of bacteria, which consume carbon dioxide but do not release oxygen.

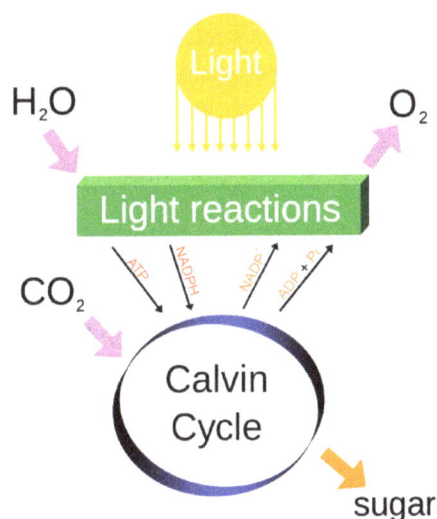

Photosynthesis changes sunlight into chemical energy, splits water to liberate O_2, and fixes CO_2 into sugar.

Carbon dioxide is converted into sugars in a process called carbon fixation. Carbon fixation is an endothermic redox reaction, so photosynthesis needs to supply both a source of energy to drive this process, and the electrons needed to convert carbon dioxide into a carbohydrate via a reduction reaction. The addition of electrons to a chemical species is called reduction. In general outline and in effect, photosynthesis is the opposite of cellular respiration, in which glucose and other compounds are oxidized to produce carbon dioxide and water, and to release chemical energy (an exothermic reaction) to drive the organism's metabolism. The two processes, reduction of carbon dioxide to carbohydrate and then later oxidation of the carbohydrate, are distinct: photosynthesis and cellular respiration take place through a different sequence of chemical reactions and in different cellular compartments.

The general equation for photosynthesis as first proposed by Cornelius van Niel is therefore:

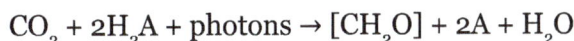

$$CO_2 + 2H_2A + photons \rightarrow [CH_2O] + 2A + H_2O$$

carbon dioxide + electron donor + light energy → carbohydrate + oxidized electron donor + water

Since water is used as the electron donor in oxygenic photosynthesis, the equation for this process is:

$$CO_2 + 2H_2O + photons \rightarrow [CH_2O] + O_2 + H_2O$$

carbon dioxide + water + light energy → carbohydrate + oxygen + water

This equation emphasizes that water is both a reactant in the light-dependent reaction and a product of the light-independent reaction, but canceling n water molecules from each side gives the net equation:

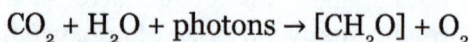

$$CO_2 + H_2O + photons \rightarrow [CH_2O] + O_2$$

carbon dioxide + water + light energy → carbohydrate + oxygen

Other processes substitute other compounds (such as arsenite) for water in the electron-supply role; for example some microbes use sunlight to oxidize arsenite to arsenate: The equation for this reaction is:

$$CO_2 + (AsO3-3) + photons \rightarrow (AsO3-4) + CO$$

carbon dioxide + arsenite + light energy → arsenate + carbon monoxide (used to build other compounds in subsequent reactions)

Photosynthesis occurs in two stages. In the first stage, *light-dependent reactions* or *light reactions* capture the energy of light and use it to make the energy-storage molecules ATP and NADPH. During the second stage, the *light-independent reactions* use these products to capture and reduce carbon dioxide.

Most organisms that utilize oxygenic photosynthesis use visible light for the light-dependent reactions, although at least three use shortwave infrared or, more specifically, far-red radiation.

Some organisms employ even more radical variants of photosynthesis. Some archea use a simpler method that employs a pigment similar to those used for vision in animals. The bacteriorhodopsin changes its configuration in response to sunlight, acting as a proton pump. This produces a proton gradient more directly, which is then converted to chemical energy. The process does not involve carbon dioxide fixation and does not release oxygen, and seems to have evolved separately from the more common types of photosynthesis.

Photosynthetic Membranes and Organelles

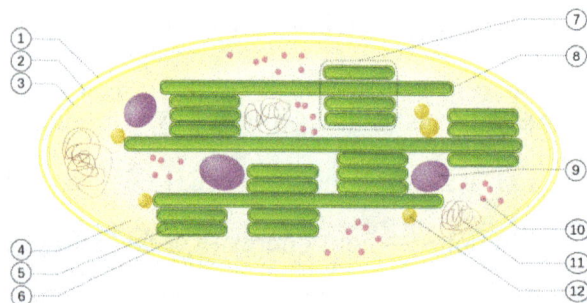

Chloroplast ultrastructure:
1. outer membrane
2. intermembrane space
3. inner membrane (1+2+3: envelope)
4. stroma (aqueous fluid)
5. thylakoid lumen (inside of thylakoid)
6. thylakoid membrane
7. granum (stack of thylakoids)
8. thylakoid (lamella)
9. starch
10. ribosome
11. plastidial DNA
12. plastoglobule (drop of lipids)

In photosynthetic bacteria, the proteins that gather light for photosynthesis are embedded in cell membranes. In its simplest form, this involves the membrane surrounding the cell itself. However, the membrane may be tightly folded into cylindrical sheets called thylakoids, or bunched up into round vesicles called *intracytoplasmic membranes*. These structures can fill most of the interior of a cell, giving the membrane a very large surface area and therefore increasing the amount of light that the bacteria can absorb.

In plants and algae, photosynthesis takes place in organelles called chloroplasts. A typical plant cell contains about 10 to 100 chloroplasts. The chloroplast is enclosed by a membrane. This membrane is composed of a phospholipid inner membrane, a phospholipid outer membrane, and an intermembrane space. Enclosed by the membrane is an aqueous fluid called the stroma. Embedded within the stroma are stacks of thylakoids (grana), which are the site of photosynthesis. The thylakoids appear as flattened disks. The thylakoid itself is enclosed by the thylakoid membrane, and within the enclosed volume is a lumen or thylakoid space. Embedded in the thylakoid membrane are integral and peripheral membrane protein complexes of the photosynthetic system.

Plants absorb light primarily using the pigment chlorophyll. The green part of the light spectrum is not absorbed but is reflected which is the reason that most plants have a green color. Besides chlorophyll, plants also use pigments such as carotenes and xanthophylls. Algae also use chlorophyll, but various other pigments are present, such as phycocyanin, carotenes, and xanthophylls in green algae, phycoerythrin in red algae (rhodophytes) and fucoxanthin in brown algae and diatoms resulting in a wide variety of colors.

These pigments are embedded in plants and algae in complexes called antenna proteins. In such proteins, the pigments are arranged to work together. Such a combination of proteins is also called a light-harvesting complex.

Although all cells in the green parts of a plant have chloroplasts, the majority of those are found in specially adapted structures called leaves. Certain species adapted to conditions of strong sunlight and aridity, such as many Euphorbia and cactus species, have their main photosynthetic organs in their stems. The cells in the interior tissues of a leaf, called the mesophyll, can contain between 450,000 and 800,000 chloroplasts for every square millimeter of leaf. The surface of the leaf is coated with a water-resistant waxy cuticle that protects the leaf from excessive evaporation of water and decreases the absorption of ultraviolet or blue light to reduce heating. The transparent epidermis layer allows light to pass through to the palisade mesophyll cells where most of the photosynthesis takes place.

Light-dependent Reactions

In the light-dependent reactions, one molecule of the pigment chlorophyll absorbs one photon and loses one electron. This electron is passed to a modified form of chlorophyll called pheophytin, which passes the electron to a quinone molecule, starting the flow of electrons down an electron transport chain that leads to the ultimate reduction of NADP to NADPH. In addition, this creates a proton gradient (energy gradient) across the chloroplast membrane, which is used by ATP synthase in the synthesis of ATP. The chlorophyll molecule ultimately regains the electron it lost when a water molecule is split in a process called photolysis, which releases a dioxygen (O_2) molecule as a waste product.

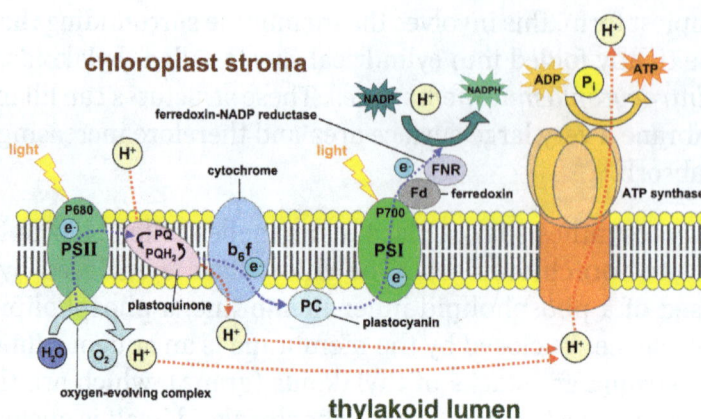

chloroplast stroma

thylakoid lumen

Light-dependent reactions of photosynthesis at the thylakoid membrane

The overall equation for the light-dependent reactions under the conditions of non-cyclic electron flow in green plants is:

$$2\ H_2O + 2\ NADP^+ + 3\ ADP + 3\ P_i + light \rightarrow 2\ NADPH + 2\ H^+ + 3\ ATP + O_2$$

Not all wavelengths of light can support photosynthesis. The photosynthetic action spectrum depends on the type of accessory pigments present. For example, in green plants, the action spectrum resembles the absorption spectrum for chlorophylls and carotenoids with peaks for violet-blue and red light. In red algae, the action spectrum is blue-green light, which allows these algae to use the blue end of the spectrum to grow in the deeper waters that filter out the longer wavelengths (red light) used by above ground green plants. The non-absorbed part of the light spectrum is what gives photosynthetic organisms their color (e.g., green plants, red algae, purple bacteria) and is the least effective for photosynthesis in the respective organisms.

Z scheme

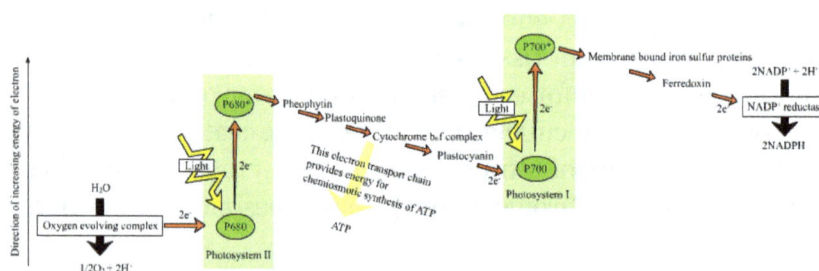

The "Z scheme"

In plants, light-dependent reactions occur in the thylakoid membranes of the chloroplasts where they drive the synthesis of ATP and NADPH. The light-dependent reactions are of two forms: cyclic and non-cyclic.

In the non-cyclic reaction, the photons are captured in the light-harvesting antenna complexes of photosystem II by chlorophyll and other accessory pigments. The absorption of a photon by the antenna complex frees an electron by a process called photoinduced charge separation. The antenna system is at the core of the chlorophyll molecule of the photosystem II reaction center. That freed electron is transferred to the primary electron-acceptor molecule, pheophytin. As the electrons are shuttled through an electron transport chain (the so-called *Z-scheme* shown in the diagram), it initially functions to generate a chemiosmotic potential by pumping proton cations (H^+) across the membrane and into the thylakoid space. An ATP synthase enzyme uses that chemiosmotic potential to make ATP during photophosphorylation, whereas NADPH is a product of the terminal redox reaction in the *Z-scheme*. The electron enters a chlorophyll molecule in Photosystem I. There it is further excited by the light absorbed by that photosystem. The electron is then passed along a chain of electron acceptors to which it transfers some of its energy. The energy delivered to the electron acceptors is used to move hydrogen ions across the thylakoid membrane into the lumen. The electron is eventually used to reduce the co-enzyme NADP with a H^+ to NADPH (which has functions in the light-independent reaction); at that point, the path of that electron ends.

The cyclic reaction is similar to that of the non-cyclic, but differs in that it generates only ATP, and no reduced NADP (NADPH) is created. The cyclic reaction takes place only at photosystem I. Once the electron is displaced from the photosystem, the electron is passed down the electron acceptor molecules and returns to photosystem I, from where it was emitted, hence the name *cyclic reaction*.

Water Photolysis

The NADPH is the main reducing agent produced by chloroplasts, which then goes on to provide a source of energetic electrons in other cellular reactions. Its production leaves chlorophyll in photosystem I with a deficit of electrons (chlorophyll has been oxidized), which must be balanced by some other reducing agent that will supply the missing electron. The excited electrons lost from chlorophyll from photosystem I are supplied from the electron transport chain by plastocyanin. However, since photosystem II is the first step of the *Z-scheme*, an external source of electrons is required to reduce its oxidized chlorophyll *a* molecules. The source of electrons in green-plant and cyanobacterial photosynthesis is water. Two water molecules are oxidized by four successive charge-separation reactions by photosystem II to yield a molecule of diatomic oxygen and four hydrogen ions; the electrons yielded are transferred to a redox-active tyrosine residue that then reduces the oxidized chlorophyll *a* (called P680) that serves as the primary light-driven electron donor in the photosystem II reaction center. That photo receptor is in effect reset and is then able to repeat the absorption of another photon and the release of another photo-dissociated electron. The oxidation of water is catalyzed in photosystem II by a redox-active structure that contains four manganese ions and a calcium ion; this oxygen-evolving complex binds two water molecules and contains the four oxidizing equivalents that are used to drive the water-oxidizing reaction. Photosystem II is the only known biological enzyme that carries out this oxidation of water. The hydrogen ions released contribute to the transmembrane chemiosmotic potential that leads to ATP synthesis. Oxygen is a waste product of light-dependent reactions, but the majority of organisms on Earth use oxygen for cellular respiration, including photosynthetic organisms.

Light-independent Reactions

Calvin Cycle

In the light-independent (or "dark") reactions, the enzyme RuBisCO captures CO_2 from the atmosphere and, in a process called the Calvin-Benson cycle, it uses the newly formed NADPH and releases three-carbon sugars, which are later combined to form sucrose and starch. The overall equation for the light-independent reactions in green plants is

$$3 \, CO_2 + 9 \, ATP + 6 \, NADPH + 6 \, H^+ \rightarrow C_3H_6O_3\text{-phosphate} + 9 \, ADP + 8 \, P_i + 6 \, NADP^+ + 3 \, H_2O$$

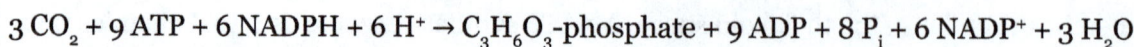

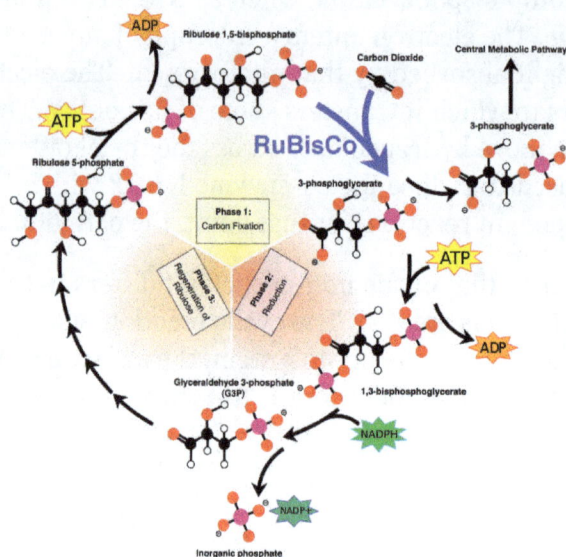

Overview of the Calvin cycle and carbon fixation

Carbon fixation produces the intermediate three-carbon sugar product, which is then converted to the final carbohydrate products. The simple carbon sugars produced by photosynthesis are then used in the forming of other organic compounds, such as the building material cellulose, the precursors for lipid and amino acid biosynthesis, or as a fuel in cellular respiration. The latter occurs not only in plants but also in animals when the energy from plants is passed through a food chain.

The fixation or reduction of carbon dioxide is a process in which carbon dioxide combines with a five-carbon sugar, ribulose 1,5-bisphosphate, to yield two molecules of a three-carbon compound, glycerate 3-phosphate, also known as 3-phosphoglycerate. Glycerate 3-phosphate, in the presence of ATP and NADPH produced during the light-dependent stages, is reduced to glyceraldehyde 3-phosphate. This product is also referred to as 3-phosphoglyceraldehyde (PGAL) or, more generically, as triose phosphate. Most (5 out of 6 molecules) of the glyceraldehyde 3-phosphate produced is used to regenerate ribulose 1,5-bisphosphate so the process can continue. The triose phosphates not thus "recycled" often condense to form hexose phosphates, which ultimately yield sucrose, starch and cellulose. The sugars produced during carbon metabolism yield carbon skeletons that can be used for other metabolic reactions like the production of amino acids and lipids.

Carbon Concentrating Mechanisms

On Land

In hot and dry conditions, plants close their stomata to prevent water loss. Under these conditions, CO_2 will decrease and oxygen gas, produced by the light reactions of photosynthesis, will increase, causing an increase of photorespiration by the oxygenase activity of ribulose-1,5-bisphosphate carboxylase/oxygenase and decrease in carbon fixation. Some plants have evolved mechanisms to increase the CO_2 concentration in the leaves under these conditions.

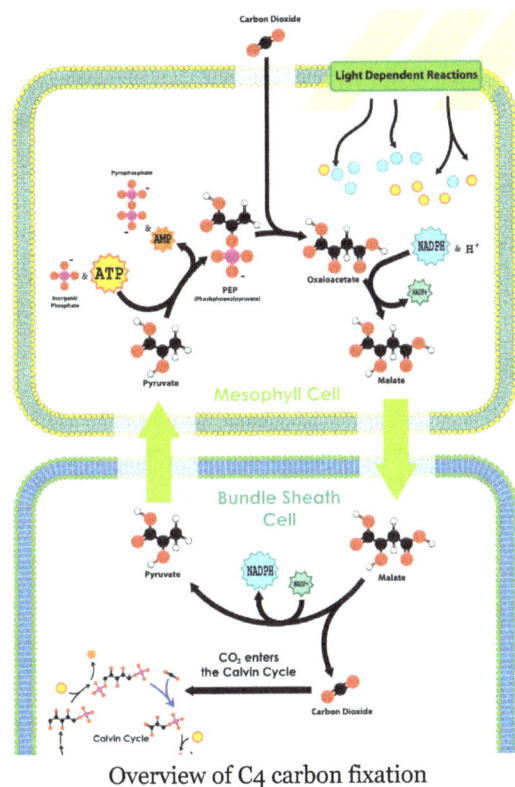
Overview of C4 carbon fixation

Plants that use the C_4 carbon fixation process chemically fix carbon dioxide in the cells of the mesophyll by adding it to the three-carbon molecule phosphoenolpyruvate (PEP), a reaction catalyzed by an enzyme called PEP carboxylase, creating the four-carbon organic acid oxaloacetic acid. Oxaloacetic acid or malate synthesized by this process is then translocated to specialized bundle sheath cells where the enzyme RuBisCO and other Calvin cycle enzymes are located, and where CO_2 released by decarboxylation of the four-carbon acids is then fixed by RuBisCO activity to the three-carbon 3-phosphoglyceric acids. The physical separation of RuBisCO from the oxygen-generating light reactions reduces photorespiration and increases CO_2 fixation and, thus, the photosynthetic capacity of the leaf. C_4 plants can produce more sugar than C_3 plants in conditions of high light and temperature. Many important crop plants are C_4 plants, including maize, sorghum, sugarcane, and millet. Plants that do not use PEP-carboxylase in carbon fixation are called C_3 plants because the primary carboxylation reaction, catalyzed by RuBisCO, produces the three-carbon 3-phosphoglyceric acids directly in the Calvin-Benson cycle. Over 90% of plants use C_3 carbon fixation, compared to 3% that use C_4 carbon fixation; however, the evolution of C_4 in over 60 plant lineages makes it a striking example of convergent evolution.

Xerophytes, such as cacti and most succulents, also use PEP carboxylase to capture carbon dioxide in a process called Crassulacean acid metabolism (CAM). In contrast to C_4 metabolism, which *spatially* separates the CO_2 fixation to PEP from the Calvin cycle, CAM *temporally* separates these two processes. CAM plants have a different leaf anatomy from C_3 plants, and fix the CO_2 at night, when their stomata are open. CAM plants store the CO_2 mostly in the form of malic acid via carboxylation of phosphoenolpyruvate to oxaloacetate, which is then reduced to malate. Decarboxylation of malate during the day releases CO_2 inside the leaves, thus allowing carbon fixation to 3-phosphoglycerate by RuBisCO. Sixteen thousand species of plants use CAM.

In Water

Cyanobacteria possess carboxysomes, which increase the concentration of CO_2 around RuBisCO to increase the rate of photosynthesis. An enzyme, carbonic anhydrase, located within the carboxysome releases CO_2 from the dissolved hydrocarbonate ions (HCO–3). Before the CO_2 diffuses out it is quickly sponged up by RuBisCO, which is concentrated within the carboxysomes. HCO–3 ions are made from CO_2 outside the cell by another carbonic anhydrase and are actively pumped into the cell by a membrane protein. They cannot cross the membrane as they are charged, and within the cytosol they turn back into CO_2 very slowly without the help of carbonic anhydrase. This causes the HCO–3 ions to accumulate within the cell from where they diffuse into the carboxysomes. Pyrenoids in algae and hornworts also act to concentrate CO_2 around rubisco.

Order and Kinetics

The overall process of photosynthesis takes place in four stages:

Stage	Description	Time scale
1	Energy transfer in antenna chlorophyll (thylakoid membranes)	femtosecond to picosecond
2	Transfer of electrons in photochemical reactions (thylakoid membranes)	picosecond to nanosecond
3	Electron transport chain and ATP synthesis (thylakoid membranes)	microsecond to millisecond
4	Carbon fixation and export of stable products	millisecond to second

Efficiency

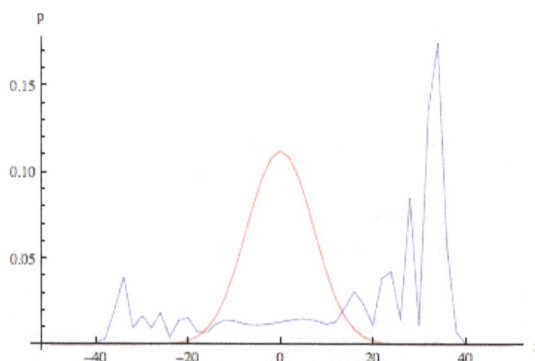

Probability distribution resulting from one-dimensional discrete time random walks.
The quantum walk created using the Hadamard coin is plotted (blue) vs a classical walk (red) after 50 time steps.

Plants usually convert light into chemical energy with a photosynthetic efficiency of 3–6%. Absorbed light that is unconverted is dissipated primarily as heat, with a small fraction (1–2%) re-emitted as chlorophyll fluorescence at longer (redder) wavelengths. A fact that allows measurement of the light reaction of photosynthesis by using chlorophyll fluorometers.

Actual plants' photosynthetic efficiency varies with the frequency of the light being converted, light intensity, temperature and proportion of carbon dioxide in the atmosphere, and can vary from 0.1% to 8%. By comparison, solar panels convert light into electric energy at an efficiency of approximately 6–20% for mass-produced panels, and above 40% in laboratory devices.

The efficiency of both light and dark reactions can be measured but the relationship between the two can be complex. For example, the ATP and NADPH energy molecules, created by the light reaction, can be used for carbon fixation or for photorespiration in C_3 plants. Electrons may also flow to other electron sinks. For this reason, it is not uncommon for authors to differentiate between work done under non-photorespiratory conditions and under photorespiratory conditions.

Chlorophyll fluorescence of photosystem II can measure the light reaction, and Infrared gas analyzers can measure the dark reaction. It is also possible to investigate both at the same time using an integrated chlorophyll fluorometer and gas exchange system, or by using two separate systems together. Infrared gas analyzers and some moisture sensors are sensitive enough to measure the photosynthetic assimilation of CO_2, and of ΔH_2O using reliable methods CO_2 is commonly measured in $\mu mols/m^2/s^{-1}$, parts per million or volume per million and H_2O is commonly measured in $mmol/m^2/s^{-1}$ or in mbars. By measuring CO_2 assimilation, ΔH_2O, leaf temperature, barometric pressure, leaf area, and photosynthetically active radiation or PAR, it becomes possible to estimate, "A" or carbon assimilation, "E" or transpiration, "gs" or stomatal conductance, and Ci or intracellular CO_2. However, it is more common to used chlorophyll fluorescence for plant stress measurement, where appropriate, because the most commonly used measuring parameters FV/FM and Y(II) or F/FM' can be made in a few seconds, allowing the measurement of larger plant populations.

Gas exchange systems that offer control of CO_2 levels, above and below ambient, allow the common practice of measurement of A/Ci curves, at different CO_2 levels, to characterize a plant's photosynthetic response.

Integrated chlorophyll fluorometer – gas exchange systems allow a more precise measure of photosynthetic response and mechanisms. While standard gas exchange photosynthesis systems can measure Ci, or substomatal CO_2 levels, the addition of integrated chlorophyll fluorescence measurements allows a more precise measurement of C_c to replace Ci. The estimation of CO_2 at the site of carboxylation in the chloroplast, or C_c, becomes possible with the measurement of mesophyll conductance or g_m using an integrated system.

Photosynthesis measurement systems are not designed to directly measure the amount of light absorbed by the leaf. But analysis of chlorophyll-fluorescence, P700- and P515-absorbance and gas exchange measurements reveal detailed information about e.g. the photosystems, quantum efficiency and the CO_2 assimilation rates. With some instruments even wavelength-dependency of the photosynthetic efficiency can be analyzed.

A phenomenon known as quantum walk increases the efficiency of the energy transport of light significantly. In the photosynthetic cell of an algae, bacterium, or plant, there are light-sensitive molecules called chromophores arranged in an antenna-shaped structure named a photocomplex. When a photon is absorbed by a chromophore, it is converted into a quasiparticle referred to as an exciton, which jumps from chromophore to chromophore towards the reaction center of the photocomplex, a collection of molecules that traps its energy in a chemical form that makes it accessible for the cell's metabolism. The exciton's wave properties enable it to cover a wider area and try out several possible paths simultaneously, allowing it to instantaneously "choose" the most efficient route, where it will have the highest probability of arriving at its destination in the minimum possible time. Because that quantum walking takes place at temperatures far higher than quantum phenomena usually occur, it is only possible over very short distances, due to obstacles in the form of destructive interference that come into play. These obstacles cause the particle to lose its wave properties for an instant before it regains them once again after it is freed from its locked position through a classic "hop". The movement of the electron towards the photo center is therefore covered in a series of conventional hops and quantum walks.

Evolution

Early photosynthetic systems, such as those in green and purple sulfur and green and purple nonsulfur bacteria, are thought to have been anoxygenic, and used various other molecules as electron donors rather than water. Green and purple sulfur bacteria are thought to have used hydrogen and sulfur as electron donors. Green nonsulfur bacteria used various amino and other organic acids as an electron donor. Purple nonsulfur bacteria used a variety of nonspecific organic molecules. The use of these molecules is consistent with the geological evidence that Earth's early atmosphere was highly reducing at that time.

Fossils of what are thought to be filamentous photosynthetic organisms have been dated at 3.4 billion years old.

The main source of oxygen in the Earth's atmosphere derives from oxygenic photosynthesis, and its first appearance is sometimes referred to as the oxygen catastrophe. Geological evidence suggests that oxygenic photosynthesis, such as that in cyanobacteria, became important during the Paleoproterozoic era around 2 billion years ago. Modern photosynthesis in plants and most photosynthetic prokaryotes is oxygenic. Oxygenic photosynthesis uses water as an electron donor, which is oxidized to molecular oxygen (O_2) in the photosynthetic reaction center.

Symbiosis and the Origin of Chloroplasts

Plant cells with visible chloroplasts (from a moss, *Plagiomnium affine*)

Several groups of animals have formed symbiotic relationships with photosynthetic algae. These are most common in corals, sponges and sea anemones. It is presumed that this is due to the particularly simple body plans and large surface areas of these animals compared to their volumes.

In addition, a few marine mollusks *Elysia viridis* and *Elysia chlorotica* also maintain a symbiotic relationship with chloroplasts they capture from the algae in their diet and then store in their bodies. This allows the mollusks to survive solely by photosynthesis for several months at a time. Some of the genes from the plant cell nucleus have even been transferred to the slugs, so that the chloroplasts can be supplied with proteins that they need to survive.

An even closer form of symbiosis may explain the origin of chloroplasts. Chloroplasts have many similarities with photosynthetic bacteria, including a circular chromosome, prokaryotic-type ribosome, and similar proteins in the photosynthetic reaction center. The endosymbiotic theory suggests that photosynthetic bacteria were acquired (by endocytosis) by early eukaryotic cells to form the first plant cells. Therefore, chloroplasts may be photosynthetic bacteria that adapted to life inside plant cells. Like mitochondria, chloroplasts possess their own DNA, separate from the nuclear DNA of their plant host cells and the genes in this chloroplast DNA resemble those found in cyanobacteria. DNA in chloroplasts codes for redox proteins such as those found in the photosynthetic reaction centers. The CoRR Hypothesis proposes that this Co-location is required for Redox Regulation.

Cyanobacteria and the Evolution of Photosynthesis

The biochemical capacity to use water as the source for electrons in photosynthesis evolved once, in a common ancestor of extant cyanobacteria. The geological record indicates that this transforming event took place early in Earth's history, at least 2450–2320 million years ago (Ma), and, it is speculated, much earlier. Because the Earth's atmosphere contained almost no oxygen during the estimated development of photosynthesis, it is believed that the first photosynthetic cyanobacteria did not generate oxygen. Available evidence from geobiological studies of Archean (>2500 Ma) sedimentary rocks indicates that life existed 3500 Ma, but the question of when oxygenic photosynthesis evolved is still unanswered. A clear paleontological window on cyanobacterial evolution opened about 2000 Ma, revealing an already-diverse biota of blue-green algae. Cyanobacteria remained the principal primary producers of oxygen throughout the Proterozoic Eon (2500–543 Ma), in part because the redox structure of the oceans favored photoautotrophs capable of nitrogen fixation. Green algae joined blue-green algae as the major primary producers of oxygen on continental shelves near the end of the Proterozoic, but it was only with the Mesozoic (251–65 Ma) radiations of dinoflagellates, coccolithophorids, and diatoms did the primary production of oxygen in marine shelf waters take modern form. Cyanobacteria remain critical to marine ecosystems as primary producers of oxygen in oceanic gyres, as agents of biological nitrogen fixation, and, in modified form, as the plastids of marine algae.

The Oriental hornet (*Vespa orientalis*) converts sunlight into electric power using a pigment called xanthopterin. This is the first evidence of a member of the animal kingdom engaging in photosynthesis.

Discovery

Although some of the steps in photosynthesis are still not completely understood, the overall photosynthetic equation has been known since the 19th century.

Jan van Helmont began the research of the process in the mid-17th century when he carefully measured the mass of the soil used by a plant and the mass of the plant as it grew. After noticing that the soil mass changed very little, he hypothesized that the mass of the growing plant must come from the water, the only substance he added to the potted plant. His hypothesis was partially accurate — much of the gained mass also comes from carbon dioxide as well as water. However, this was a signaling point to the idea that the bulk of a plant's biomass comes from the inputs of photosynthesis, not the soil itself.

Joseph Priestley, a chemist and minister, discovered that, when he isolated a volume of air under an inverted jar, and burned a candle in it, the candle would burn out very quickly, much before it ran out of wax. He further discovered that a mouse could similarly "injure" air. He then showed that the air that had been "injured" by the candle and the mouse could be restored by a plant.

In 1778, Jan Ingenhousz, repeated Priestley's experiments. He discovered that it was the influence of sunlight on the plant that could cause it to revive a mouse in a matter of hours.

In 1796, Jean Senebier, a Swiss pastor, botanist, and naturalist, demonstrated that green plants consume carbon dioxide and release oxygen under the influence of light. Soon afterward, Nicolas-Théodore de Saussure showed that the increase in mass of the plant as it grows could not be due only to uptake of CO_2 but also to the incorporation of water. Thus, the basic reaction by which photosynthesis is used to produce food (such as glucose) was outlined.

Cornelis Van Niel made key discoveries explaining the chemistry of photosynthesis. By studying purple sulfur bacteria and green bacteria he was the first to demonstrate that photosynthesis is a light-dependent redox reaction, in which hydrogen reduces carbon dioxide.

Robert Emerson discovered two light reactions by testing plant productivity using different wavelengths of light. With the red alone, the light reactions were suppressed. When blue and red were combined, the output was much more substantial. Thus, there were two photosystems, one absorbing up to 600 nm wavelengths, the other up to 700 nm. The former is known as PSII, the latter is PSI. PSI contains only chlorophyll "a", PSII contains primarily chlorophyll "a" with most of the available chlorophyll "b", among other pigment. These include phycobilins, which are the red and blue pigments of red and blue algae respectively, and fucoxanthol for brown algae and diatoms. The process is most productive when the absorption of quanta are equal in both the PSII and PSI, assuring that input energy from the antenna complex is divided between the PSI and PSII system, which in turn powers the photochemistry.

Melvin Calvin works in his photosynthesis laboratory.

Robert Hill thought that a complex of reactions consisting of an intermediate to cytochrome b_6 (now a plastoquinone), another is from cytochrome f to a step in the carbohydrate-generating mechanisms. These are linked by plastoquinone, which does require energy to reduce cytochrome f for it is a sufficient reductant. Further experiments to prove that the oxygen developed during the photosynthesis of green plants came from water, were performed by Hill in 1937 and 1939. He showed that isolated chloroplasts give off oxygen in the presence of unnatural reducing agents like iron oxalate, ferricyanide or benzoquinone after exposure to light. The Hill reaction is as follows:

$$2\ H_2O + 2\ A + (\text{light, chloroplasts}) \rightarrow 2\ AH_2 + O_2$$

where A is the electron acceptor. Therefore, in light, the electron acceptor is reduced and oxygen is evolved.

Samuel Ruben and Martin Kamen used radioactive isotopes to determine that the oxygen liberated in photosynthesis came from the water.

Melvin Calvin and Andrew Benson, along with James Bassham, elucidated the path of carbon assimilation (the photosynthetic carbon reduction cycle) in plants. The carbon reduction cycle is known as the Calvin cycle, which ignores the contribution of Bassham and Benson. Many scientists refer to the cycle as the Calvin-Benson Cycle, Benson-Calvin, and some even call it the Calvin-Benson-Bassham (or CBB) Cycle.

Nobel Prize-winning scientist Rudolph A. Marcus was able to discover the function and significance of the electron transport chain.

Otto Heinrich Warburg and Dean Burk discovered the I-quantum photosynthesis reaction that splits the CO_2, activated by the respiration.

Louis N.M. Duysens and Jan Amesz discovered that chlorophyll a will absorb one light, oxidize cytochrome f, chlorophyll a (and other pigments) will absorb another light, but will reduce this same oxidized cytochrome, stating the two light reactions are in series.

Development of the Concept

In 1893, Charles Reid Barnes proposed two terms, *photosyntax* and *photosynthesis*, for the biological process of *synthesis of complex carbon compounds out of carbonic acid, in the presence of chlorophyll, under the influence of light*. Over time, the term *photosynthesis* came into common usage as the term of choice. Later discovery of anoxygenic photosynthetic bacteria and photophosphorylation necessitated redefinition of the term.

C3 : C4 Photosynthesis Research

After WWII at late 1940 at the University of California, Berkeley, the details of photosynthetic carbon metabolism were sorted out by the chemists Melvin Calvin, Andrew Benson, James Bassham and a score of students and researchers utilizing the carbon-14 isotope and paper chromatography techniques. The pathway of CO2 fixation by the algae *Chlorella* in a fraction of a second in light resulted in a 3 carbon molecule called phosphoglyceric acid (PGA). For that original and ground-breaking work, a Nobel Prize in Chemistry was awarded to Melvin Calvin 1961. In parallel, plant physiologists studied leaf gas exchanges using the new method of infrared gas analysis and a leaf chamber where the net photosynthetic rates ranged from 10 to 13 u mole CO_2/square metere. sec., with the conclusion that all terrestrial plants having the same photosynthetic capacities that were light saturated at less than 50% of sunlight. These rates were determined in potted plants grown indoors under low light intensity.

Later in 1958-1963 at Cornell University, field grown maize was reported to have much greater leaf photosynthetic rates of 40 u mol CO_2/square meter.sec and was not saturated at near full sunlight. This higher rate in maize was almost double those observed in other species such as wheat and soybean, indicating that large differences in photosynthesis exist among higher plants. At the University of Arizona, detailed gas exchange research on more than 15 species of monocot and dicot uncovered for the first time that differences in leaf anatomy are crucial factors in differentiating photosynthetic capacities among species. In tropical grasses, including maize, sorghum, sugarcane, Bermuda grass and in the dicot amaranthus, leaf photosynthetic rates were around 38–40 u mol CO_2/square meter.sec., and the leaves have two types of green cells, i. e. outer layer of mesophyll cells surrounding a tightly packed cholorophyllous vascular bundle sheath cells. This type of anatomy was termed Kranz anatomy in the 19th century by the botanist Gottlieb Haberlandt while studying leaf anatomy of sugarcane. Plant species with the greatest photosynthetic rates and Kranz anatomy showed no apparent photorespiration, very low CO_2 compensation point, high optimum temperature, high stomatal resistances and lower mesophyll resistances for gas diffusion and rates never saturated at full sun light. The research at Arizona was designated Citation Classic by the ISI 1986. These species was later termed C4 plants as the first stable compound of CO_2 fixation in light has 4 carbon as malate and aspartate. Other species that lack Kranz anatomy were termed C3 type such as cotton and sunflower, as the first stable carbon compound is the 3-carbon PGA acid. At 1000 ppm CO_2 in measuring air, both the C3 and C4 plants had similar leaf photosynthetic rates around 60 u mole CO_2/square meter.sec. indicating the suppression of phototorespiration in C3 plants.

Factors

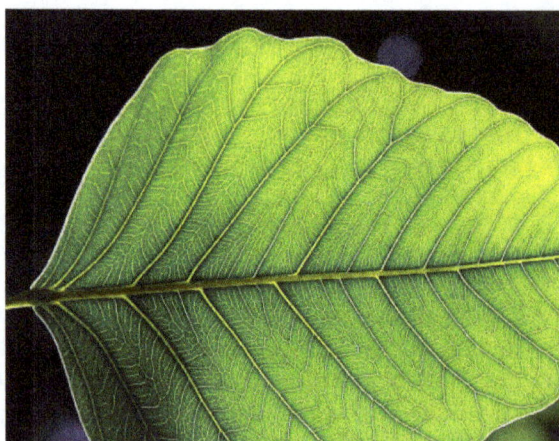

The leaf is the primary site of photosynthesis in plants.

There are three main factors affecting photosynthesis and several corollary factors. The three main are:

- Light irradiance and wavelength

- Carbon dioxide concentration

- Temperature.

Light Intensity (Irradiance), Wavelength and Temperature

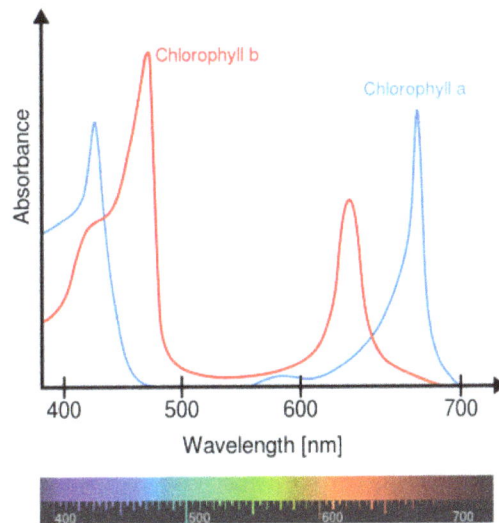

Absorbance spectra of free chlorophyll a (green) and b (red) in a solvent. The action spectra of chlorophyll molecules are slightly modified *in vivo* depending on specific pigment-protein interactions.

The process of photosynthesis provides the main input of free energy into the biosphere, and is one of four main ways in which radiation is important for plant life.

The radiation climate within plant communities is extremely variable, with both time and space.

In the early 20th century, Frederick Blackman and Gabrielle Matthaei investigated the effects of light intensity (irradiance) and temperature on the rate of carbon assimilation.

- At constant temperature, the rate of carbon assimilation varies with irradiance, increasing as the irradiance increases, but reaching a plateau at higher irradiance.

- At low irradiance, increasing the temperature has little influence on the rate of carbon assimilation. At constant high irradiance, the rate of carbon assimilation increases as the temperature is increased.

These two experiments illustrate several important points: First, it is known that, in general, photochemical reactions are not affected by temperature. However, these experiments clearly show that temperature affects the rate of carbon assimilation, so there must be two sets of reactions in the full process of carbon assimilation. These are, of course, the light-dependent 'photochemical' temperature-independent stage, and the light-independent, temperature-dependent stage. Second, Blackman's experiments illustrate the concept of limiting factors. Another limiting factor is the wavelength of light. Cyanobacteria, which reside several meters underwater, cannot receive the correct wavelengths required to cause photoinduced charge separation in conventional photosynthetic pigments. To combat this problem, a series of proteins with different pigments surround the reaction center. This unit is called a phycobilisome.

Carbon Dioxide Levels and Photorespiration

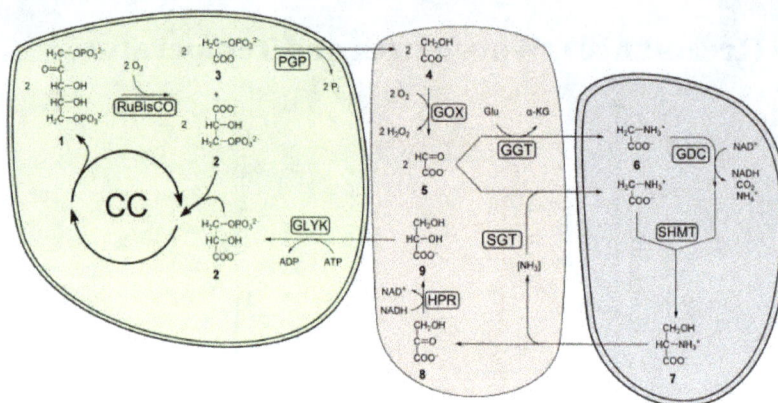

Photorespiration

As carbon dioxide concentrations rise, the rate at which sugars are made by the light-independent reactions increases until limited by other factors. RuBisCO, the enzyme that captures carbon dioxide in the light-independent reactions, has a binding affinity for both carbon dioxide and oxygen. When the concentration of carbon dioxide is high, RuBisCO will fix carbon dioxide. However, if the carbon dioxide concentration is low, RuBisCO will bind oxygen instead of carbon dioxide. This process, called photorespiration, uses energy, but does not produce sugars.

RuBisCO oxygenase activity is disadvantageous to plants for several reasons:

1. One product of oxygenase activity is phosphoglycolate (2 carbon) instead of 3-phosphoglycerate (3 carbon). Phosphoglycolate cannot be metabolized by the Calvin-Benson cycle and represents carbon lost from the cycle. A high oxygenase activity, therefore, drains the sugars that are required to recycle ribulose 5-bisphosphate and for the continuation of the Calvin-Benson cycle.

2. Phosphoglycolate is quickly metabolized to glycolate that is toxic to a plant at a high concentration; it inhibits photosynthesis.

3. Salvaging glycolate is an energetically expensive process that uses the glycolate pathway, and only 75% of the carbon is returned to the Calvin-Benson cycle as 3-phosphoglycerate. The reactions also produce ammonia (NH_3), which is able to diffuse out of the plant, leading to a loss of nitrogen.

   A highly simplified summary is:

   2 glycolate + ATP → 3-phosphoglycerate + carbon dioxide + ADP + NH_3

The salvaging pathway for the products of RuBisCO oxygenase activity is more commonly known as photorespiration, since it is characterized by light-dependent oxygen consumption and the release of carbon dioxide.

Chloroplast

Chloroplasts are organelles, specialized subunits, in plant and algal cells. Their discovery inside plant cells is usually credited to Julius von Sachs (1832–1897), an influential botanist and author of standard botanical textbooks – sometimes called "The Father of Plant Physiology".

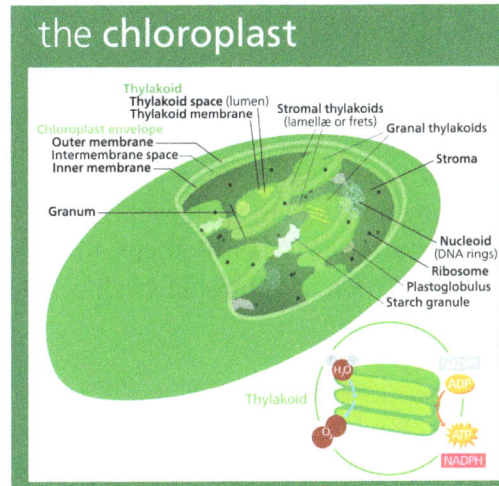

Structure of a typical higher-plant chloroplast

Chloroplasts' main role is to conduct photosynthesis, where the photosynthetic pigment chlorophyll captures the energy from sunlight and converts it and stores it in the energy-storage molecules ATP and NADPH while freeing oxygen from water. They then use the ATP and NADPH to make organic molecules from carbon dioxide in a process known as the Calvin cycle. Chloroplasts carry out a number of other functions, including fatty acid synthesis, much amino acid synthesis, and the immune response in plants. The number of chloroplasts per cell varies from 1 in algae up to 100 in plants like Arabidopsis and wheat.

A chloroplast is one of three types of plastids, characterized by its high concentration of chlorophyll, the other two types, the leucoplast and the chromoplast, contain little chlorophyll and do not carry out photosynthesis.

Chloroplasts are highly dynamic—they circulate and are moved around within plant cells, and occasionally pinch in two to reproduce. Their behavior is strongly influenced by environmental factors like light color and intensity. Chloroplasts, like mitochondria, contain their own DNA, which is thought to be inherited from their ancestor—a photosynthetic cyanobacterium that was engulfed by an early eukaryotic cell. Chloroplasts cannot be made by the plant cell and must be inherited by each daughter cell during cell division.

With one exception (the amoeboid *Paulinella chromatophora*), all chloroplasts can probably be traced back to a single endosymbiotic event, when a cyanobacterium was engulfed by the eukaryote. Despite this, chloroplasts can be found in an extremely wide set of organisms, some not even directly related to each other—a consequence of many secondary and even tertiary endosymbiotic events.

The word *chloroplast* is derived from the Greek words *chloros*, which means green, and *plastes*, which means «the one who forms».

Discovery

The first definitive description of a chloroplast (*Chlorophyllkörnen*, "grain of chlorophyll") was given by Hugo von Mohl in 1837 as discrete bodies within the green plant cell. In 1883, A. F. W. Schimper would name these bodies as "chloroplastids" (*Chloroplastiden*). In 1884, Eduard Strasburger adopted the term "chloroplasts" (*Chloroplasten*).

Chloroplast Lineages and Evolution

Chloroplasts are one of many types of organelles in the plant cell. They are considered to have originated from cyanobacteria through endosymbiosis—when a eukaryotic cell engulfed a photosynthesizing cyanobacterium that became a permanent resident in the cell. Mitochondria are thought to have come from a similar event, where an aerobic prokaryote was engulfed. This origin of chloroplasts was first suggested by the Russian biologist Konstantin Mereschkowski in 1905 after Andreas Schimper observed in 1883 that chloroplasts closely resemble cyanobacteria. Chloroplasts are only found in plants and algae.

Cyanobacterial Ancestor

Cyanobacteria are considered the ancestors of chloroplasts. They are sometimes called blue-green algae even though they are prokaryotes. They are a diverse phylum of bacteria capable of carrying out photosynthesis, and are gram-negative, meaning that they have two cell membranes. Cyanobacteria also contain a peptidoglycan cell wall, which is thicker than in other gram-negative bacteria, and which is located between their two cell membranes. Like chloroplasts, they have thylakoids within. On the thylakoid membranes are photosynthetic pigments, including chlorophyll *a*. Phycobilins are also common cyanobacterial pigments, usually organized into hemispherical phycobilisomes attached to the outside of the thylakoid membranes (phycobilins are not shared with all chloroplasts though).

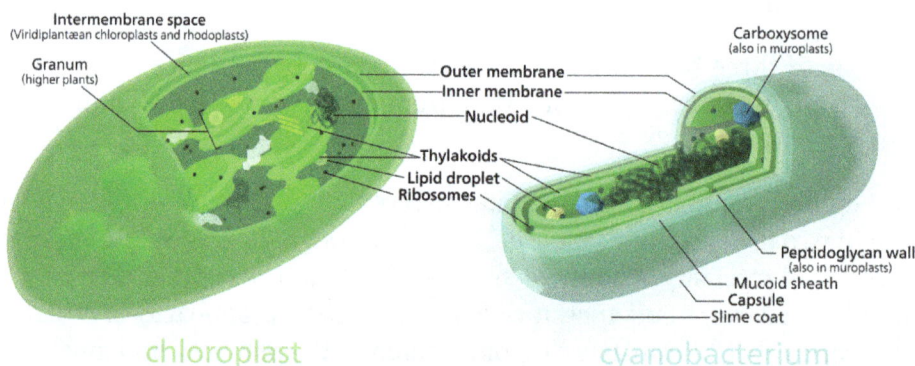

Both chloroplasts and cyanobacteria have a double membrane, DNA, ribosomes, and thylakoids. Both the chloroplast and cyanobacterium depicted are idealized versions (the chloroplast is that of a higher plant)—a lot of diversity exists among chloroplasts and cyanobacteria.

Primary Endosymbiosis

Somewhere around a billion years ago, a free-living cyanobacterium entered an early eukaryotic cell, either as food or as an internal parasite, but managed to escape the phagocytic vacuole it was contained in. The two innermost lipid-bilayer membranes that surround all chloroplasts correspond to the outer and inner membranes of the ancestral cyanobacterium's gram negative cell wall, and not the phagosomal membrane from the host, which was probably lost. The new cellular resident quickly became an advantage, providing food for the eukaryotic host, which allowed it to live within it. Over time, the cyanobacterium was assimilated, and many of its genes were lost or transferred to the nucleus of the host. Some of its proteins were then synthesized in the cytoplasm of the host cell, and imported back into the chloroplast (formerly the cyanobacterium).

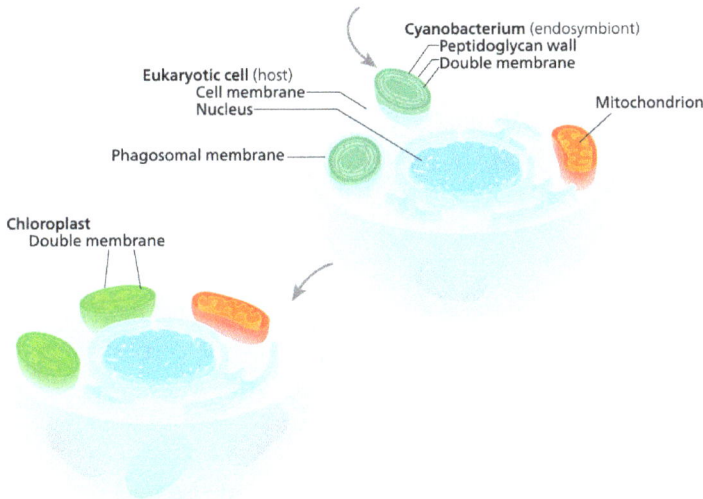

Primary endosymbiosis A eukaryote with mitochondria engulfed a cyanobacterium in an event of serial primary endosymbiosis, creating a lineage of cells with both organelles. It is important to note that the cyanobacterial endosymbiont already had a double membrane—the phagosomal vacuole-derived membrane was lost.

This event is called *endosymbiosis*, or "cell living inside another cell". The cell living inside the other cell is called the *endosymbiont*; the endosymbiont is found inside the *host cell*.

Chloroplasts are believed to have arisen after mitochondria, since all eukaryotes contain mitochondria, but not all have chloroplasts. This is called *serial endosymbiosis*—an early eukaryote engulfing the mitochondrion ancestor, and some descendants of it then engulfing the chloroplast ancestor, creating a cell with both chloroplasts and mitochondria.

Whether or not chloroplasts came from a single endosymbiotic event, or many independent engulfments across various eukaryotic lineages, has been long debated, but it is now generally held that all organisms with chloroplasts either share a single ancestor or obtained their chloroplast from organisms that share a common ancestor that took in a cyanobacterium 600–1600 million years ago.

These chloroplasts, which can be traced back directly to a cyanobacterial ancestor, are known as *primary plastids* ("*plastid*" in this context means almost the same thing as chloroplast). All primary chloroplasts belong to one of three chloroplast lineages—the glaucophyte chloroplast lineage, the rhodophyte, or red algal chloroplast lineage, or the chloroplastidan, or green chloroplast lineage. The second two are the largest, and the green chloroplast lineage is the one that contains the land plants.

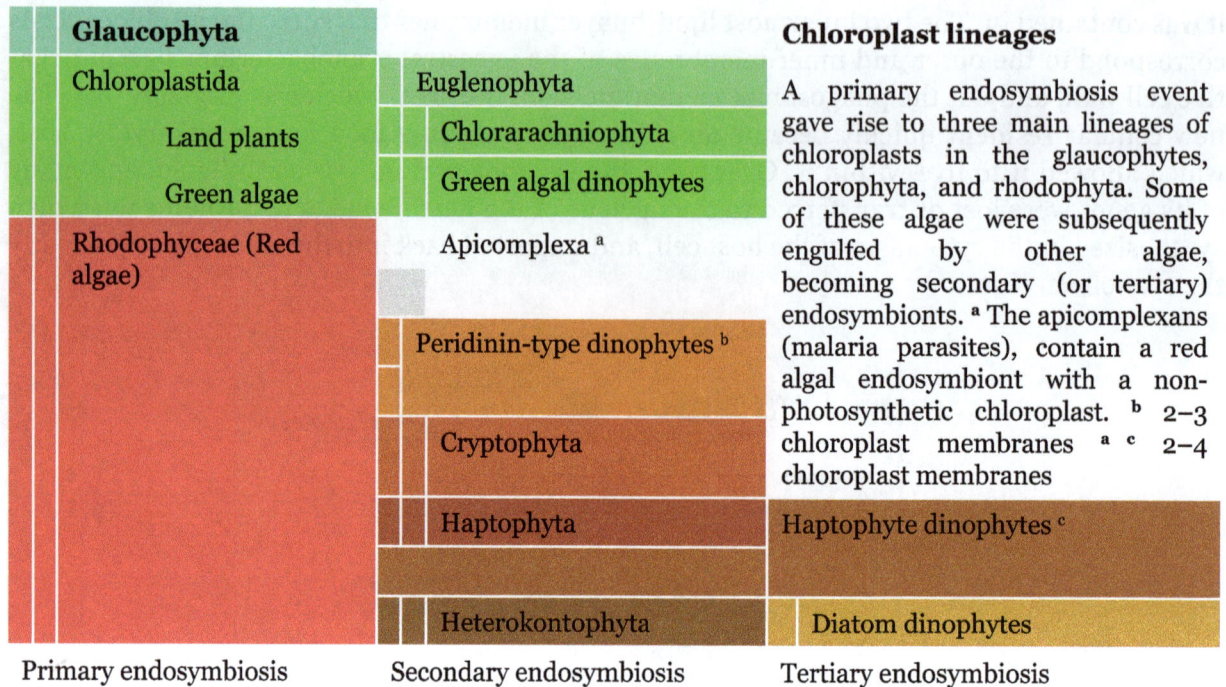

		Chloroplast lineages
Glaucophyta		A primary endosymbiosis event gave rise to three main lineages of chloroplasts in the glaucophytes, chlorophyta, and rhodophyta. Some of these algae were subsequently engulfed by other algae, becoming secondary (or tertiary) endosymbionts. [a] The apicomplexans (malaria parasites), contain a red algal endosymbiont with a non-photosynthetic chloroplast. [b] 2–3 chloroplast membranes [a] [c] 2–4 chloroplast membranes
Chloroplastida	Euglenophyta	
Land plants	Chlorarachniophyta	
Green algae	Green algal dinophytes	
Rhodophyceae (Red algae)	Apicomplexa [a]	
	Peridinin-type dinophytes [b]	
	Cryptophyta	
	Haptophyta	Haptophyte dinophytes [c]
	Heterokontophyta	Diatom dinophytes

Primary endosymbiosis   Secondary endosymbiosis   Tertiary endosymbiosis

Glaucophyta

The alga *Cyanophora*, a glaucophyte, is thought to be one of the first organisms to contain a chloroplast. The glaucophyte chloroplast group is the smallest of the three primary chloroplast lineages, being found in only 13 species, and is thought to be the one that branched off the earliest. Glaucophytes have chloroplasts that retain a peptidoglycan wall between their double membranes, like their cyanobacterial parent. For this reason, glaucophyte chloroplasts are also known as *muroplasts*. Glaucophyte chloroplasts also contain concentric unstacked thylakoids, which surround a carboxysome - an icosahedral structure that glaucophyte chloroplasts and cyanobacteria keep their carbon fixation enzyme rubisco in. The starch that they synthesize collects outside the chloroplast. Like cyanobacteria, glaucophyte chloroplast thylakoids are studded with light collecting structures called phycobilisomes. For these reasons, glaucophyte chloroplasts are considered a primitive intermediate between cyanobacteria and the more evolved chloroplasts in red algae and plants.

Diversity of red algae Clockwise from top left: *Bornetia secundiflora*, *Peyssonnelia squamaria*, *Cyanidium*, *Laurencia*, *Callophyllis laciniata*. Red algal chloroplasts are characterized by phycobilin pigments which often give them their reddish color.

Rhodophyceae (Red Algae)

The rhodophyte, or red algal chloroplast group is another large and diverse chloroplast lineage. Rhodophyte chloroplasts are also called *rhodoplasts*, literally "red chloroplasts".

Rhodoplasts have a double membrane with an intermembrane space and phycobilin pigments organized into phycobilisomes on the thylakoid membranes, preventing their thylakoids from stacking. Some contain pyrenoids. Rhodoplasts have chlorophyll *a* and phycobilins for photosynthetic pigments; the phycobilin phycoerytherin is responsible for giving many red algae their distinctive red color. However, since they also contain the blue-green chlorophyll *a* and other pigments, many are reddish to purple from the combination. The red phycoerytherin pigment is an adaptation to help red algae catch more sunlight in deep water—as such, some red algae that live in shallow water have less phycoerytherin in their rhodoplasts, and can appear more greenish. Rhodoplasts synthesize a form of starch called floridean starch, which collects into granules outside the rhodoplast, in the cytoplasm of the red alga.

Chloroplastida (Green Algae and Plants)

Diversity of green algae Clockwise from top left: *Scenedesmus*, *Micrasterias*, *Hydrodictyon*, *Stigeoclonium*, *Volvox*. Green algal chloroplasts are characterized by their pigments chlorophyll *a* and chlorophyll *b* which give them their green color.

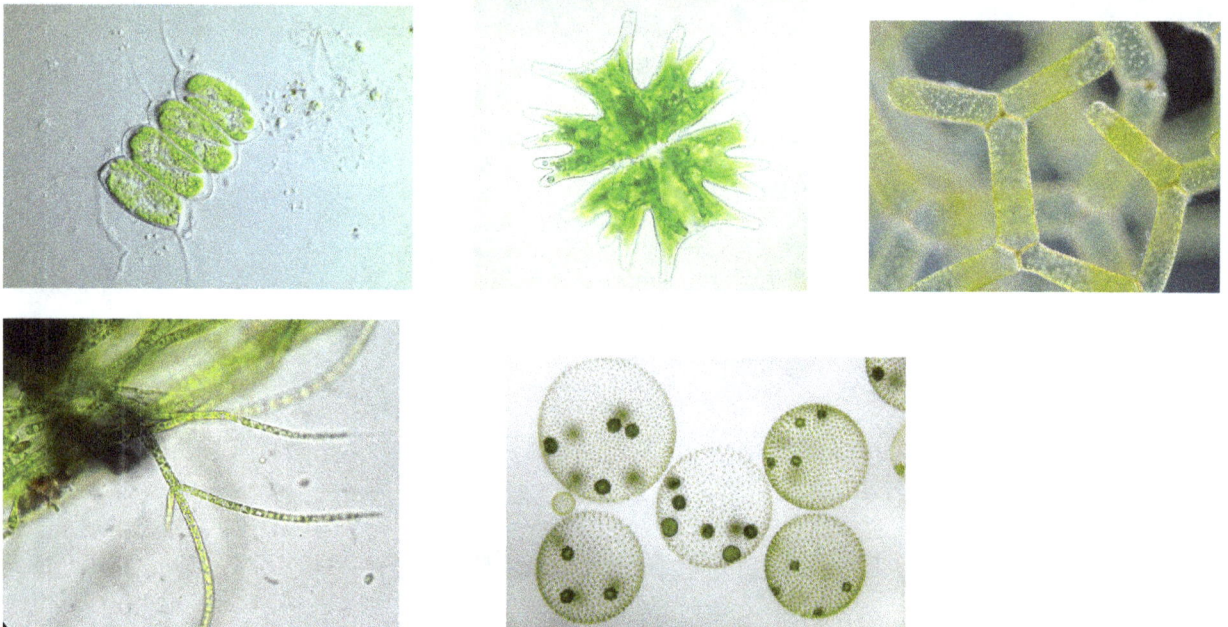

The chloroplastidan chloroplasts, or green chloroplasts, are another large, highly diverse primary chloroplast lineage. Their host organisms are commonly known as the green algae and land plants. They differ from glaucophyte and red algal chloroplasts in that they have lost their phycobilisomes, and contain chlorophyll *b* instead. Most green chloroplasts are (obviously) green, though some aren't, like some forms of *Hæmatococcus pluvialis*, due to accessory pigments that override the chlorophylls' green colors. Chloroplastidan chloroplasts have lost the peptidoglycan wall between their double membrane, and have replaced it with an intermembrane space. Some plants seem to have kept the genes for the synthesis of the peptidoglycan layer, though they've been repurposed for use in chloroplast division instead.

Most of the chloroplasts depicted in this article are green chloroplasts.

Green algae and plants keep their starch *inside* their chloroplasts, and in plants and some algae, the chloroplast thylakoids are arranged in grana stacks. Some green algal chloroplasts contain a structure called a pyrenoid, which is functionally similar to the glaucophyte carboxysome in that it is where rubisco and CO_2 are concentrated in the chloroplast.

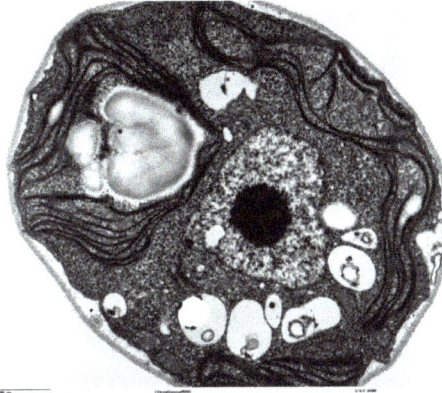

Transmission electron micrograph of *Chlamydomonas reinhardtii*, a green alga that contains a pyrenoid surrounded by starch.

Helicosporidium

Helicosporidium is a genus of nonphotosynthetic parasitic green algae that is thought to contain a vestigial chloroplast. Genes from a chloroplast and nuclear genes indicating the presence of a chloroplast have been found in Helicosporidium even if nobody's seen the chloroplast itself.

Secondary and Tertiary Endosymbiosis

Many other organisms obtained chloroplasts from the primary chloroplast lineages through secondary endosymbiosis—engulfing a red or green alga that contained a chloroplast. These chloroplasts are known as secondary plastids.

While primary chloroplasts have a double membrane from their cyanobacterial ancestor, secondary chloroplasts have additional membranes outside of the original two, as a result of the secondary endosymbiotic event, when a nonphotosynthetic eukaryote engulfed a chloroplast-containing alga but failed to digest it—much like the cyanobacterium at the beginning of this story. The engulfed alga was broken down, leaving only its chloroplast, and sometimes its cell membrane and nucleus, forming a chloroplast with three or four membranes—the two cyanobacterial membranes, sometimes the eaten alga's cell membrane, and the phagosomal vacuole from the host's cell membrane.

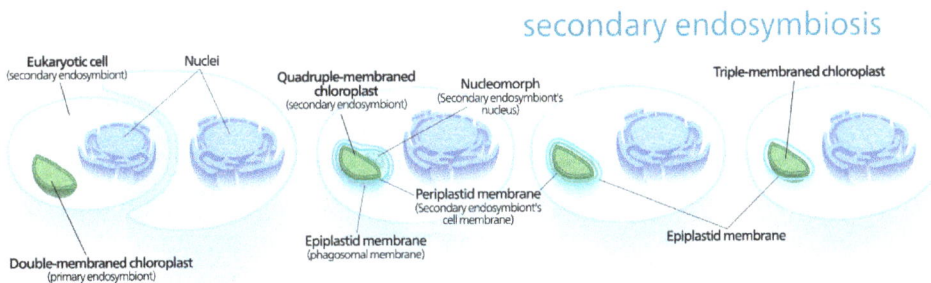

Secondary endosymbiosis consisted of a eukaryotic alga being engulfed by another eukaryote, forming a chloroplast with three or four membranes.

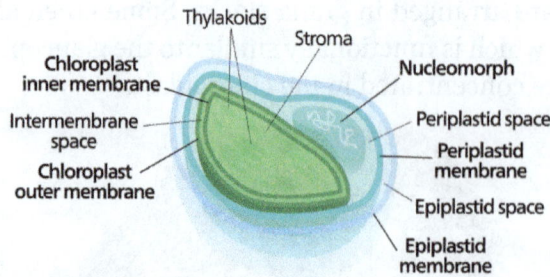

Diagram of a four membraned chloroplast containing a nucleomorph.

The genes in the phagocytosed eukaryote's nucleus are often transferred to the secondary host's nucleus. Cryptomonads and chlorarachniophytes retain the phagocytosed eukaryote's nucleus, an object called a nucleomorph, located between the second and third membranes of the chloroplast.

All secondary chloroplasts come from green and red algae—no secondary chloroplasts from glaucophytes have been observed, probably because glaucophytes are relatively rare in nature, making them less likely to have been taken up by another eukaryote.

Green Algal Derived Chloroplasts

Green algae have been taken up by the euglenids, chlorarachniophytes, a lineage of dinoflagellates, and possibly the ancestor of the chromalveolates in three or four separate engulfments. Many green algal derived chloroplasts contain pyrenoids, but unlike chloroplasts in their green algal ancestors, starch collects in granules outside the chloroplast.

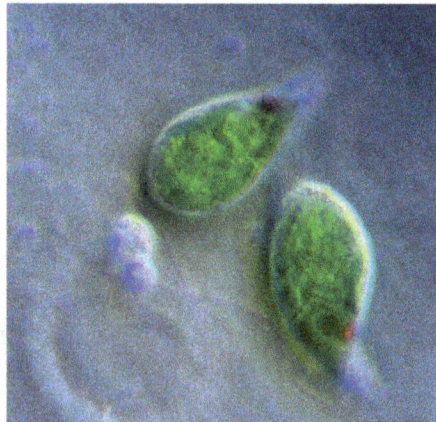

Euglena, a euglenophyte, contains secondary chloroplasts from green algae.

Euglenophytes

Euglenophytes are a group of common flagellated protists that contain chloroplasts derived from a green alga. Euglenophyte chloroplasts have three membranes—it is thought that the membrane of the primary endosymbiont was lost, leaving the cyanobacterial membranes, and the secondary host's phagosomal membrane. Euglenophyte chloroplasts have a pyrenoid and thylakoids stacked in groups of three. Starch is stored in the form of paramylon, which is contained in membrane-bound granules in the cytoplasm of the euglenophyte.

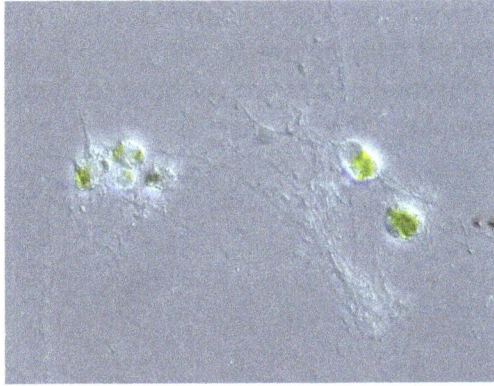

Chlorarachnion reptans is a chlorarachniophyte. Chlorarachniophytes replaced their original red algal endosymbiont with a green alga.

Chlorarachniophytes

Chlorarachniophytes are a rare group of organisms that also contain chloroplasts derived from green algae, though their story is more complicated than that of the euglenophytes. The ancestor of chlorarachniophytes is thought to have been a chromalveolate, a eukaryote with a *red* algal derived chloroplast. It is then thought to have lost its first red algal chloroplast, and later engulfed a green alga, giving it its second, green algal derived chloroplast.

Chlorarachniophyte chloroplasts are bounded by four membranes, except near the cell membrane, where the chloroplast membranes fuse into a double membrane. Their thylakoids are arranged in loose stacks of three. Chlorarachniophytes have a form of starch called chrysolaminarin, which they store in the cytoplasm, often collected around the chloroplast pyrenoid, which bulges into the cytoplasm.

Chlorarachniophyte chloroplasts are notable because the green alga they are derived from has not been completely broken down—its nucleus still persists as a nucleomorph found between the second and third chloroplast membranes—the periplastid space, which corresponds to the green alga's cytoplasm.

Early Chromalveolates

Recent research has suggested that the ancestor of the chromalveolates acquired a green algal prasinophyte endosymbiont. The green algal derived chloroplast was lost and replaced with a red algal derived chloroplast, but not before contributing some of its genes to the early chromalveolate's nucleus. The presence of both green algal and red algal genes in chromalveolates probably helps them thrive under fluctuating light conditions.

Red Algal Derived Chloroplasts (Chromalveolate Chloroplasts)

Like green algae, red algae have also been taken up in secondary endosymbiosis, though it is thought that all red algal derived chloroplasts are descended from a single red alga that was engulfed by an early chromalveolate, giving rise to the chromalveolates, some of which, like the ciliates, subsequently lost the chloroplast. This is still debated though.

Pyrenoids and stacked thylakoids are common in chromalveolate chloroplasts, and the outermost membrane of many are continuous with the rough endoplasmic reticulum and studded with ribosomes. They have lost their phycobilisomes and exchanged them for chlorophyll c, which isn't found in primary red algal chloroplasts themselves.

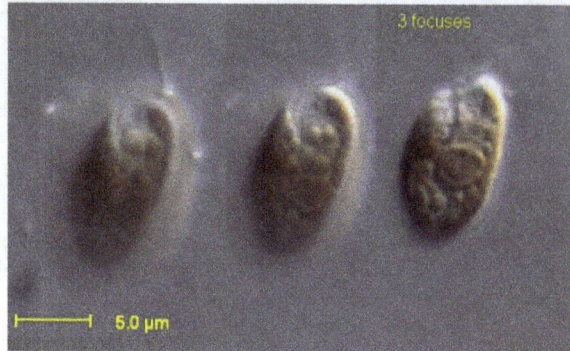

Rhodomonas salina is a cryptophyte.

Cryptophytes

Cryptophytes, or cryptomonads are a group of algae that contain a red-algal derived chloroplast. Cryptophyte chloroplasts contain a nucleomorph that superficially resembles that of the chlorarachniophytes. Cryptophyte chloroplasts have four membranes, the outermost of which is continuous with the rough endoplasmic reticulum. They synthesize ordinary starch, which is stored in granules found in the periplastid space—outside the original double membrane, in the place that corresponds to the red alga's cytoplasm. Inside cryptophyte chloroplasts is a pyrenoid and thylakoids in stacks of two.

Their chloroplasts do not have phycobilisomes, but they do have phycobilin pigments which they keep in their thylakoid space, rather than anchored on the outside of their thylakoid membranes.

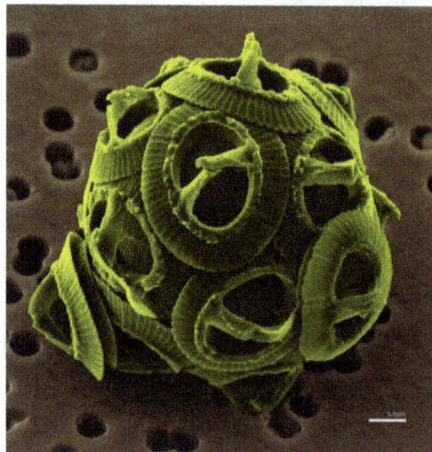

Scanning electron micrograph of *Gephyrocapsa oceanica*, a haptophyte.

Haptophytes

Haptophytes are similar and closely related to cryptophytes, and are thought to be the first chromalveolates to branch off. Their chloroplasts lack a nucleomorph, their thylakoids are in stacks of three, and they synthesize chrysolaminarin sugar, which they store completely outside of the chloroplast, in the cytoplasm of the haptophyte.

Heterokontophytes (Stramenopiles)

The photosynthetic pigments present in their chloroplasts give diatoms a greenish-brown color.

The heterokontophytes, also known as the stramenopiles, are a very large and diverse group of algae that also contain red algal derived chloroplasts. Heterokonts include the diatoms and the brown algae, golden algae, and yellow-green algae.

Heterokont chloroplasts are very similar to haptophyte chloroplasts, containing a pyrenoid, triplet thylakoids, and with some exceptions, having an epiplastid membrane connected to the endoplasmic reticulum. Like haptophytes, heterokontophytes store sugar in chrysolaminarin granules in the cytoplasm. Heterokontophyte chloroplasts contain chlorophyll *a* and with a few exceptions chlorophyll *c*, but also have carotenoids which give them their many colors.

Possible cladogram of chloroplast evolution Circles represent endosymbiotic events. For clarity, dinophyte tertiary endosymbioses and many nonphotosynthetic lineages have been omitted.

[a] It is now suspected that Chromalveolata is paraphyletic to Rhizaria.

Apicomplexans

Apicomplexans are another group of chromalveolates. Like the helicosproidia, they're parasitic, and have a nonphotosynthetic chloroplast. They were once thought to be related to the helicosproidia, but it is now known that the helicosproida are green algae rather than chromalveolates. The apicomplexans include *Plasmodium*, the malaria parasite. Many apicomplexans keep a vestigial red algal derived chloroplast called an apicoplast, which they inherited from their ancestors. Other apicomplexans like *Cryptosporidium* have lost the chloroplast completely. Apicomplexans store their energy in amylopectin starch granules that are located in their cytoplasm, even though they are nonphotosynthetic.

Apicoplasts have lost all photosynthetic function, and contain no photosynthetic pigments or true thylakoids. They are bounded by four membranes, but the membranes are not connected to the endoplasmic reticulum. The fact that apicomplexans still keep their nonphotosynthetic chloro-plast around demonstrates how the chloroplast carries out important functions other than pho-tosynthesis. Plant chloroplasts provide plant cells with many important things besides sugar, and apicoplasts are no different—they synthesize fatty acids, isopentenyl pyrophosphate, iron-sulfur clusters, and carry out part of the heme pathway. This makes the apicoplast an attractive target for drugs to cure apicomplexan-related diseases. The most important apicoplast function is isopen-tenyl pyrophosphate synthesis—in fact, apicomplexans die when something interferes with this apicoplast function, and when apicomplexans are grown in an isopentenyl pyrophosphate-rich medium, they dump the organelle.

Dinophytes

The dinoflagellates are yet another very large and diverse group of protists, around half of which are (at least partially) photosynthetic.

Most dinophyte chloroplasts are secondary red algal derived chloroplasts, like other chromalveolate chloroplasts. Many other dinophytes have lost the chloroplast (becoming the nonphotosynthetic kind of dinoflagellate), or replaced it though *tertiary* endosymbiosis—the engulfment of another chromal-veolate containing a red algal derived chloroplast. Others replaced their original chloroplast with a green algal derived one.

Most dinophyte chloroplasts contain at least the photosynthetic pigments chlorophyll *a*, chloro-phyll c_2, *beta*-carotene, and at least one dinophyte-unique xanthophyll (peridinin, dinoxanthin, or diadinoxanthin), giving many a golden-brown color. All dinophytes store starch in their cyto-plasm, and most have chloroplasts with thylakoids arranged in stacks of three.

Peridinin-containing Dinophyte Chloroplast

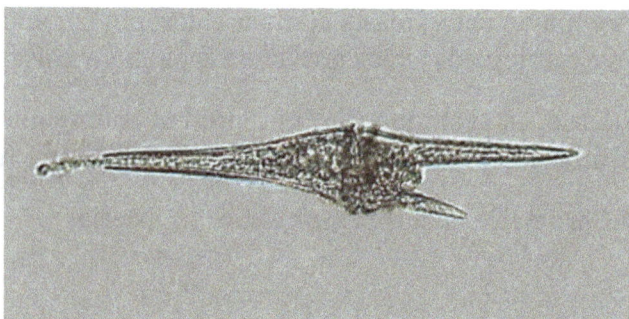

Ceratium furca, a peridinin-containing dinophyte

The most common dinophyte chloroplast is the peridinin-type chloroplast, characterized by the carotenoid pigment peridinin in their chloroplasts, along with chlorophyll *a* and chlorophyll c_2. Peridinin is not found in any other group of chloroplasts. The peridinin chloroplast is bounded by three membranes (occasionally two), having lost the red algal endosymbiont's original cell mem-brane. The outermost membrane is not connected to the endoplasmic reticulum. They contain a pyrenoid, and have triplet-stacked thylakoids. Starch is found outside the chloroplast. An import-ant feature of these chloroplasts is that their chloroplast DNA is highly reduced and fragmented into many small circles. Most of the genome has migrated to the nucleus, and only critical photo-synthesis-related genes remain in the chloroplast.

The peridinin chloroplast is thought to be the dinophytes' "original" chloroplast, which has been lost, reduced, replaced, or has company in several other dinophyte lineages.

Fucoxanthin-containing Dinophyte Chloroplasts (Haptophyte Endosymbionts)

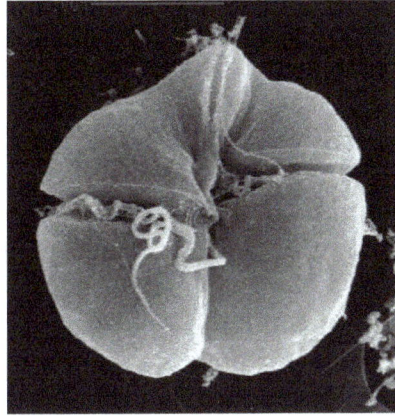

Karenia brevis is a fucoxanthin-containing dynophyte responsible for algal blooms called "red tides".

The fucoxanthin dinophyte lineages (including *Karlodinium* and *Karenia*) lost their original red algal derived chloroplast, and replaced it with a new chloroplast derived from a haptophyte endosymbiont. *Karlodinium* and *Karenia* probably took up different heterokontophytes. Because the haptophyte chloroplast has four membranes, tertiary endosymbiosis would be expected to create a six membraned chloroplast, adding the haptophyte's cell membrane and the dinophyte's phagosomal vacuole. However, the haptophyte was heavily reduced, stripped of a few membranes and its nucleus, leaving only its chloroplast (with its original double membrane), and possibly one or two additional membranes around it.

Fucoxanthin-containing chloroplasts are characterized by having the pigment fucoxanthin (actually 19'-hexanoyloxy-fucoxanthin and/or 19'-butanoyloxy-fucoxanthin) and no peridinin. Fucoxanthin is also found in haptophyte chloroplasts, providing evidence of ancestry.

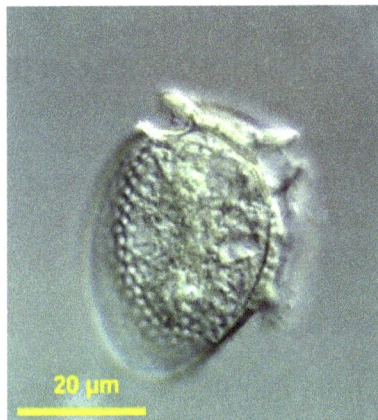

Dinophysis acuminata has chloroplasts taken from a cryptophyte.

Cryptophyte Derived Dinophyte Chloroplast

Members of the genus *Dinophysis* have a phycobilin-containing chloroplast taken from a cryptophyte. However, the cryptophyte is not an endosymbiont—only the chloroplast seems to have been taken, and the chloroplast has been stripped of its nucleomorph and outermost two membranes, leaving just a two-membraned chloroplast. Cryptophyte chloroplasts require their nucleomorph to maintain themselves, and *Dinophysis* species grown in cell culture alone cannot survive, so it is possible (but not confirmed) that the *Dinophysis* chloroplast is a kleptoplast—if so, *Dinophysis* chloroplasts wear out and *Dinophysis* species must continually engulf cryptophytes to obtain new chloroplasts to replace the old ones.

Diatom Derived Dinophyte Chloroplasts

Some dinophytes, like *Kryptoperidinium* and *Durinskia* have a diatom (heterokontophyte) derived chloroplast. These chloroplasts are bounded by up to *five* membranes, (depending on whether you count the entire diatom endosymbiont as the chloroplast, or just the red algal derived chloroplast inside it). The diatom endosymbiont has been reduced relatively little—it still retains its original mitochondria, and has endoplasmic reticulum, ribosomes, a nucleus, and of course, red algal derived chloroplasts—practically a complete cell, all inside the host's endoplasmic reticulum lumen. However the diatom endosymbiont can't store its own food—its starch is found in granules in the dinophyte host's cytoplasm instead. The diatom endosymbiont's nucleus is present, but it probably can't be called a nucleomorph because it shows no sign of genome reduction, and might have even been *expanded*. Diatoms have been engulfed by dinoflagellates at least three times.

The diatom endosymbiont is bounded by a single membrane, inside it are chloroplasts with four membranes. Like the diatom endosymbiont's diatom ancestor, the chloroplasts have triplet thylakoids and pyrenoids.

In some of these genera, the diatom endosymbiont's chloroplasts aren't the only chloroplasts in the dinophyte. The original three-membraned peridinin chloroplast is still around, converted to an eyespot.

Prasinophyte (Green Algal) Derived Dinophyte Chloroplast

Lepidodinium viride and its close relatives are dinophytes that lost their original peridinin chloroplast and replaced it with a green algal derived chloroplast (more specifically, a prasinophyte). *Lepidodinium* is the only dinophyte that has a chloroplast that's not from the rhodoplast lineage. The chloroplast is surrounded by two membranes and has no nucleomorph—all the nucleomorph genes have been transferred to the dinophyte nucleus. The endosymbiotic event that led to this chloroplast was serial secondary endosymbiosis rather than tertiary endosymbiosis—the endosymbiont was a green alga containing a primary chloroplast (making a secondary chloroplast).

Chromatophores

While most chloroplasts originate from that first set of endosymbiotic events, *Paulinella chromatophora* is an exception that acquired a photosynthetic cyanobacterial endosymbiont more recently. It is not clear whether that symbiont is closely related to the ancestral chloroplast of other eukaryotes. Being in the early stages of endosymbiosis, *Paulinella chromatophora* can offer some insights into how chloroplasts evolved. *Paulinella* cells contain one or two sausage shaped blue-green photosynthesizing structures called chromatophores, descended from the cyanobacterium

Synechococcus. Chromatophores cannot survive outside their host. Chromatophore DNA is about a million base pairs long, containing around 850 protein encoding genes—far less than the three million base pair *Synechococcus* genome, but much larger than the approximately 150,000 base pair genome of the more assimilated chloroplast. Chromatophores have transferred much less of their DNA to the nucleus of their host. About 0.3–0.8% of the nuclear DNA in *Paulinella* is from the chromatophore, compared with 11–14% from the chloroplast in plants.

Kleptoplastidy

In some groups of mixotrophic protists, like some dinoflagellates, chloroplasts are separated from a captured alga or diatom and used temporarily. These klepto chloroplasts may only have a lifetime of a few days and are then replaced.

Chloroplast DNA

Chloroplasts have their own DNA, often abbreviated as ctDNA, or cpDNA. It is also known as the plastome. Its existence was first proved in 1962, and first sequenced in 1986—when two Japanese research teams sequenced the chloroplast DNA of liverwort and tobacco. Since then, hundreds of chloroplast DNAs from various species have been sequenced, but they're mostly those of land plants and green algae—glaucophytes, red algae, and other algal groups are extremely underrepresented, potentially introducing some bias in views of "typical" chloroplast DNA structure and content.

Molecular Structure

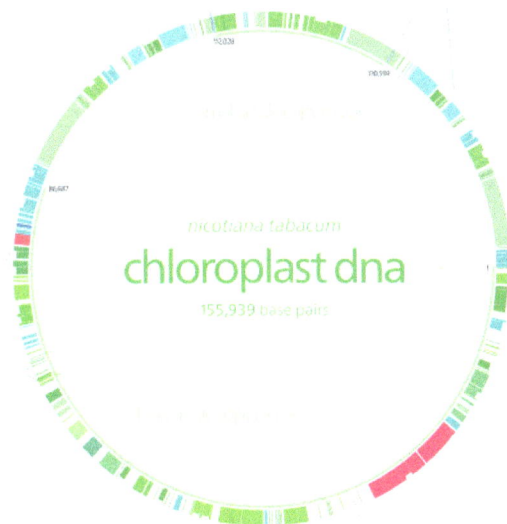

With few exceptions, most chloroplasts have their entire chloroplast genome combined into a single large circular DNA molecule, typically 120,000–170,000 base pairs long. They can have a contour length of around 30–60 micrometers, and have a mass of about 80–130 million daltons.

While usually thought of as a circular molecule, there is some evidence that chloroplast DNA molecules more often take on a linear shape.

Inverted Repeats

Many chloroplast DNAs contain two *inverted repeats*, which separate a long single copy section (LSC) from a short single copy section (SSC). While a given pair of inverted repeats are rarely completely identical, they are always very similar to each other, apparently resulting from concerted evolution.

The inverted repeats vary wildly in length, ranging from 4,000 to 25,000 base pairs long each and containing as few as four or as many as over 150 genes. Inverted repeats in plants tend to be at the upper end of this range, each being 20,000–25,000 base pairs long.

The inverted repeat regions are highly conserved among land plants, and accumulate few mutations. Similar inverted repeats exist in the genomes of cyanobacteria and the other two chloroplast lineages (glaucophyta and rhodophyceae), suggesting that they predate the chloroplast, though some chloroplast DNAs have since lost or flipped the inverted repeats (making them direct repeats). It is possible that the inverted repeats help stabilize the rest of the chloroplast genome, as chloroplast DNAs which have lost some of the inverted repeat segments tend to get rearranged more.

Nucleoids

New chloroplasts may contain up to 100 copies of their DNA, though the number of chloroplast DNA copies decreases to about 15–20 as the chloroplasts age. They are usually packed into nucleoids, which can contain several identical chloroplast DNA rings. Many nucleoids can be found in each chloroplast. In primitive red algae, the chloroplast DNA nucleoids are clustered in the center of the chloroplast, while in green plants and green algae, the nucleoids are dispersed throughout the stroma.

Though chloroplast DNA is not associated with true histones, in red algae, similar proteins that tightly pack each chloroplast DNA ring into a nucleoid have been found.

DNA Replication

The Leading Model of cpDNA Replication

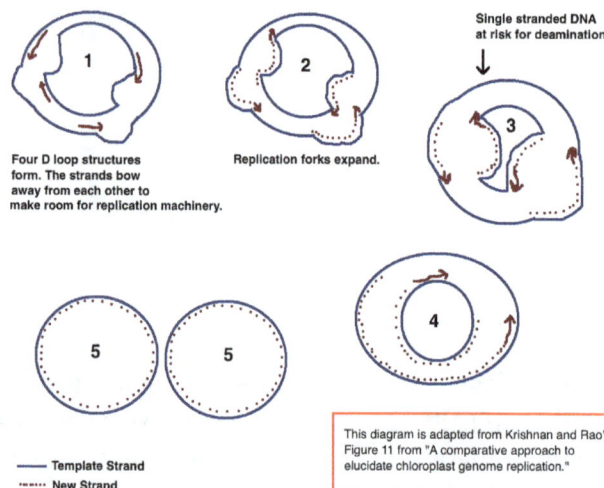

Chloroplast DNA replication via multiple D loop mechanisms. Adapted from Krishnan NM, Rao BJ's paper "A comparative approach to elucidate chloroplast genome replication."

The mechanism for chloroplast DNA (cpDNA) replication has not been conclusively determined, but two main models have been proposed. Scientists have attempted to observe chloroplast replication via electron microscopy since the 1970s. The results of the microscopy experiments led to the idea that chloroplast DNA replicates using a double displacement loop (D-loop). As the D-loop moves through the circular DNA, it adopts a theta intermediary form, also known as a Cairns replication intermediate, and completes replication with a rolling circle mechanism. Transcription starts at specific points of origin. Multiple replication forks open up, allowing replication machinery to transcribe the DNA. As replication continues, the forks grow and eventually converge. The new cpDNA structures separate, creating daughter cpDNA chromosomes.

In addition to the early microscopy experiments, this model is also supported by the amounts of deamination seen in cpDNA. Deamination occurs when an amino group is lost and is a mutation that often results in base changes. When adenine is deaminated, it becomes hypoxanthine. Hypoxanthine can bind to cytosine, and when the XC base pair is replicated, it becomes a GC (thus, an A → G base change).

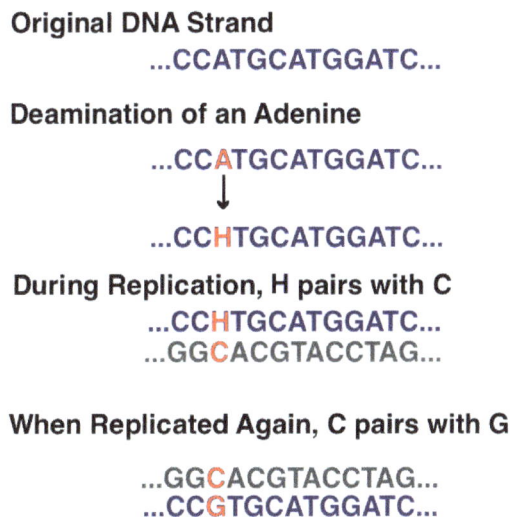

Original DNA Strand
...CCATGCATGGATC...

Deamination of an Adenine
...CCATGCATGGATC...
↓
...CCHTGCATGGATC...

During Replication, H pairs with C
...CCHTGCATGGATC...
...GGCACGTACCTAG...

When Replicated Again, C pairs with G
...GGCACGTACCTAG...
...CCGTGCATGGATC...

Over time, base changes in the DNA sequence can arise from deamination mutations. When adenine is deaminated, it becomes hypoxanthine, which can pair with cytosine. During replication, the cytosine will pair with guanine, causing an A --> G base change.

Deamination

In cpDNA, there are several A → G deamination gradients. DNA becomes susceptible to deamination events when it is single stranded. When replication forks form, the strand not being copied is single stranded, and thus at risk for A → G deamination. Therefore, gradients in deamination indicate that replication forks were most likely present and the direction that they initially opened (the highest gradient is most likely nearest the start site because it was single stranded for the longest amount of time). This mechanism is still the leading theory today; however, a second theory suggests that most cpDNA is actually linear and replicates through homologous recombination. It further contends that only a minority of the genetic material is kept in circular chromosomes while the rest is in branched, linear, or other complex structures.

Alternative Model of Replication

One of competing model for cpDNA replication asserts that most cpDNA is linear and participates in homologous recombination and replication structures similar to bacteriophage T4. It has been established that some plants have linear cpDNA, such as maize, and that more species still contain complex structures that scientists do not yet understand. When the original experiments on cpDNA were performed, scientists did notice linear structures; however, they attributed these linear forms to broken circles. If the branched and complex structures seen in cpDNA experiments are real and not artifacts of concatenated circular DNA or broken circles, then a D-loop mechanism of replication is insufficient to explain how those structures would replicate. At the same time, homologous recombination does not expand the multiple A --> G gradients seen in plastomes. Because of the failure to explain the deamination gradient as well as the numerous plant species that have been shown to have circular cpDNA, the predominant theory continues to hold that most cpDNA is circular and most likely replicates via a D loop mechanism.

Gene Content and Protein Synthesis

The chloroplast genome most commonly includes around 100 genes that code for a variety of things, mostly to do with the protein pipeline and photosynthesis. As in prokaryotes, genes in chloroplast DNA are organized into operons. Interestingly though, unlike prokaryotic DNA molecules, chloroplast DNA molecules contain introns (plant mitochondrial DNAs do too, but not human mtDNAs).

Among land plants, the contents of the chloroplast genome are fairly similar.

Chloroplast Genome Reduction and Gene Transfer

Over time, many parts of the chloroplast genome were transferred to the nuclear genome of the host, a process called *endosymbiotic gene transfer*. As a result, the chloroplast genome is heavily reduced compared to that of free-living cyanobacteria. Chloroplasts may contain 60–100 genes whereas cyanobacteria often have more than 1500 genes in their genome. Recently, a plastid without a genome was found, demonstrating chloroplasts can lose their genome during endosymbiotic the gene transfer process.

Endosymbiotic gene transfer is how we know about the lost chloroplasts in many chromalveolate lineages. Even if a chloroplast is eventually lost, the genes it donated to the former host's nucleus persist, providing evidence for the lost chloroplast's existence. For example, while diatoms (a heterokontophyte) now have a red algal derived chloroplast, the presence of many green algal genes in the diatom nucleus provide evidence that the diatom ancestor (probably the ancestor of all chromalveolates too) had a green algal derived chloroplast at some point, which was subsequently replaced by the red chloroplast.

In land plants, some 11–14% of the DNA in their nuclei can be traced back to the chloroplast, up to 18% in *Arabidopsis*, corresponding to about 4,500 protein-coding genes. There have been a few recent transfers of genes from the chloroplast DNA to the nuclear genome in land plants.

Of the approximately 3000 proteins found in chloroplasts, some 95% of them are encoded by nuclear genes. Many of the chloroplast's protein complexes consist of subunits from both the chloroplast genome and the host's nuclear genome. As a result, protein synthesis must be coordinated between the chloroplast and the nucleus. The chloroplast is mostly under nuclear control, though chloroplasts can also give out signals regulating gene expression in the nucleus, called *retrograde signaling*.

Protein Synthesis

Protein synthesis within chloroplasts relies on two RNA polymerases. One is coded by the chloroplast DNA, the other is of nuclear origin. The two RNA polymerases may recognize and bind to different kinds of promoters within the chloroplast genome. The ribosomes in chloroplasts are similar to bacterial ribosomes.

Protein Targeting and Import

Because so many chloroplast genes have been moved to the nucleus, many proteins that would originally have been translated in the chloroplast are now synthesized in the cytoplasm of the plant cell. These proteins must be directed back to the chloroplast, and imported through at least two chloroplast membranes.

Curiously, around half of the protein products of transferred genes aren't even targeted back to the chloroplast. Many became exaptations, taking on new functions like participating in cell division, protein routing, and even disease resistance. A few chloroplast genes found new homes in the mitochondrial genome—most became nonfunctional pseudogenes, though a few tRNA genes still work in the mitochondrion. Some transferred chloroplast DNA protein products get directed to the secretory pathway (though it should be noted that many secondary plastids are bounded by an outermost membrane derived from the host's cell membrane, and therefore topologically outside of the cell, because to reach the chloroplast from the cytosol, you have to cross the cell membrane, just like if you were headed for the extracellular space. In those cases, chloroplast-targeted proteins do initially travel along the secretory pathway).

Because the cell acquiring a chloroplast already had mitochondria (and peroxisomes, and a cell membrane for secretion), the new chloroplast host had to develop a unique protein targeting system to avoid having chloroplast proteins being sent to the wrong organelle, and the wrong proteins being sent to the chloroplast.

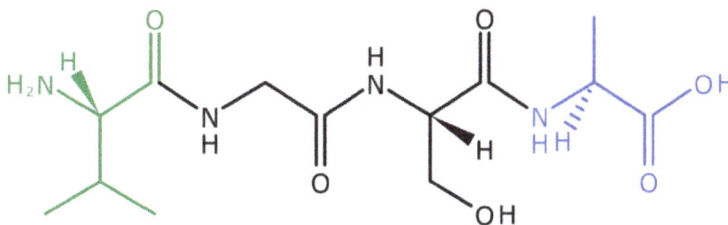

The two ends of a polypeptide are called the N-terminus, or *amino end*, and the C-terminus, or *carboxyl end*. This polypeptide has four amino acids linked together. At the left is the N-terminus, with its amino (H_2N) group in green. The blue C-terminus, with its carboxyl group (CO_2H) is at the right.

In most, but not all cases, nuclear-encoded chloroplast proteins are translated with a *cleavable transit peptide* that's added to the N-terminus of the protein precursor. Sometimes the transit sequence is found on the C-terminus of the protein, or within the functional part of the protein.

Transport Proteins and Membrane Translocons

After a chloroplast polypeptide is synthesized on a ribosome in the cytosol, an enzyme specific to chloroplast proteins phosphorylates, or adds a phosphate group to many (but not all) of them in their transit sequences. Phosphorylation helps many proteins bind the polypeptide, keeping it from folding prematurely. This is important because it prevents chloroplast proteins from assuming their active form and carrying out their chloroplast functions in the wrong place—the cytosol. At the same time, they have to keep just enough shape so that they can be recognized by the chloroplast. These proteins also help the polypeptide get imported into the chloroplast.

From here, chloroplast proteins bound for the stroma must pass through two protein complexes—the TOC complex, or *translocon on the outer chloroplast membrane*, and the TIC translocon, or *translocon on the inner chloroplast membrane translocon*. Chloroplast polypeptide chains probably often travel through the two complexes at the same time, but the TIC complex can also retrieve preproteins lost in the intermembrane space.

Structure

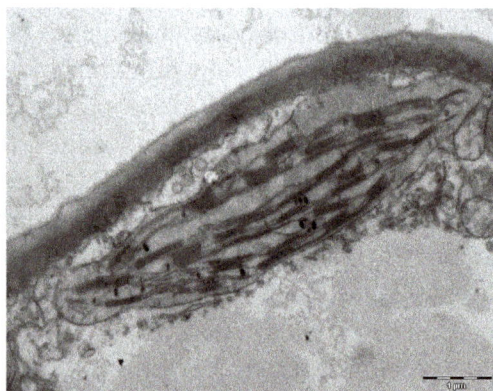

Transmission electron microscope image of a chloroplast.
Grana of thylakoids and their connecting lamellae are clearly visible.

In land plants, chloroplasts are generally lens-shaped, 5–8 μm in diameter and 1–3 μm thick. Greater diversity in chloroplast shapes exists among the algae, which often contain a single chloroplast that can be shaped like a net (e.g., *Oedogonium*), a cup (e.g., *Chlamydomonas*), a ribbon-like spiral around the edges of the cell (e.g., *Spirogyra*), or slightly twisted bands at the cell edges (e.g., *Sirogonium*). Some algae have two chloroplasts in each cell; they are star-shaped in *Zygnema*, or may follow the shape of half the cell in order Desmidiales. In some algae, the chloroplast takes up most of the cell, with pockets for the nucleus and other organelles (for example some species of *Chlorella* have a cup-shaped chloroplast that occupies much of the cell).

All chloroplasts have at least three membrane systems—the outer chloroplast membrane, the inner chloroplast membrane, and the thylakoid system. Chloroplasts that are the product of secondary endosymbiosis may have additional membranes surrounding these three. Inside the outer and inner chloroplast membranes is the chloroplast stroma, a semi-gel-like fluid that makes up much of a chloroplast's volume, and in which the thylakoid system floats.

3 Thylakoid
  3.1 Thylakoid space (lumen) **4** Stromal thylakoids
  3.2 Thylakoid membrane   (lamellæ or frets)

2 Chloroplast envelope   **5 Granal thylakoids**
  2.1 Outer membrane   **6 Stroma**
  2.2 Intermembrane space
  2.3 Inner membrane

1 Granum

7 Nucleoid
  (DNA rings)
8 Ribosome

9 Plastoglobulus

10 Starch granule

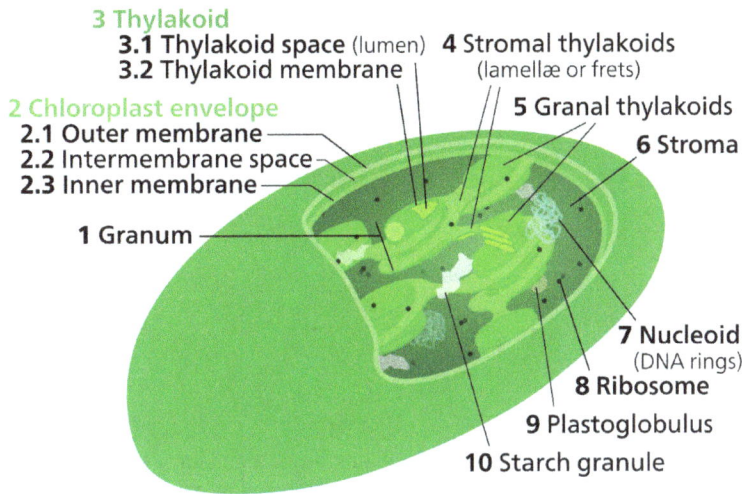

There are some common misconceptions about the outer and inner chloroplast membranes. The fact that chloroplasts are surrounded by a double membrane is often cited as evidence that they are the descendants of endosymbiotic cyanobacteria. This is often interpreted as meaning the outer chloroplast membrane is the product of the host's cell membrane infolding to form a vesicle to surround the ancestral cyanobacterium—which is not true—both chloroplast membranes are homologous to the cyanobacterium's original double membranes.

The chloroplast double membrane is also often compared to the mitochondrial double membrane. This is not a valid comparison—the inner mitochondria membrane is used to run proton pumps and carry out oxidative phosphorylation across to generate ATP energy. The only chloroplast structure that can considered analogous to it is the internal thylakoid system. Even so, in terms of "in-out", the direction of chloroplast H^+ ion flow is in the opposite direction compared to oxidative phosphorylation in mitochondria. In addition, in terms of function, the inner chloroplast membrane, which regulates metabolite passage and synthesizes some materials, has no counterpart in the mitochondrion.

Outer Chloroplast Membrane

The outer chloroplast membrane is a semi-porous membrane that small molecules and ions can easily diffuse across. However, it is not permeable to larger proteins, so chloroplast polypeptides being synthesized in the cell cytoplasm must be transported across the outer chloroplast membrane by the TOC complex, or *translocon on the outer chloroplast* membrane.

The chloroplast membranes sometimes protrude out into the cytoplasm, forming a stromule, or stroma-containing tubule. Stromules are very rare in chloroplasts, and are much more common in other plastids like chromoplasts and amyloplasts in petals and roots, respectively. They may exist to increase the chloroplast's surface area for cross-membrane transport, because they are often branched and tangled with the endoplasmic reticulum. When they were first observed in 1962, some plant biologists dismissed the structures as artifactual, claiming that stromules were just oddly shaped chloroplasts with constricted regions or dividing chloroplasts. However, there is a growing body of evidence that stromules are functional, integral features of plant cell plastids, not merely artifacts.

Intermembrane Space and Peptidoglycan Wall

Instead of an intermembrane space, glaucophyte algae have a peptidoglycan
wall between their inner and outer chloroplast membranes.

Usually, a thin intermembrane space about 10–20 nanometers thick exists between the outer and inner chloroplast membranes.

Glaucophyte algal chloroplasts have a peptidoglycan layer between the chloroplast membranes. It corresponds to the peptidoglycan cell wall of their cyanobacterial ancestors, which is located between their two cell membranes. These chloroplasts are called muroplasts. Other chloroplasts have lost the cyanobacterial wall, leaving an intermembrane space between the two chloroplast envelope membranes.

Inner Chloroplast Membrane

The inner chloroplast membrane borders the stroma and regulates passage of materials in and out of the chloroplast. After passing through the TOC complex in the outer chloroplast membrane, polypeptides must pass through the TIC complex *(translocon on the inner chloroplast membrane)* which is located in the inner chloroplast membrane.

In addition to regulating the passage of materials, the inner chloroplast membrane is where fatty acids, lipids, and carotenoids are synthesized.

Peripheral Reticulum

Some chloroplasts contain a structure called the chloroplast peripheral reticulum. It is often found in the chloroplasts of C_4 plants, though it has also been found in some C_3 angiosperms, and even some gymnosperms. The chloroplast peripheral reticulum consists of a maze of membranous tubes and vesicles continuous with the inner chloroplast membrane that extends into the internal stromal fluid of the chloroplast. Its purpose is thought to be to increase the chloroplast's surface area for cross-membrane transport between its stroma and the cell cytoplasm. The small vesicles sometimes observed may serve as transport vesicles to shuttle stuff between the thylakoids and intermembrane space.

Stroma

The protein-rich, alkaline, aqueous fluid within the inner chloroplast membrane and outside of the thylakoid space is called the stroma, which corresponds to the cytosol of the original cyanobacterium. Nucleoids of chloroplast DNA, chloroplast ribosomes, the thylakoid system with plastoglobuli, starch granules, and many proteins can be found floating around in it. The Calvin cycle, which fixes CO_2 into sugar takes place in the stroma.

Chloroplast Ribosomes

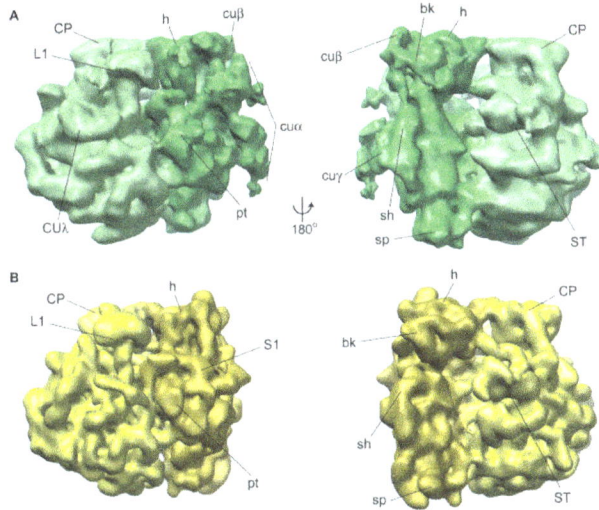

Chloroplast ribosomes Comparison of a chloroplast ribosome (green) and a bacterial ribosome (yellow). Important features common to both ribosomes and chloroplast-unique features are labeled.

Chloroplasts have their own ribosomes, which they use to synthesize a small fraction of their proteins. Chloroplast ribosomes are about two-thirds the size of cytoplasmic ribosomes (around 17 nm vs 25 nm). They take mRNAs transcribed from the chloroplast DNA and translate them into protein. While similar to bacterial ribosomes, chloroplast translation is more complex than in bacteria, so chloroplast ribosomes include some chloroplast-unique features. Small subunit ribosomal RNAs in several Chlorophyta and euglenid chloroplasts lack motifs for shine-dalgarno sequence recognition, which is considered essential for translation initiation in most chloroplasts and prokaryotes. Such loss is also rarely observed in other plastids and prokaryotes.

Plastoglobuli

Plastoglobuli (singular *plastoglobulus*, sometimes spelled *plastoglobule(s)*), are spherical bubbles of lipids and proteins about 45–60 nanometers across. They are surrounded by a lipid monolayer. Plastoglobuli are found in all chloroplasts, but become more common when the chloroplast is under oxidative stress, or when it ages and transitions into a gerontoplast. Plastoglobuli also exhibit a greater size variation under these conditions. They are also common in etioplasts, but decrease in number as the etioplasts mature into chloroplasts.

Plastoglubuli contain both structural proteins and enzymes involved in lipid synthesis and metabolism. They contain many types of lipids including plastoquinone, vitamin E, carotenoids and chlorophylls.

Plastoglobuli were once thought to be free-floating in the stroma, but it is now thought that they are permanently attached either to a thylakoid or to another plastoglobulus attached to a thylakoid, a configuration that allows a plastoglobulus to exchange its contents with the thylakoid network. In normal green chloroplasts, the vast majority of plastoglobuli occur singularly, attached directly to their parent thylakoid. In old or stressed chloroplasts, plastoglobuli tend to occur in linked groups or chains, still always anchored to a thylakoid.

Plastoglobuli form when a bubble appears between the layers of the lipid bilayer of the thylakoid membrane, or bud from existing plastoglubuli—though they never detach and float off into the stroma. Practically all plastoglobuli form on or near the highly curved edges of the thylakoid disks or sheets. They are also more common on stromal thylakoids than on granal ones.

Starch Granules

Starch granules are very common in chloroplasts, typically taking up 15% of the organelle's volume, though in some other plastids like amyloplasts, they can be big enough to distort the shape of the organelle. Starch granules are simply accumulations of starch in the stroma, and are not bounded by a membrane.

Starch granules appear and grow throughout the day, as the chloroplast synthesizes sugars, and are consumed at night to fuel respiration and continue sugar export into the phloem, though in mature chloroplasts, it is rare for a starch granule to be completely consumed or for a new granule to accumulate.

Starch granules vary in composition and location across different chloroplast lineages. In red algae, starch granules are found in the cytoplasm rather than in the chloroplast. In C_4 plants, mesophyll chloroplasts, which do not synthesize sugars, lack starch granules.

Rubisco

Rubisco, shown here in a space-filling model, is the main enzyme responsible for carbon fixation in chloroplasts.

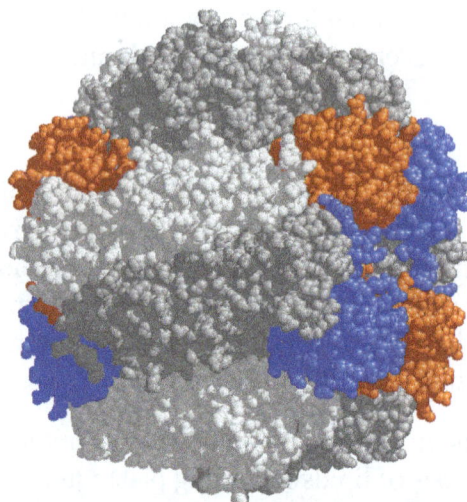

The chloroplast stroma contains many proteins, though the most common and important is Rubisco, which is probably also the most abundant protein on the planet. Rubisco is the enzyme that fixes CO_2 into sugar molecules. In C_3 plants, rubisco is abundant in all chloroplasts, though in C_4 plants, it is confined to the bundle sheath chloroplasts, where the Calvin cycle is carried out in C_4 plants.

Pyrenoids

The chloroplasts of some hornworts and algae contain structures called pyrenoids. They are not found in higher plants. Pyrenoids are roughly spherical and highly refractive bodies which are a site of starch accumulation in plants that contain them. They consist of a matrix opaque to electrons, surrounded by two hemispherical starch plates. The starch is accumulated as the pyrenoids mature. In algae with carbon concentrating mechanisms, the enzyme rubisco is found in the pyrenoids. Starch can also accumulate around the pyrenoids when CO_2 is scarce. Pyrenoids can divide to form new pyrenoids, or be produced "de novo".

Thylakoid System

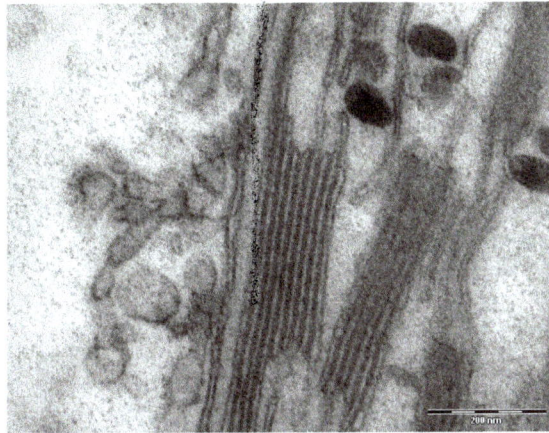

Transmission electron microscope image of some thylakoids arranged in grana stacks and lamellæ. Plastoglobuli (dark blobs) are also present.

Suspended within the chloroplast stroma is the thylakoid system, a highly dynamic collection of membranous sacks called thylakoids where chlorophyll is found and the light reactions of photosynthesis happen. In most vascular plant chloroplasts, the thylakoids are arranged in stacks called grana, though in certain C_4 plant chloroplasts and some algal chloroplasts, the thylakoids are free floating.

Granal Structure

Using a light microscope, it is just barely possible to see tiny green granules—which were named grana. With electron microscopy, it became possible to see the thylakoid system in more detail, revealing it to consist of stacks of flat thylakoids which made up the grana, and long interconnecting stromal thylakoids which linked different grana. In the transmission electron microscope, thylakoid membranes appear as alternating light-and-dark bands, 8.5 nanometers thick.

For a long time, the three-dimensional structure of the thylakoid system has been unknown or disputed. One model has the granum as a stack of thylakoids linked by helical stromal thylakoids; the other has the granum as a single folded thylakoid connected in a "hub and spoke" way to other grana by stromal thylakoids. While the thylakoid system is still commonly depicted according to the folded thylakoid model, it was determined in 2011 that the stacked and helical thylakoids model is correct.

Granum structure The prevailing model for granal structure is a stack of granal thylakoids linked by helical stromal thylakoids that wrap around the grana stacks and form large sheets that connect different grana.

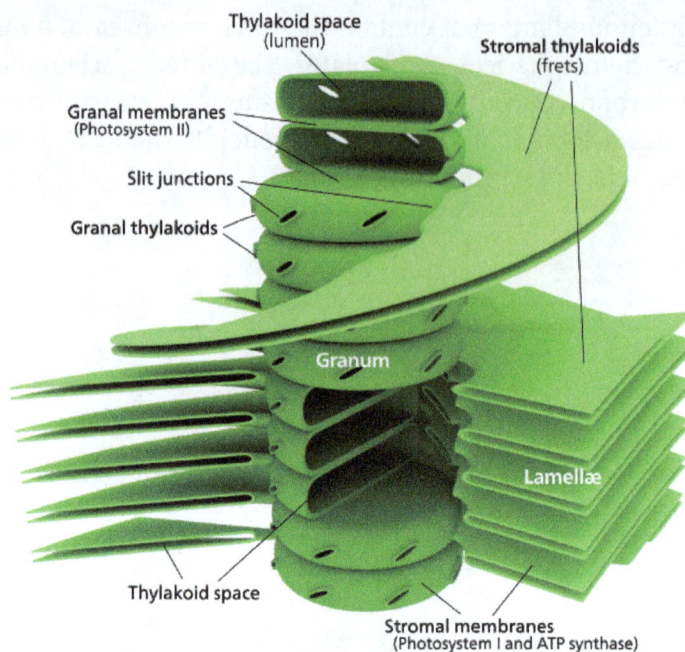

In the helical thylakoid model, grana consist of a stack of flattened circular granal thylakoids that resemble pancakes. Each granum can contain anywhere from two to a hundred thylakoids, though grana with 10–20 thylakoids are most common. Wrapped around the grana are helicoid stromal thylakoids, also known as frets or lamellar thylakoids. The helices ascend at an angle of 20–25°, connecting to each granal thylakoid at a bridge-like slit junction. The helicoids may extend as large sheets that link multiple grana, or narrow to tube-like bridges between grana. While different parts of the thylakoid system contain different membrane proteins, the thylakoid membranes are continuous and the thylakoid space they enclose form a single continuous labyrinth.

Thylakoids

Thylakoids (sometimes spelled *thylakoïds*), are small interconnected sacks which contain the membranes that the light reactions of photosynthesis take place on. The word *thylakoid* comes from the Greek word *thylakos* which means "sack".

Embedded in the thylakoid membranes are important protein complexes which carry out the light reactions of photosynthesis. Photosystem II and photosystem I contain light-harvesting complexes with chlorophyll and carotenoids that absorb light energy and use it to energize electrons. Molecules in the thylakoid membrane use the energized electrons to pump hydrogen ions into the thylakoid space, decreasing the pH and turning it acidic. ATP synthase is a large protein complex that harnesses the concentration gradient of the hydrogen ions in the thylakoid space to generate ATP energy as the hydrogen ions flow back out into the stroma—much like a dam turbine.

There are two types of thylakoids—granal thylakoids, which are arranged in grana, and stromal thylakoids, which are in contact with the stroma. Granal thylakoids are pancake-shaped circular disks about 300–600 nanometers in diameter. Stromal thylakoids are helicoid sheets that spiral around grana. The flat tops and bottoms of granal thylakoids contain only the relatively flat photosystem II protein complex. This allows them to stack tightly, forming grana with many layers of tightly appressed membrane, called granal membrane, increasing stability and surface area for light capture.

In contrast, photosystem I and ATP synthase are large protein complexes which jut out into the stroma. They can't fit in the appressed granal membranes, and so are found in the stromal thylakoid membrane—the edges of the granal thylakoid disks and the stromal thylakoids. These large protein complexes may act as spacers between the sheets of stromal thylakoids.

The number of thylakoids and the total thylakoid area of a chloroplast is influenced by light exposure. Shaded chloroplasts contain larger and more grana with more thylakoid membrane area than chloroplasts exposed to bright light, which have smaller and fewer grana and less thylakoid area. Thylakoid extent can change within minutes of light exposure or removal.

Pigments and Chloroplast Colors

Inside the photosystems embedded in chloroplast thylakoid membranes are various photosynthetic pigments, which absorb and transfer light energy. The types of pigments found are different in various groups of chloroplasts, and are responsible for a wide variety of chloroplast colorations.

Paper chroma-tography of some spinach leaf extract shows the various pigments present in their chloroplasts.

Xanthophylls
Chlorophyll *a*
Chlorophyll *b*

Chlorophylls

Chlorophyll *a* is found in all chloroplasts, as well as their cyanobacterial ancestors. Chlorophyll *a* is a blue-green pigment partially responsible for giving most cyanobacteria and chloroplasts their color. Other forms of chlorophyll exist, such as the accessory pigments chlorophyll *b*, chlorophyll *c*, chlorophyll *d*, and chlorophyll *f*.

Chlorophyll *b* is an olive green pigment found only in the chloroplasts of plants, green algae, any secondary chloroplasts obtained through the secondary endosymbiosis of a green alga, and a few cyanobacteria. It is the chlorophylls *a* and *b* together that make most plant and green algal chloroplasts green.

Chlorophyll *c* is mainly found in secondary endosymbiotic chloroplasts that originated from a red alga, although it is not found in chloroplasts of red algae themselves. Chlorophyll *c* is also found in some green algae and cyanobacteria.

Chlorophylls *d* and *f* are pigments found only in some cyanobacteria.

Carotenoids

Delesseria sanguinea, a red alga, has chloroplasts that contain red pigments like phycoerytherin that mask their blue-green chlorophyll *a*.

In addition to chlorophylls, another group of yellow–orange pigments called carotenoids are also found in the photosystems. There are about thirty photosynthetic carotenoids. They help transfer and dissipate excess energy, and their bright colors sometimes override the chlorophyll green, like during the fall, when the leaves of some land plants change color. β-carotene is a bright red-orange carotenoid found in nearly all chloroplasts, like chlorophyll *a*. Xanthophylls, especially the orange-red zeaxanthin, are also common. Many other forms of carotenoids exist that are only found in certain groups of chloroplasts.

Phycobilins

Phycobilins are a third group of pigments found in cyanobacteria, and glaucophyte, red algal, and cryptophyte chloroplasts. Phycobilins come in all colors, though phycoerytherin is one of the pigments that makes many red algae red. Phycobilins often organize into relatively large protein complexes about 40 nanometers across called phycobilisomes. Like photosystem I and ATP synthase, phycobilisomes jut into the stroma, preventing thylakoid stacking in red algal chloroplasts. Cryptophyte chloroplasts and some cyanobacteria don't have their phycobilin pigments organized into phycobilisomes, and keep them in their thylakoid space instead.

Photosynthetic pigments Table of the presence of various pigments across chloroplast groups. Colored cells represent pigment presence.

Specialized Chloroplasts in C$_4$ Plants

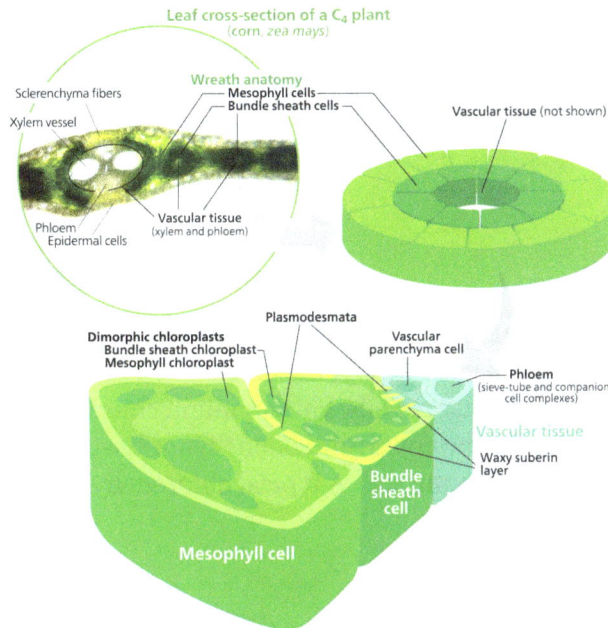

Many C$_4$ plants have their mesophyll cells and bundle sheath cells arranged radially around their leaf veins. The two types of cells contain different types of chloroplasts specialized for a particular part of photosynthesis.

To fix carbon dioxide into sugar molecules in the process of photosynthesis, chloroplasts use an enzyme called rubisco. Rubisco has a problem—it has trouble distinguishing between carbon dioxide and oxygen, so at high oxygen concentrations, rubisco starts accidentally adding oxygen to sugar precursors. This has the end result of ATP energy being wasted and CO$_2$ being released, all with no sugar being produced. This is a big problem, since O$_2$ is produced by the initial light reactions of photosynthesis, causing issues down the line in the Calvin cycle which uses rubisco.

C$_4$ plants evolved a way to solve this—by spatially separating the light reactions and the Calvin cycle. The light reactions, which store light energy in ATP and NADPH, are done in the mesophyll cells of a C$_4$ leaf. The Calvin cycle, which uses the stored energy to make sugar using rubisco, is done in the bundle sheath cells, a layer of cells surrounding a vein in a leaf.

As a result, chloroplasts in C$_4$ mesophyll cells and bundle sheath cells are specialized for each stage of photosynthesis. In mesophyll cells, chloroplasts are specialized for the light reactions, so they lack rubisco, and have normal grana and thylakoids, which they use to make ATP and NADPH, as well as oxygen. They store CO$_2$ in a four-carbon compound, which is why the process is called *C$_4$ photosynthesis*. The four-carbon compound is then transported to the bundle sheath chloroplasts, where it drops off CO$_2$ and returns to the mesophyll. Bundle sheath chloroplasts do not carry out the light reactions, preventing oxygen from building up in them and disrupting rubisco activity. Because of this, they lack thylakoids organized into grana stacks—though bundle sheath chloroplasts still have free-floating thylakoids in the stroma where they still carry out cyclic electron flow, a light-driven method of synthesizing ATP to power the Calvin cycle without generating oxygen. They lack photosystem II, and only have photosystem I—the only protein complex needed for cyclic electron flow. Because the job of bundle sheath chloroplasts is to carry out the Calvin cycle and make sugar, they often contain large starch grains.

Both types of chloroplast contain large amounts of chloroplast peripheral reticulum, which they use to get more surface area to transport stuff in and out of them. Mesophyll chloroplasts have a little more peripheral reticulum than bundle sheath chloroplasts.

Location

Distribution in a Plant

Not all cells in a multicellular plant contain chloroplasts. All green parts of a plant contain chloroplasts—the chloroplasts, or more specifically, the chlorophyll in them are what make the photosynthetic parts of a plant green. The plant cells which contain chloroplasts are usually parenchyma cells, though chloroplasts can also be found in collenchyma tissue. A plant cell which contains chloroplasts is known as a chlorenchyma cell. A typical chlorenchyma cell of a land plant contains about 10 to 100 chloroplasts.

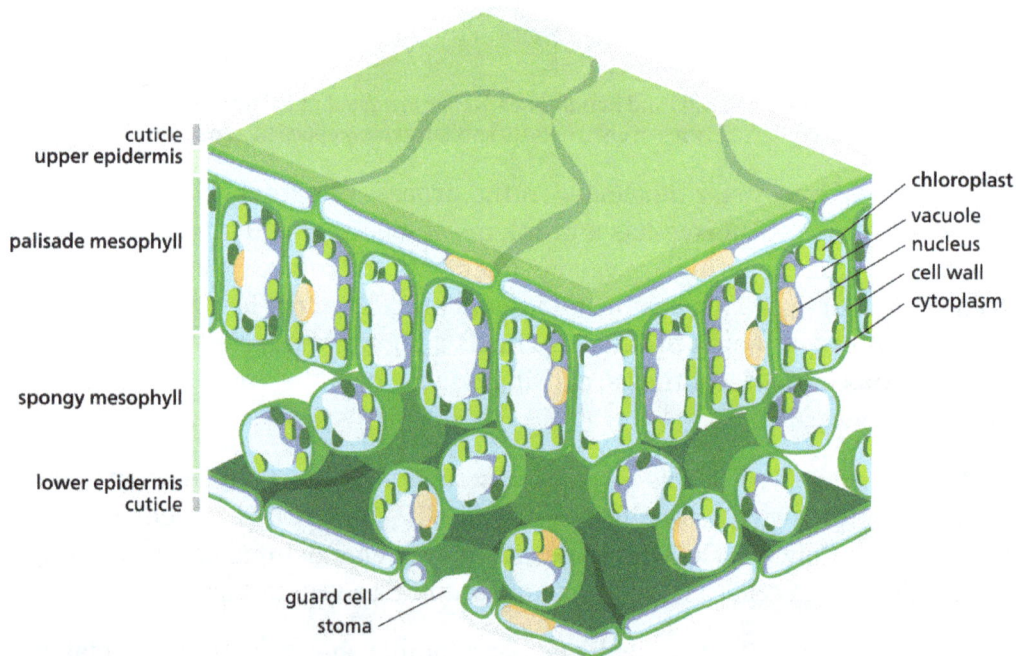

A cross section of a leaf, showing chloroplasts in its mesophyll cells.
Stomal guard cells also have chloroplasts, though much fewer than mesophyll cells.

In some plants such as cacti, chloroplasts are found in the stems, though in most plants, chloroplasts are concentrated in the leaves. One square millimeter of leaf tissue can contain half a million chloroplasts. Within a leaf, chloroplasts are mainly found in the mesophyll layers of a leaf, and the guard cells of stomata. Palisade mesophyll cells can contain 30–70 chloroplasts per cell, while stomatal guard cells contain only around 8–15 per cell, as well as much less chlorophyll. Chloroplasts can also be found in the bundle sheath cells of a leaf, especially in

C_4 plants, which carry out the Calvin cycle in their bundle sheath cells. They are often absent from the epidermis of a leaf.

Cellular Location

Chloroplast Movement

The chloroplasts of plant and algal cells can orient themselves to best suit the available light. In low-light conditions, they will spread out in a sheet—maximizing the surface area to absorb light. Under intense light, they will seek shelter by aligning in vertical columns along the plant cell's cell wall or turning sideways so that light strikes them edge-on. This reduces exposure and protects them from photooxidative damage. This ability to distribute chloroplasts so that they can take shelter behind each other or spread out may be the reason why land plants evolved to have many small chloroplasts instead of a few big ones. Chloroplast movement is considered one of the most closely regulated stimulus-response systems that can be found in plants. Mitochondria have also been observed to follow chloroplasts as they move.

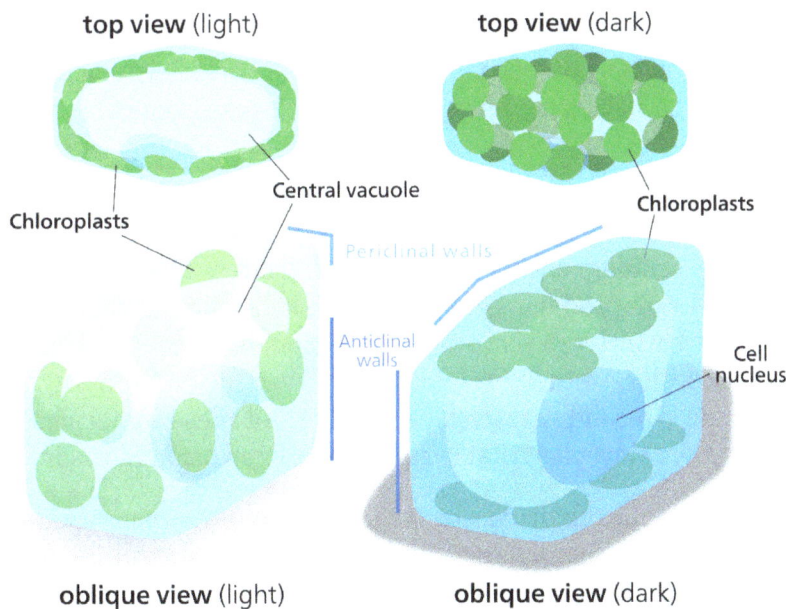

When chloroplasts are exposed to direct sunlight, they stack along the anticlinal cell walls to minimize exposure. In the dark they spread out in sheets along the periclinal walls to maximize light absorption.

In higher plants, chloroplast movement is run by phototropins, blue light photoreceptors also responsible for plant phototropism. In some algae, mosses, ferns, and flowering plants, chloroplast movement is influenced by red light in addition to blue light, though very long red wavelengths inhibit movement rather than speeding it up. Blue light generally causes chloroplasts to seek shelter, while red light draws them out to maximize light absorption.

Studies of *Vallisneria gigantea*, an aquatic flowering plant, have shown that chloroplasts can get moving within five minutes of light exposure, though they don't initially show any net directionality. They may move along microfilament tracks, and the fact that the microfilament mesh changes shape to form a honeycomb structure surrounding the chloroplasts after they have moved suggests that microfilaments may help to anchor chloroplasts in place.

Function and Chemistry

Guard Cell Chloroplasts

Unlike most epidermal cells, the guard cells of plant stomata contain relatively well-developed chloroplasts. However, exactly what they do is controversial.

Plant Innate Immunity

Plants lack specialized immune cells—all plant cells participate in the plant immune response. Chloroplasts, along with the nucleus, cell membrane, and endoplasmic reticulum, are key players in pathogen defense. Due to its role in a plant cell's immune response, pathogens frequently target the chloroplast.

Plants have two main immune responses—the hypersensitive response, in which infected cells seal themselves off and undergo programmed cell death, and systemic acquired resistance, where infected cells release signals warning the rest of the plant of a pathogen's presence. Chloroplasts stimulate both responses by purposely damaging their photosynthetic system, producing reactive oxygen species. High levels of reactive oxygen species will cause the hypersensitive response. The reactive oxygen species also directly kill any pathogens within the cell. Lower levels of reactive oxygen species initiate systemic acquired resistance, triggering defense-molecule production in the rest of the plant.

In some plants, chloroplasts are known to move closer to the infection site and the nucleus during an infection.

Chloroplasts can serve as cellular sensors. After detecting stress in a cell, which might be due to a pathogen, chloroplasts begin producing molecules like salicylic acid, jasmonic acid, nitric oxide and reactive oxygen species which can serve as defense-signals. As cellular signals, reactive oxygen species are unstable molecules, so they probably don't leave the chloroplast, but instead pass on their signal to an unknown second messenger molecule. All these molecules initiate retrograde signaling—signals from the chloroplast that regulate gene expression in the nucleus.

In addition to defense signaling, chloroplasts, with the help of the peroxisomes, help synthesize an important defense molecule, jasmonate. Chloroplasts synthesize all the fatty acids in a plant cell—linoleic acid, a fatty acid, is a precursor to jasmonate.

Photosynthesis

One of the main functions of the chloroplast is its role in photosynthesis, the process by which light is transformed into chemical energy, to subsequently produce food in the form of sugars. Water (H_2O) and carbon dioxide (CO_2) are used in photosynthesis, and sugar and oxygen (O_2) is made, using light energy. Photosynthesis is divided into two stages—the light reactions, where water is split to produce oxygen, and the dark reactions, or Calvin cycle, which builds sugar molecules from carbon dioxide. The two phases are linked by the energy carriers adenosine triphosphate (ATP) and nicotinamide adenine dinucleotide phosphate ($NADP^+$).

Light Reactions

The light reactions of photosynthesis take place across the thylakoid membranes.

The light reactions take place on the thylakoid membranes. They take light energy and store it in NADPH, a form of NADP⁺, and ATP to fuel the dark reactions.

Energy Carriers

ATP is the phosphorylated version of adenosine diphosphate (ADP), which stores energy in a cell and powers most cellular activities. ATP is the energized form, while ADP is the (partially) depleted form. NADP⁺ is an electron carrier which ferries high energy electrons. In the light reactions, it gets reduced, meaning it picks up electrons, becoming NADPH.

Photophosphorylation

Like mitochondria, chloroplasts use the potential energy stored in an H^+, or hydrogen ion gradient to generate ATP energy. The two photosystems capture light energy to energize electrons taken from water, and release them down an electron transport chain. The molecules between the photosystems harness the electrons' energy to pump hydrogen ions into the thylakoid space, creating a concentration gradient, with more hydrogen ions (up to a thousand times as many) inside the thylakoid system than in the stroma. The hydrogen ions in the thylakoid space then diffuse back down their concentration gradient, flowing back out into the stroma through ATP synthase. ATP synthase uses the energy from the flowing hydrogen ions to phosphorylate adenosine diphosphate into adenosine triphosphate, or ATP. Because chloroplast ATP synthase projects out into the stroma, the ATP is synthesized there, in position to be used in the dark reactions.

NADP⁺ Reduction

Electrons are often removed from the electron transport chains to charge NADP⁺ with electrons, reducing it to NADPH. Like ATP synthase, ferredoxin-NADP⁺ reductase, the enzyme that reduces NADP⁺, releases the NADPH it makes into the stroma, right where it is needed for the dark reactions.

Because NADP⁺ reduction removes electrons from the electron transport chains, they must be replaced—the job of photosystem II, which splits water molecules (H_2O) to obtain the electrons from its hydrogen atoms.

Cyclic Photophosphorylation

While photosystem II photolyzes water to obtain and energize new electrons, photosystem I simply reenergizes depleted electrons at the end of an electron transport chain. Normally, the reenergized electrons are taken by NADP⁺, though sometimes they can flow back down more H^+-pumping electron transport chains to transport more hydrogen ions into the thylakoid space to generate more ATP. This is termed cyclic photophosphorylation because the electrons are recycled. Cyclic photophosphorylation is common in C_4 plants, which need more ATP than NADPH.

Dark Reactions

The Calvin cycle *(Interactive diagram)* The Calvin cycle incorporates carbon dioxide into sugar molecules.

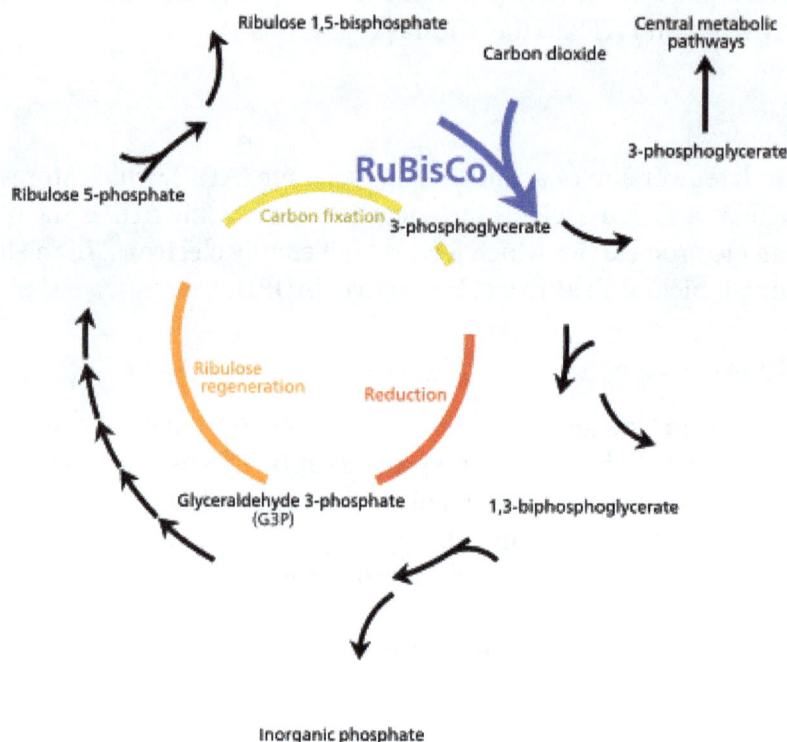

The Calvin cycle, also known as the dark reactions, is a series of biochemical reactions that fixes CO_2 into G3P sugar molecules and uses the energy and electrons from the ATP and NADPH made in the light reactions. The Calvin cycle takes place in the stroma of the chloroplast.

While named *"the dark reactions"*, in most plants, they take place in the light, since the dark reactions are dependent on the products of the light reactions.

Carbon Fixation and G3P Synthesis

The Calvin cycle starts by using the enzyme Rubisco to fix CO_2 into five-carbon Ribulose bisphosphate (RuBP) molecules. The result is unstable six-carbon molecules that immediately break down into three-carbon molecules called 3-phosphoglyceric acid, or 3-PGA. The ATP and NADPH made in the light reactions is used to convert the 3-PGA into glyceraldehyde-3-phosphate, or G3P sugar molecules. Most of the G3P molecules are recycled back into RuBP using energy from more ATP, but one out of every six produced leaves the cycle—the end product of the dark reactions.

Sugars and Starches

Glyceraldehyde-3-phosphate can double up to form larger sugar molecules like glucose and fructose. These molecules are processed, and from them, the still larger sucrose, a disaccharide commonly known as table sugar, is made, though this process takes place outside of the chloroplast, in the cytoplasm.

Sucrose is made up of a glucose monomer (left), and a fructose monomer (right).

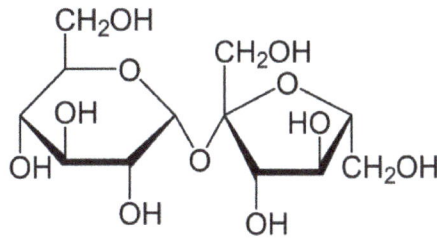

Alternatively, glucose monomers in the chloroplast can be linked together to make starch, which accumulates into the starch grains found in the chloroplast. Under conditions such as high atmospheric CO_2 concentrations, these starch grains may grow very large, distorting the grana and thylakoids. The starch granules displace the thylakoids, but leave them intact. Waterlogged roots can also cause starch buildup in the chloroplasts, possibly due to less sucrose being exported out of the chloroplast (or more accurately, the plant cell). This depletes a plant's free phosphate supply, which indirectly stimulates chloroplast starch synthesis. While linked to low photosynthesis rates, the starch grains themselves may not necessarily interfere significantly with the efficiency of photosynthesis, and might simply be a side effect of another photosynthesis-depressing factor.

Photorespiration

Photorespiration can occur when the oxygen concentration is too high. Rubisco cannot distinguish between oxygen and carbon dioxide very well, so it can accidentally add O_2 instead of CO_2 to RuBP. This process reduces the efficiency of photosynthesis—it consumes ATP and oxygen, releases CO_2, and produces no sugar. It can waste up to half the carbon fixed by the Calvin cycle. Several mechanisms have evolved in different lineages that raise the carbon dioxide concentration relative to oxygen within the chloroplast, increasing the efficiency of photosynthesis. These mechanisms are called carbon dioxide concentrating mechanisms, or CCMs. These include Crassulacean acid metabolism, C_4 carbon fixation, and pyrenoids. Chloroplasts in C_4 plants are notable as they exhibit a distinct chloroplast dimorphism.

pH

Because of the H^+ gradient across the thylakoid membrane, the interior of the thylakoid is acidic, with a pH around 4, while the stroma is slightly basic, with a pH of around 8. The optimal stroma pH for the Calvin cycle is 8.1, with the reaction nearly stopping when the pH falls below 7.3.

CO_2 in water can form carbonic acid, which can disturb the pH of isolated chloroplasts, interfering with photosynthesis, even though CO_2 is used in photosynthesis. However, chloroplasts in living plant cells are not affected by this as much.

Chloroplasts can pump K^+ and H^+ ions in and out of themselves using a poorly understood light-driven transport system.

In the presence of light, the pH of the thylakoid lumen can drop up to 1.5 pH units, while the pH of the stroma can rise by nearly one pH unit.

Amino Acid Synthesis

Chloroplasts alone make almost all of a plant cell's amino acids in their stroma except the sulfur-containing ones like cysteine and methionine. Cysteine is made in the chloroplast (the proplastid too) but it is also synthesized in the cytosol and mitochondria, probably because it has trouble crossing membranes to get to where it is needed. The chloroplast is known to make the precursors to methionine but it is unclear whether the organelle carries out the last leg of the pathway or if it happens in the cytosol.

Other Nitrogen Compounds

Chloroplasts make all of a cell's purines and pyrimidines—the nitrogenous bases found in DNA and RNA. They also convert nitrite (NO_2^-) into ammonia (NH_3) which supplies the plant with nitrogen to make its amino acids and nucleotides.

Other Chemical Products

Chloroplasts are the site of complex lipid metabolism.

Differentiation, Replication, and Inheritance

Chloroplasts are a special type of a plant cell organelle called a plastid, though the two terms are sometimes used interchangeably. There are many other types of plastids, which carry out various functions. All chloroplasts in a plant are descended from undifferentiated proplastids found in the zygote, or fertilized egg. Proplastids are commonly found in an adult plant's apical meristems. Chloroplasts do not normally develop from proplastids in root tip meristems—instead, the formation of starch-storing amyloplasts is more common.

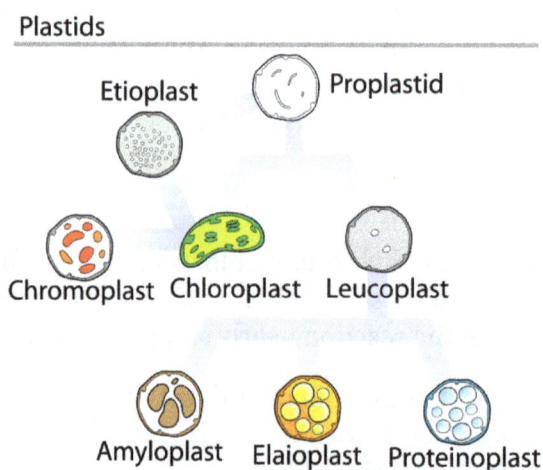

Plastid types *(Interactive diagram)* Plants contain many different kinds of plastids in their cells.

In shoots, proplastids from shoot apical meristems can gradually develop into chloroplasts in photosynthetic leaf tissues as the leaf matures, if exposed to the required light. This process involves invaginations of the inner plastid membrane, forming sheets of membrane that project into the internal stroma. These membrane sheets then fold to form thylakoids and grana.

If angiosperm shoots are not exposed to the required light for chloroplast formation, proplastids may develop into an etioplast stage before becoming chloroplasts. An etioplast is a plastid that lacks chlorophyll, and has inner membrane invaginations that form a lattice of tubes in their stroma, called a prolamellar body. While etioplasts lack chlorophyll, they have a yellow chlorophyll precursor stocked. Within a few minutes of light exposure, the prolamellar body begins to reorganize into stacks of thylakoids, and chlorophyll starts to be produced. This process, where the etioplast becomes a chloroplast, takes several hours. Gymnosperms do not require light to form chloroplasts.

Light, however, does not guarantee that a proplastid will develop into a chloroplast. Whether a proplastid develops into a chloroplast some other kind of plastid is mostly controlled by the nucleus and is largely influenced by the kind of cell it resides in.

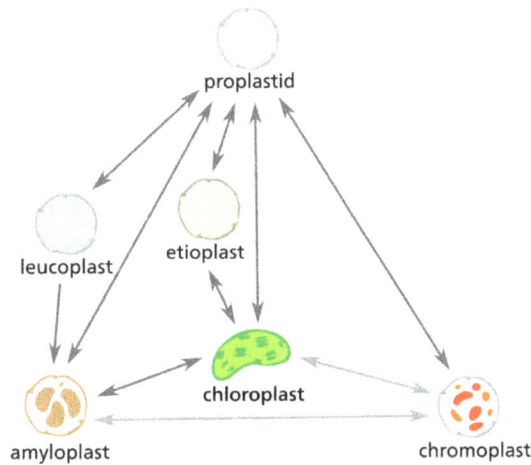

possible plastid interconversions

Many plastid interconversions are possible.

Plastid Interconversion

Plastid differentiation is not permanent, in fact many interconversions are possible. Chloroplasts may be converted to chromoplasts, which are pigment-filled plastids responsible for the bright colors seen in flowers and ripe fruit. Starch storing amyloplasts can also be converted to chromoplasts, and it is possible for proplastids to develop straight into chromoplasts. Chromoplasts and amyloplasts can also become chloroplasts, like what happens when a carrot or a potato is illuminated. If a plant is injured, or something else causes a plant cell to revert to a meristematic state, chloroplasts and other plastids can turn back into proplastids. Chloroplast, amyloplast, chromoplast, proplast, etc., are not absolute states—intermediate forms are common.

Chloroplast Division

Most chloroplasts in a photosynthetic cell do not develop directly from proplastids or etioplasts. In fact, a typical shoot meristematic plant cell contains only 7–20 proplastids. These proplastids differentiate into chloroplasts, which divide to create the 30–70 chloroplasts found in a mature photosynthetic plant cell. If the cell divides, chloroplast division provides the additional chloroplasts to partition between the two daughter cells.

In single-celled algae, chloroplast division is the only way new chloroplasts are formed. There is no proplastid differentiation—when an algal cell divides, its chloroplast divides along with it, and each daughter cell receives a mature chloroplast.

Almost all chloroplasts in a cell divide, rather than a small group of rapidly dividing chloroplasts. Chloroplasts have no definite S-phase—their DNA replication is not synchronized or limited to that of their host cells. Much of what we know about chloroplast division comes from studying organisms like *Arabidopsis* and the red alga *Cyanidioschyzon merolæ*.

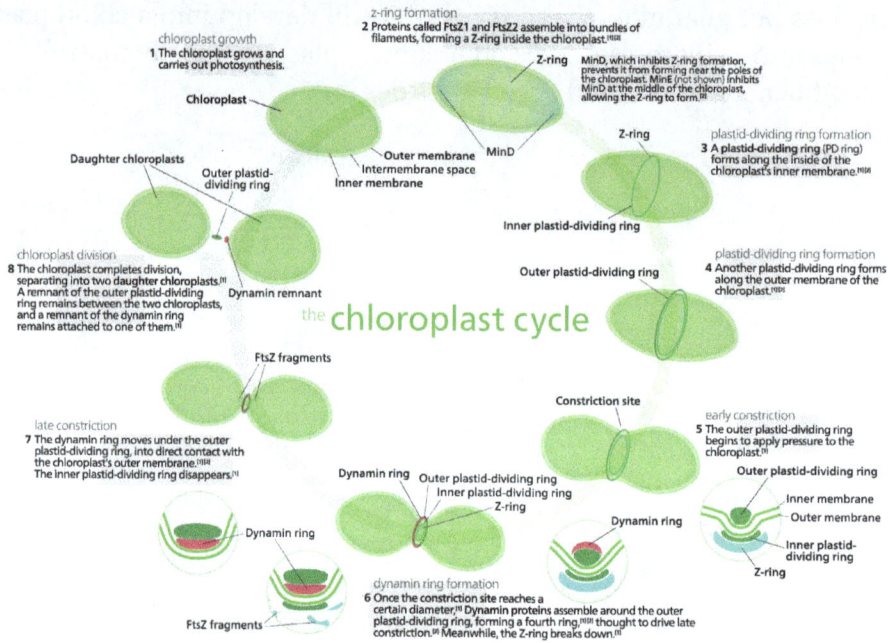

the chloroplast cycle

chloroplast growth
1 The chloroplast grows and carries out photosynthesis.

Chloroplast

Daughter chloroplasts
Outer plastid-dividing ring
Outer membrane
Intermembrane space
Inner membrane
MinD

z-ring formation
2 Proteins called FtsZ1 and FtsZ2 assemble into bundles of filaments, forming a Z-ring inside the chloroplast.[II]

Z-ring

MinD, which inhibits Z-ring formation, prevents it from forming near the poles of the chloroplast. MinE (not shown) inhibits MinD at the middle of the chloroplast, allowing the Z-ring to form.[II]

Z-ring

plastid-dividing ring formation
3 A plastid-dividing ring (PD ring) forms along the inside of the chloroplast's inner membrane.[III][IV]

Inner plastid-dividing ring

Outer plastid-dividing ring

plastid-dividing ring formation
4 Another plastid-dividing ring forms along the outer membrane of the chloroplast.[III][IV]

chloroplast division
8 The chloroplast completes division, separating into two daughter chloroplasts.[I] A remnant of the outer plastid-dividing ring remains between the two chloroplasts, and a remnant of the dynamin ring remains attached to one of them.[I]

Dynamin remnant

FtsZ fragments

late constriction
7 The dynamin ring moves under the outer plastid-dividing ring, into direct contact with the chloroplast's outer membrane.[I][II] The inner plastid-dividing ring disappears.[V]

Dynamin ring

FtsZ fragments

Dynamin ring
Outer plastid-dividing ring
Inner plastid-dividing ring
Z-ring

dynamin ring formation
6 Once the constriction site reaches a certain diameter,[I] Dynamin proteins assemble around the outer plastid-dividing ring, forming a fourth ring,[II][III] thought to drive late constriction.[II] Meanwhile, the Z-ring breaks down.[I]

Constriction site

Dynamin ring

early constriction
5 The outer plastid-dividing ring begins to apply pressure to the chloroplast.[I]

Outer plastid-dividing ring

Inner membrane
Outer membrane
Inner plastid-dividing ring
Z-ring

Most chloroplasts in plant cells, and all chloroplasts in algae arise from chloroplast division. *Picture references,*

The division process starts when the proteins FtsZ1 and FtsZ2 assemble into filaments, and with the help of a protein ARC6, form a structure called a Z-ring within the chloroplast's stroma. The Min system manages the placement of the Z-ring, ensuring that the chloroplast is cleaved more or less evenly. The protein MinD prevents FtsZ from linking up and forming filaments. Another protein ARC3 may also be involved, but it is not very well understood. These proteins are active at the poles of the chloroplast, preventing Z-ring formation there, but near the center of the chloroplast, MinE inhibits them, allowing the Z-ring to form.

Next, the two plastid-dividing rings, or PD rings form. The inner plastid-dividing ring is located in the inner side of the chloroplast's inner membrane, and is formed first. The outer plastid-dividing ring is found wrapped around the outer chloroplast membrane. It consists of filaments about 5 nanometers across, arranged in rows 6.4 nanometers apart, and shrinks to squeeze the chloroplast. This is when chloroplast constriction begins. In a few species like *Cyanidioschyzon merolæ*, chloroplasts have a third plastid-dividing ring located in the chloroplast's intermembrane space.

Late into the constriction phase, dynamin proteins assemble around the outer plastid-dividing ring, helping provide force to squeeze the chloroplast. Meanwhile, the Z-ring and the inner plastid-dividing ring break down. During this stage, the many chloroplast DNA plasmids floating around in the stroma are partitioned and distributed to the two forming daughter chloroplasts.

Later, the dynamins migrate under the outer plastid dividing ring, into direct contact with the chloroplast's outer membrane, to cleave the chloroplast in two daughter chloroplasts.

A remnant of the outer plastid dividing ring remains floating between the two daughter chloroplasts, and a remnant of the dynamin ring remains attached to one of the daughter chloroplasts.

Of the five or six rings involved in chloroplast division, only the outer plastid-dividing ring is present for the entire constriction and division phase—while the Z-ring forms first, constriction does not begin until the outer plastid-dividing ring forms.

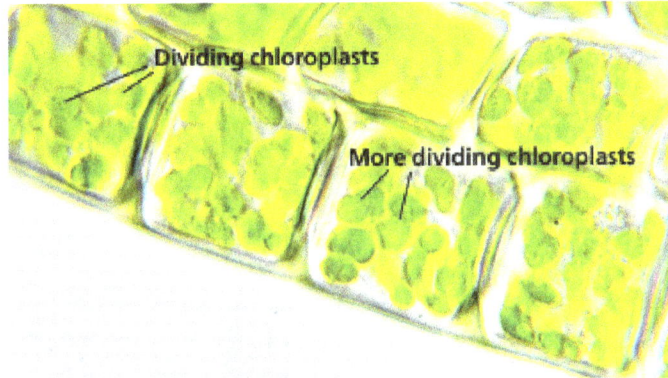

Chloroplast division In this light micrograph of some moss chloroplasts, many dumbbell-shaped chloroplasts can be seen dividing. Grana are also just barely visible as small granules.

Regulation

In species of algae that contain a single chloroplast, regulation of chloroplast division is extremely important to ensure that each daughter cell receives a chloroplast—chloroplasts can't be made from scratch. In organisms like plants, whose cells contain multiple chloroplasts, coordination is looser and less important. It is likely that chloroplast and cell division are somewhat synchronized, though the mechanisms for it are mostly unknown.

Light has been shown to be a requirement for chloroplast division. Chloroplasts can grow and progress through some of the constriction stages under poor quality green light, but are slow to complete division—they require exposure to bright white light to complete division. Spinach leaves grown under green light have been observed to contain many large dumbbell-shaped chloroplasts. Exposure to white light can stimulate these chloroplasts to divide and reduce the population of dumbbell-shaped chloroplasts.

Chloroplast Inheritance

Like mitochondria, chloroplasts are usually inherited from a single parent. Biparental chloroplast inheritance—where plastid genes are inherited from both parent plants—occurs in very low levels in some flowering plants.

Many mechanisms prevent biparental chloroplast DNA inheritance, including selective destruction of chloroplasts or their genes within the gamete or zygote, and chloroplasts from one parent being excluded from the embryo. Parental chloroplasts can be sorted so that only one type is present in each offspring.

Gymnosperms, such as pine trees, mostly pass on chloroplasts paternally, while flowering plants often inherit chloroplasts maternally. Flowering plants were once thought to only inherit chloroplasts maternally. However, there are now many documented cases of angiosperms inheriting chloroplasts paternally.

Angiosperms, which pass on chloroplasts maternally, have many ways to prevent paternal inheritance. Most of them produce sperm cells that do not contain any plastids. There are many other documented mechanisms that prevent paternal inheritance in these flowering plants, such as different rates of chloroplast replication within the embryo.

Among angiosperms, paternal chloroplast inheritance is observed more often in hybrids than in offspring from parents of the same species. This suggests that incompatible hybrid genes might interfere with the mechanisms that prevent paternal inheritance.

Transplastomic Plants

Recently, chloroplasts have caught attention by developers of genetically modified crops. Since, in most flowering plants, chloroplasts are not inherited from the male parent, transgenes in these plastids cannot be disseminated by pollen. This makes plastid transformation a valuable tool for the creation and cultivation of genetically modified plants that are biologically contained, thus posing significantly lower environmental risks. This biological containment strategy is therefore suitable for establishing the coexistence of conventional and organic agriculture. While the reliability of this mechanism has not yet been studied for all relevant crop species, recent results in tobacco plants are promising, showing a failed containment rate of transplastomic plants at 3 in 1,000,000.

Light-dependent Reactions

In photosynthesis, the light-dependent reactions take place on the thylakoid membranes. The inside of the thylakoid membrane is called the lumen, and outside the thylakoid membrane is the stroma, where the light-independent reactions take place. The thylakoid membrane contains some integral membrane protein complexes that catalyze the light reactions. There are four major protein complexes in the thylakoid membrane: Photosystem II (PSII), Cytochrome b6f complex, Photosystem I (PSI), and ATP synthase. These four complexes work together to ultimately create the products ATP and NADPH.

The two photosystems absorb light energy through pigments - primarily the chlorophylls, which are responsible for the green color of leaves. The light-dependent reactions begin in photosystem II. When a chlorophyll *a* molecule within the reaction center of PSII absorbs a photon, an electron in this molecule attains a higher energy level. Because this state of an electron is very unstable, the electron is transferred from one to another molecule creating a chain of redox reactions, called an electron transport chain (ETC). The electron flow goes from PSII to cytochrome b6f to PSI. In PSI, the electron gets the energy from another photon. The final electron acceptor is NADP. In oxygenic photosynthesis, the first electron donor is water, creating oxygen as a waste product. In anoxygenic photosynthesis various electron donors are used.

Cytochrome b6f and ATP synthase work together to create ATP. This process is called photophosphorylation, which occurs in two different ways. In non-cyclic photophosphorylation, cytochrome b6f uses the energy of electrons from PSII to pump protons from the stroma to the lumen. The proton gradient across the thylakoid membrane creates a proton-motive force, used by ATP synthase to form ATP. In cyclic photophosphorylation, cytochrome b6f uses the energy of electrons from not only PSII but also PSI to create more ATP and to stop the production of NADPH. Cyclic phosphorylation is important to create ATP and maintain NADPH in the right proportion for the light-independent reactions.

The net-reaction of all light-dependent reactions in oxygenic photosynthesis is:

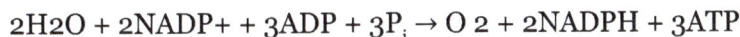

$$2H_2O + 2NADP+ + 3ADP + 3P_i \rightarrow O_2 + 2NADPH + 3ATP$$

The two photosystems are protein complexes that absorb photons and are able to use this energy to create an electron transport chain. Photosystem I and II are very similar in structure and function. They use special proteins, called light-harvesting complexes, to absorb the photons with very high effectiveness. If a special pigment molecule in a photosynthetic reaction center absorbs a photon, an electron in this pigment attains the excited state and then is transferred to another molecule in the reaction center. This reaction, called photoinduced charge separation, is the start of the electron flow and is unique because it transforms light energy into chemical forms.

The Reaction Center

The reaction center is in the thylakoid membrane. It transfers light energy to a dimer of chlorophyll pigment molecules near the periplasmic (or thylakoid lumen) side of the membrane. This dimer is called a special pair because of its fundamental role in photosynthesis. This special pair is slightly different in PSI and PSII reaction center. In PSII, it absorbs photons with a wavelength of 680 nm, and it is therefore called P680. In PSI, it absorbs photons at 700 nm, and it is called P700. In bacteria, the special pair is called P760, P840, P870, or P960.

If an electron of the special pair in the reaction center becomes excited, it cannot transfer this energy to another pigment using resonance energy transfer. In normal circumstances, the electron should return to the ground state, but, because the reaction center is arranged so that a suitable electron acceptor is nearby, the excited electron can move from the initial molecule to the acceptor. This process results in the formation of a positive charge on the special pair (due to the loss of an electron) and a negative charge on the acceptor and is, hence, referred to as photoinduced charge separation. In other words, electrons in pigment molecules can exist at specific energy levels. Under normal circumstances, they exist at the lowest possible energy level they can. However, if there is enough energy to move them into the next energy level, they can absorb that energy and occupy that higher energy level. The light they absorb contains the necessary amount of energy needed to push them into the next level. Any light that does not have enough or has too much energy cannot be absorbed and is reflected. The electron in the higher energy level, however, does not want to be there; the electron is unstable and must return to its normal lower energy level. To do this, it must release the energy that has put it into the higher energy state to begin with. This can happen various ways. The extra energy can be converted into molecular motion and lost as heat. Some of the extra energy can be lost as heat energy, while the rest is lost as light. This re-emission of light energy is called fluorescence. The energy, but not the e- itself, can be passed onto another molecule. This is called resonance.

The energy and the e- can be transferred to another molecule. Plant pigments usually utilize the last two of these reactions to convert the sun's energy into their own.

This initial charge separation occurs in less than 10 picoseconds (10^{-11} seconds). In their high-energy states, the special pigment and the acceptor could undergo charge recombination; that is, the electron on the acceptor could move back to neutralize the positive charge on the special pair. Its return to the special pair would waste a valuable high-energy electron and simply convert the absorbed light energy into heat. In the case of PSII, this backflow of electrons can produce reactive oxygen species leading to photoinhibition. Three factors in the structure of the reaction center work together to suppress charge recombination nearly completely.

- Another electron acceptor is less than 10 Å away from the first acceptor, and so the electron is rapidly transferred farther away from the special pair.

- An electron donor is less than 10 Å away from the special pair, and so the positive charge is neutralized by the transfer of another electron

- The electron transfer back from the electron acceptor to the positively charged special pair is especially slow. The rate of an of electron transfer reaction increases with its thermodynamic favorability up to a point and then decreases. The back transfer is so favourable that it takes place in the inverted region where electron-transfer rates become slower.

Thus, electron transfer proceeds efficiently from the first electron acceptor to the next, creating an electron transport chain that ends if it has reached NADPH.

Photosynthetic Electron Transport Chains in Chloroplasts

The photosynthesis process in chloroplasts begins when an electron of P680 of PSII attains a higher-energy level. This energy is used to reduce a chain of electron acceptors that have subsequently lowered redox-potentials. This chain of electron acceptors is known as an electron transport chain. When this chain reaches PS I, an electron is again excited, creating a high redox-potential. The electron transport chain of photosynthesis is often put in a diagram called the z-scheme, because the redox diagram from P680 to P700 resembles the letter z.

The final product of PSII is plastoquinol, a mobile electron carrier in the membrane. Plastoquinol transfers the electron from PSII to the proton pump, cytochrome b6f. The ultimate electron donor of PSII is water. Cytochrome b6f proceeds the electron chain to PSI through plastocyanin molecules. PSI is able to continue the electron transfer in two different ways. It can transfer the electrons either to plastoquinol again, creating a cyclic electron flow, or to an enzyme called FNR (Ferredoxin—NADP(+) reductase), creating a non-cyclic electron flow. PSI releases FNR into the stroma, where it reduces NADP+to NADPH.

Activities of the electron transport chain, especially from cytochrome b6f, lead to pumping of protons from the stroma to the lumen. The resulting transmembrane proton gradient is used to make ATP via ATP synthase.

The overall process of the photosynthetic electron transport chain in chloroplasts is:

$$H\ 2O \rightarrow PS\ II \rightarrow plastoquinone \rightarrow cytb\ 6 \rightarrow plastocyanin \rightarrow PS\ I \rightarrow NADPH$$

Photosystem II

PS II is an extremely complex, highly organized transmembrane structure that contains a *water-splitting complex*, chlorophylls and carotenoid pigments, a *reaction center* (P680), pheophytin (a pigment similar to chlorophyll), and two quinones. It uses the energy of sunlight to transfer electrons from water to a mobile electron carrier in the membrane called *plastoquinone*:

$$H_2O \rightarrow P680 \rightarrow P680^* \rightarrow plastoquinone$$

Plastoquinone, in turn, transfers electrons to *cytb 6*, which feeds them into PS I.

The Water-splitting Complex

The step $H_2O \rightarrow P680$ is performed by a poorly understood structure embedded within PS II called the *water-splitting complex* or the *oxygen-evolving complex*. It catalyzes a reaction that splits water into electrons, protons and oxygen:

$$2H_2O \rightarrow 4H^+ + 4e^- + O_2$$

The electrons are transferred to special chlorophyll molecules (embedded in PS II) that are promoted to a higher-energy state by the energy of photons.

The Reaction Center

The excitation $P680 \rightarrow P680^*$ of the reaction center pigment P680 occurs here. These special chlorophyll molecules embedded in PS II absorb the energy of photons, with maximal absorption at 680 nm. Electrons within these molecules are promoted to a higher-energy state. This is one of two core processes in photosynthesis, and it occurs with astonishing efficiency (greater than 90%) because, in addition to direct excitation by light at 680 nm, the energy of light first harvested by *antenna proteins* at other wavelengths in the light-harvesting system is also transferred to these special chlorophyll molecules.

This is followed by the step $P680^* \rightarrow$ pheophytin, and then on to plastoquinone, which occurs within the reaction center of PS II. High-energy electrons are transferred to plastoquinone before it subsequently picks up two protons to become plastoquinol. Plastoquinol is then released into the membrane as a mobile electron carrier.

This is the second core process in photosynthesis. The initial stages occur within *picoseconds*, with an efficiency of 100%. The seemingly impossible efficiency is due to the precise positioning of molecules within the reaction center. This is a solid-state process, not a chemical reaction. It occurs within an essentially crystalline environment created by the macromolecular structure of PS II. The usual rules of chemistry (which involve random collisions and random energy distributions) do not apply in solid-state environments.

Link of Water-splitting Complex and Chlorophyll Excitation

When the chlorophyll passes the electron to pheophytin, it obtains an electron from P_{680}^*. In turn, P_{680}^* can oxidize the Z (or Y_Z) molecule. Once oxidized, the Z molecule can derive electrons from the oxygen-evolving complex.

Summary

PS II is a transmembrane structure found in all chloroplasts. It splits water into electrons, protons and molecular oxygen. The electrons are transferred to plastoquinone, which carries them to a proton pump. Molecular oxygen is released into the atmosphere.

The emergence of such an incredibly complex structure, a macromolecule that converts the energy of sunlight into potentially useful work with efficiencies that are impossible in ordinary experience, seems almost magical at first glance. Thus, it is of considerable interest that, in essence, the same structure is found in purple bacteria.

Cytochrome B6

PS II and PS I are connected by a transmembrane proton pump, cytochrome *b6* complex (plastoquinol—plastocyanin reductase; EC 1.10.99.1). Electrons from PS II are carried by plastoquinol to cyt*b6*, where they are removed in a stepwise fashion (reforming plastoquinone) and transferred to a water-soluble electron carrier called *plastocyanin*. This redox process is coupled to the pumping of four protons across the membrane. The resulting proton gradient (together with the proton gradient produced by the water-splitting complex in PS II) is used to make ATP via ATP synthase.

The similarity in structure and function between cytochrome *b6* (in chloroplasts) and cytochrome *bc1* (*Complex III* in mitochondria) is striking. Both are transmembrane structures that remove electrons from a mobile, lipid-soluble electron carrier (plastoquinone in chloroplasts; ubiquinone in mitochondria) and transfer them to a mobile, water-soluble electron carrier (plastocyanin in chloroplasts; cytochrome *c* in mitochondria). Both are proton pumps that produce a transmembrane proton gradient.

Photosystem I

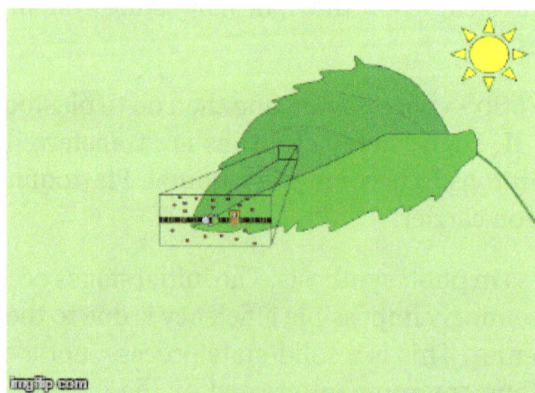

The cyclic light-dependent reactions occur when only the sole photosystem being used is photosystem 1. Photosystem 1 excites electrons which then cycle from the transport protein, ferredoxin (Fd), to the cytochrome complex, b6f, to another transport protein, plastocyanin (Pc), and back to photosystem I. A proton gradient is created across the thylakoid membrane (6) as protons (3) are transported from the chloroplast stroma (4) to the thylakoid lumen (5). Through chemiosmosis, ATP (9) is produced where ATP synthase (1) binds an inorganic phosphate group (8) to an ADP molecule (7).

PS I accepts electrons from plastocyanin and transfers them either to NADPH (*noncyclic electron transport*) or back to cytochrome *b6* (*cyclic electron transport*):

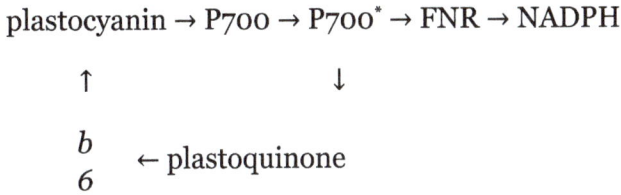

$$\text{plastocyanin} \rightarrow \text{P700} \rightarrow \text{P700}^* \rightarrow \text{FNR} \rightarrow \text{NADPH}$$

$$\uparrow \qquad\qquad\qquad \downarrow$$

$$\begin{matrix} b \\ 6 \end{matrix} \quad \leftarrow \text{plastoquinone}$$

PS I, like PS II, is a complex, highly organized transmembrane structure that contains antenna chlorophylls, a reaction center (P700), phylloquinine, and a number of iron-sulfur proteins that serve as intermediate redox carriers.

The light-harvesting system of PS I uses multiple copies of the same transmembrane proteins used by PS II. The energy of absorbed light (in the form of delocalized, high-energy electrons) is funneled into the reaction center, where it excites special chlorophyll molecules (P700, maximum light absorption at 700 nm) to a higher energy level. The process occurs with astonishingly high efficiency.

Electrons are removed from excited chlorophyll molecules and transferred through a series of intermediate carriers to *ferredoxin*, a water-soluble electron carrier. As in PS II, this is a solid-state process that operates with 100% efficiency.

There are two different pathways of electron transport in PS I. In *noncyclic electron transport*, ferredoxin carries the electron to the enzyme ferredoxin NADP+oxidoreductase (FNR) that reduces NADP+to NADPH. In *cyclic electron transport*, electrons from ferredoxin are transferred (via plastoquinone) to a proton pump, cytochrome *b6*. They are then returned (via plastocyanin) to P700.

NADPH and ATP are used to synthesize organic molecules from CO_2. The ratio of NADPH to ATP production can be adjusted by adjusting the balance between cyclic and noncyclic electron transport.

It is noteworthy that PS I closely resembles photosynthetic structures found in green sulfur bacteria, just as PS II resembles structures found in purple bacteria.

Photosynthetic Electron Transport Chains in Bacteria

PS II, PS I, and cytochrome*b6* are found in chloroplasts. All plants and all photosynthetic algae contain chloroplasts, which produce NADPH and ATP by the mechanisms described above. In essence, the same transmembrane structures are also found in *cyanobacteria*.

Unlike plants and algae, cyanobacteria are prokaryotes. They do not contain chloroplasts. Rather, they bear a striking resemblance to chloroplasts themselves. This suggests that organisms resembling cyanobacteria were the evolutionary precursors of chloroplasts. One imagines primitive eukaryotic cells taking up cyanobacteria as intracellular symbionts in a process known as endosymbiosis.

Cyanobacteria

Cyanobacteria contain structures similar to PS II and PS I in chloroplasts. Their light-harvesting system is different from that found in plants (they use *phycobilins*, rather than chlorophylls, as antenna pigments), but their electron transport chain

$$H_2O \rightarrow PS\ II \rightarrow plastoquinone \rightarrow b_6 \rightarrow cytochrome\ c_6 \rightarrow PS\ I \rightarrow ferredoxin \rightarrow NADPH$$

$$\uparrow \qquad\qquad\qquad\qquad \downarrow$$

$$b_6 \qquad \leftarrow \qquad plastoquinone$$

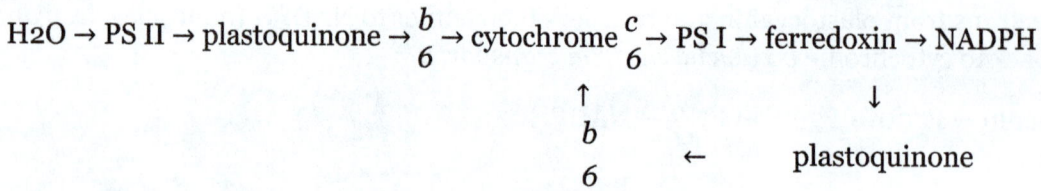

is, in essence, the same as the electron transport chain in chloroplasts. The mobile water-soluble electron carrier is cytochrome $c6$ in cyanobacteria, plastocyanin in plants.

Cyanobacteria can also synthesize ATP by oxidative phosphorylation, in the manner of other bacteria. The electron transport chain is

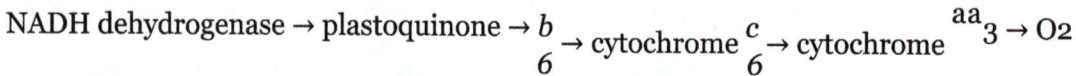

$$NADH\ dehydrogenase \rightarrow plastoquinone \rightarrow b_6 \rightarrow cytochrome\ c_6 \rightarrow cytochrome\ aa_3 \rightarrow O_2$$

where the mobile electron carriers are plastoquinone and cytochrome $c6$, while the proton pumps are NADH dehydrogenase, $b6$ and cytochrome $aa3$.

Cyanobacteria are the only bacteria that produce oxygen during photosynthesis. Earth's primordial atmosphere was anoxic. Organisms like cyanobacteria produced our present-day oxygen-containing atmosphere.

The other two major groups of photosynthetic bacteria, purple bacteria and green sulfur bacteria, contain only a single photosystem and do not produce oxygen.

Purple Bacteria

Purple bacteria contain a single photosystem that is structurally related to PS II in cyanobacteria and chloroplasts:

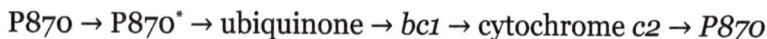

$$P870 \rightarrow P870^* \rightarrow ubiquinone \rightarrow bc1 \rightarrow cytochrome\ c2 \rightarrow P870$$

This is a *cyclic* process in which electrons are removed from an excited chlorophyll molecule (*bacteriochlorophyll*; P870), passed through an electron transport chain to a proton pump (cytochrome *bc1* complex, similar but not identical to cytochrome *bc1* in chloroplasts), and then returned to the chlorophyll molecule. The result is a proton gradient, which is used to make ATP via ATP synthase. As in cyanobacteria and chloroplasts, this is a solid-state process that depends on the precise orientation of various functional groups within a complex transmembrane macromolecular structure.

To make NADPH, purple bacteria use an external electron donor (hydrogen, hydrogen sulfide, sulfur, sulfite, or organic molecules such as succinate and lactate) to feed electrons into a reverse electron transport chain.

Green Sulfur Bacteria

Green sulfur bacteria contain a photosystem that is analogous to PS I in chloroplasts:

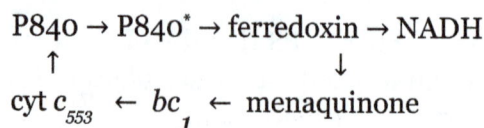

$$P840 \rightarrow P840^* \rightarrow ferredoxin \rightarrow NADH$$

$$\uparrow \qquad\qquad\qquad \downarrow$$

$$cyt\ c_{553} \leftarrow bc_1 \leftarrow menaquinone$$

There are two pathways of electron transfer. In *cyclic electron transfer*, electrons are removed from an excited chlorophyll molecule, passed through an electron transport chain to a proton pump, and then returned to the chlorophyll. The mobile electron carriers are, as usual, a lipid-soluble quinone and a water-soluble cytochrome. The resulting proton gradient is used to make ATP.

In *noncyclic electron transfer*, electrons are removed from an excited chlorophyll molecule and used to reduce NAD^+ to NADH. The electrons removed from P840 must be replaced. This is accomplished by removing electrons from H_2S, which is oxidized to sulfur (hence the name "green *sulfur* bacteria").

Purple bacteria and green sulfur bacteria occupy relatively minor ecological niches in the present day biosphere. They are of interest because of their importance in precambrian ecologies, and because their methods of photosynthesis were the likely evolutionary precursors of those in modern plants.

History

The first ideas about light being used in photosynthesis were proposed by Colin Flannery in 1779 who recognized it was sunlight falling on plants that was required, although Joseph Priestley had noted the production of oxygen without the association with light in 1772. Cornelius Van Niel proposed in 1931 that photosynthesis is a case of general mechanism where a photon of light is used to photo decompose a hydrogen donor and the hydrogen being used to reduce CO 2. Then in 1939, Robin Hill showed that isolated chloroplasts would make oxygen, but not fix CO 2 showing the light and dark reactions occurred in different places. Although they are referred to as light and dark reactions, both of them take place only in the presence of light. This led later to the discovery of photosystems 1 and 2.

Light-independent Reactions

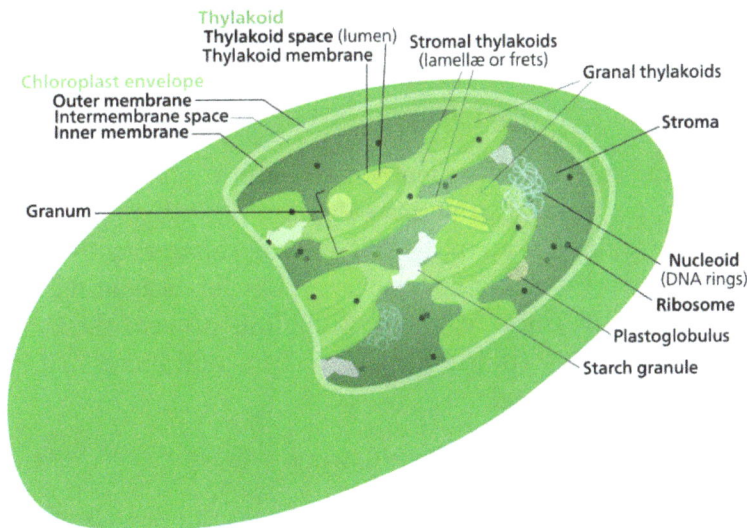

The internal structure of a chloroplast

The light-independent reactions of photosynthesis are chemical reactions that convert carbon dioxide and other compounds into glucose. These reactions occur in the stroma, the fluid-filled area of a chloroplast outside of the thylakoid membranes. These reactions take the products (ATP and NADPH) of light-dependent reactions and perform further chemical processes on them. There are three phases to the light-independent reactions, collectively called the Calvin cycle: carbon fixation, reduction reactions, and ribulose 1,5-bisphosphate (RuBP) regeneration.

Despite its name, this process occurs only when light is available. Plants do not carry out the Calvin cycle during nighttime. They instead release sucrose into the phloem from their starch reserves. This process happens when light is available independent of the kind of photosynthesis (C3 carbon fixation, C4 carbon fixation, and Crassulacean acid metabolism); CAM plants store malic acid in their vacuoles every night and release it by day in order to make this process work.

Coupling to other Metabolic Pathways

These reactions are closely coupled to the thylakoid electron transport chain as reducing power provided by NADPH produced in the photosystem I is actively needed. The process of photorespiration, also known as C2 cycle, is also coupled to the dark reactions, as it results from an alternative reaction of the Rubisco enzyme, and its final byproduct is also another glyceraldehyde-3-P.

The Calvin cycle, Calvin–Benson–Bassham (CBB) cycle, reductive pentose phosphate cycle or C3 cycle is a series of biochemical redox reactions that take place in the stroma of chloroplast in photosynthetic organisms. It is also known as the light-independent reactions.

The cycle was discovered by Melvin Calvin, James Bassham, and Andrew Benson at the University of California, Berkeley by using the radioactive isotope carbon-14. It is one of the light-independent reactions used for carbon fixation.

Photosynthesis occurs in two stages in a cell. In the first stage, light-dependent reactions capture the energy of light and use it to make the energy-storage and transport molecules ATP and NADPH. The light-independent Calvin cycle uses the energy from short-lived electronically excited carriers to convert carbon dioxide and water into organic compounds that can be used by the organism (and by animals that feed on it). This set of reactions is also called *carbon fixation*. The key enzyme of the cycle is called RuBisCO. In the following biochemical equations, the chemical species (phosphates and carboxylic acids) exist in equilibria among their various ionized states as governed by the pH.

The enzymes in the Calvin cycle are functionally equivalent to most enzymes used in other metabolic pathways such as gluconeogenesis and the pentose phosphate pathway, but they are to be found in the chloroplast stroma instead of the cell cytosol, separating the reactions. They are activated in the light (which is why the name "dark reaction" is misleading), and also by products of the light-dependent reaction. These regulatory functions prevent the Calvin cycle from being respired to carbon dioxide. Energy (in the form of ATP) would be wasted in carrying out these reactions that have no net productivity.

The sum of reactions in the Calvin cycle is the following:

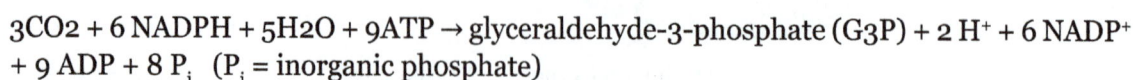

$$3CO_2 + 6\,NADPH + 5H_2O + 9ATP \rightarrow \text{glyceraldehyde-3-phosphate (G3P)} + 2\,H^+ + 6\,NADP^+ + 9\,ADP + 8\,P_i \quad (P_i = \text{inorganic phosphate})$$

Hexose (six-carbon) sugars are not a product of the Calvin cycle. Although many texts list a product of photosynthesis as C6H12O6, this is mainly a convenience to counter the equation of respiration, where six-carbon sugars are oxidized in mitochondria. The carbohydrate products of the Calvin cycle are three-carbon sugar phosphate molecules, or "triose phosphates," namely, glyceraldehyde-3-phosphate (G3P).

Steps

In the first stage of the Calvin cycle, a CO_2 molecule is incorporated into one of two three-carbon molecules (glyceraldehyde 3-phosphate or G3P), where it uses up two molecules of ATP and two molecules of NADPH, which had been produced in the light-dependent stage. The three steps involved are:

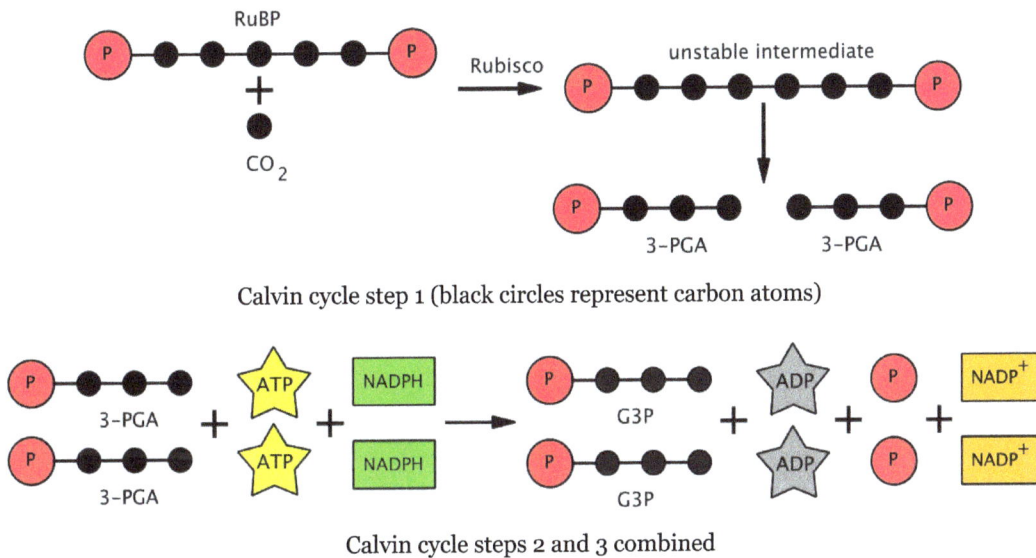

Calvin cycle step 1 (black circles represent carbon atoms)

Calvin cycle steps 2 and 3 combined

1. The enzyme RuBisCO catalyses the carboxylation of ribulose-1,5-bisphosphate, RuBP, a 5-carbon compound, by carbon dioxide (a total of 6 carbons) in a two-step reaction. The product of the first step is enediol-enzyme complex that can capture CO2 or O2. Thus, enediol-enzyme complex is the real carboxylase/oxygenase. The CO2 that is captured by enediol in second step produces a six-carbon intermediate initially that immediately splits in half, forming two molecules of 3-phosphoglycerate, or 3-PGA, a 3-carbon compound (also: 3-phosphoglyceric acid, PGA, 3PGA).

2. The enzyme phosphoglycerate kinase catalyses the phosphorylation of 3-PGA by ATP (which was produced in the light-dependent stage). 1,3-Bisphosphoglycerate (1,3BPGA, glycerate-1,3-bisphosphate) and ADP are the products. (However, note that two 3-PGAs are produced for every CO2 that enters the cycle, so this step utilizes two ATP per CO2 fixed.)

3. The enzyme glyceraldehyde 3-phosphate dehydrogenase catalyses the reduction of 1,3BPGA by NADPH (which is another product of the light-dependent stage). Glyceraldehyde 3-phosphate (also called G3P, GP, TP, PGAL, GAP) is produced, and the NADPH itself is oxidized and becomes $NADP^+$. Again, two NADPH are utilized per CO2 fixed.

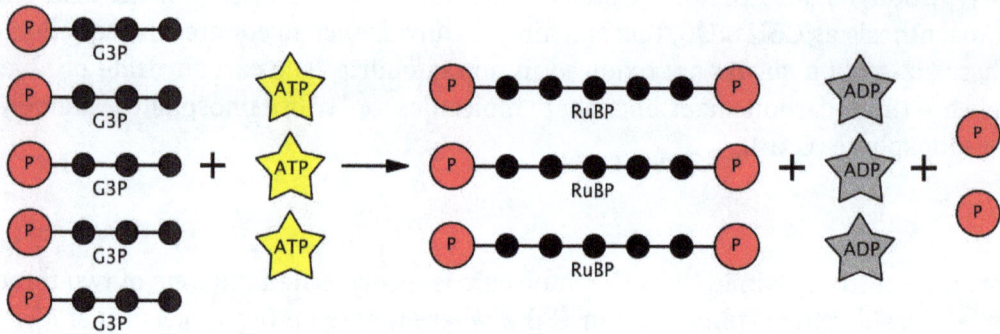

Regeneration stage of the Calvin cycle

The next stage in the Calvin cycle is to regenerate RuBP. Five G3P molecules produce three RuBP molecules, using up three molecules of ATP. Since each CO_2 molecule produces two G3P molecules, three CO_2 molecules produce six G3P molecules, of which five are used to regenerate RuBP, leaving a net gain of one G3P molecule per three CO_2 molecules (as would be expected from the number of carbon atoms involved).

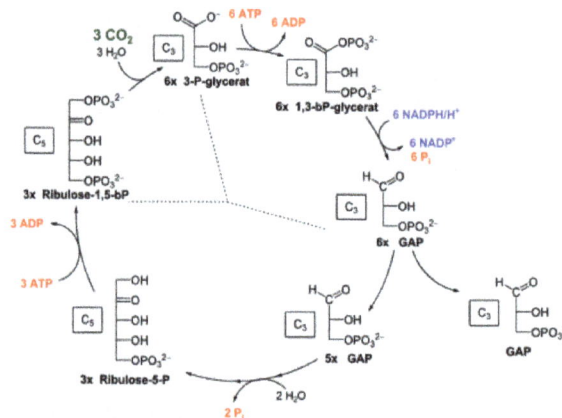

Simplified C3 cycle with structural formulas

The regeneration stage can be broken down into steps.

1. Triose phosphate isomerase converts all of the G3P reversibly into dihydroxyacetone phosphate (DHAP), also a 3-carbon molecule.

2. Aldolase and fructose-1,6-bisphosphatase convert a G3P and a DHAP into fructose 6-phosphate (6C). A phosphate ion is lost into solution.

3. Then fixation of another CO2 generates two more G3P.

4. F6P has two carbons removed by transketolase, giving erythrose-4-phosphate. The two carbons on transketolase are added to a G3P, giving the ketose xylulose-5-phosphate (Xu5P).

5. E4P and a DHAP (formed from one of the G3P from the second CO2 fixation) are converted into sedoheptulose-1,7-bisphosphate (7C) by aldolase enzyme.

6. Sedoheptulose-1,7-bisphosphatase (one of only three enzymes of the Calvin cycle that are

unique to plants) cleaves sedoheptulose-1,7-bisphosphate into sedoheptulose-7-phosphate, releasing an inorganic phosphate ion into solution.

7. Fixation of a third CO_2 generates two more G3P. The ketose S7P has two carbons removed by transketolase, giving ribose-5-phosphate (R5P), and the two carbons remaining on transketolase are transferred to one of the G3P, giving another Xu5P. This leaves one G3P as the product of fixation of 3 CO_2, with generation of three pentoses that can be converted to Ru5P.

8. R5P is converted into ribulose-5-phosphate (Ru5P, RuP) by phosphopentose isomerase. Xu5P is converted into RuP by phosphopentose epimerase.

9. Finally, phosphoribulokinase (another plant-unique enzyme of the pathway) phosphorylates RuP into RuBP, ribulose-1,5-bisphosphate, completing the Calvin *cycle*. This requires the input of one ATP.

Thus, of six G3P produced, five are used to make three RuBP (5C) molecules (totaling 15 carbons), with only one G3P available for subsequent conversion to hexose. This requires nine ATP molecules and six NADPH molecules per three CO_2 molecules. The equation of the overall Calvin cycle is shown diagrammatically below.

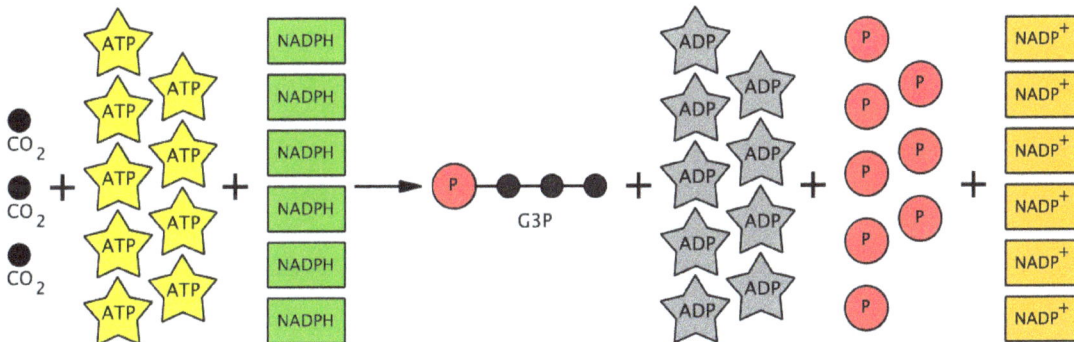

The overall equation of the Calvin cycle (black circles represent carbon atoms)

RuBisCO also reacts competitively with O_2 instead of CO_2 in photorespiration. The rate of photorespiration is higher at high temperatures. Photorespiration turns RuBP into 3-PGA and 2-phosphoglycolate, a 2-carbon molecule that can be converted via glycolate and glyoxalate to glycine. Via the glycine cleavage system and tetrahydrofolate, two glycines are converted into serine +CO_2. Serine can be converted back to 3-phosphoglycerate. Thus, only 3 of 4 carbons from two phosphoglycolates can be converted back to 3-PGA. It can be seen that photorespiration has very negative consequences for the plant, because, rather than fixing CO_2, this process leads to loss of CO_2. C4 carbon fixation evolved to circumvent photorespiration, but can occur only in certain plants native to very warm or tropical climates—corn, for example.

Products

The immediate products of one turn of the Calvin cycle are 2 glyceraldehyde-3-phosphate (G3P) molecules, 3 ADP, and 2 NADP+. (ADP and NADP+ are not really "products." They are regenerated and later used again in the Light-dependent reactions). Each G3P molecule is composed of

3 carbons. In order for the Calvin cycle to continue, RuBP (ribulose 1,5-bisphosphate) must be regenerated. So, 5 out of 6 carbons from the 2 G3P molecules are used for this purpose. Therefore, there is only 1 net carbon produced to play with for each turn. To create 1 surplus G3P requires 3 carbons, and therefore 3 turns of the Calvin cycle. To make one glucose molecule (which can be created from 2 G3P molecules) would require 6 turns of the Calvin cycle. Surplus G3P can also be used to form other carbohydrates such as starch, sucrose, and cellulose, depending on what the plant needs.

Light-dependent Regulation

Despite its widespread names (both light-independent and dark reactions), these reactions do not occur in the dark or at night. There is a light-dependent regulation of the cycle enzymes, as the third step requires reduced NADP; and this process would be a waste of energy, as there is no electron flow in the dark.

There are 2 regulation systems at work when the cycle needs to be turned on or off: thioredoxin/ferredoxin activation system, which activates some of the cycle enzymes, and the RuBisCo enzyme activation, active in the Calvin cycle, which involves its own activase.

The thioredoxin/ferredoxin system activates the enzymes glyceraldehyde-3-P dehydrogenase, glyceraldehyde-3-P phosphatase, fructose-1,6-bisphosphatase, sedoheptulose-1,7-bisphosphatase, and ribulose-5-phosphatase kinase, which are key points of the process. This happens when light is available, as the ferredoxin protein is reduced in the photosystem I complex of the thylakoid electron chain when electrons are circulating through it. Ferredoxin then binds to and reduces the thioredoxin protein, which activates the cycle enzymes by severing a cystine bond found in all these enzymes. This is a dynamic process as the same bond is formed again by other proteins that deactivate the enzymes. The implications of this process are that the enzymes remain mostly activated by day and are deactivated in the dark when there is no more reduced ferredoxin available.

The enzyme RuBisCo has its own activation process, which involves a more complex process. It is necessary that a specific lysine amino acid be carbamylated in order to activate the enzyme. This lysine binds to RuBP and leads to a non-functional state if left uncarbamylated. A specific activase enzyme, called RuBisCo activase, helps this carbamylation process by removing one proton from the lysine and making the binding of the carbon dioxide molecule possible. Even then the RuBisCo enzyme is not yet functional, as it needs a magnesium ion to be bound to the lysine in order to function. This magnesium ion is released from the thylakoid lumen when the inner pH drops due to the active pumping of protons from the electron flow. RuBisCo activase itself is activated by increased concentrations of ATP in the stroma caused by its phosphorylation.

Photosynthetic Reaction Centre

A photosynthetic reaction centre is a complex of several proteins, pigments and other co-factors that together execute the primary energy conversion reactions of photosynthesis. Molecular excitations, either originating directly from sunlight or transferred as excitation energy via light-harvesting antenna systems, give rise to electron transfer reactions along the path of a series of protein-bound co-factors. These co-factors are light-absorbing molecules (also named chromophores or pigments) such as chlorophyll and phaeophytin, as well as quinones. The energy of the photon is used to excite an electron of a pigment. The free energy created is then used to reduce a chain of nearby electron acceptors, which have subsequently higher redox-potentials. These electron transfer steps are the initial phase of a series of energy conversion reactions, ultimately resulting in the conversion of the energy of photons to the storage of that energy by the production of chemical bonds.

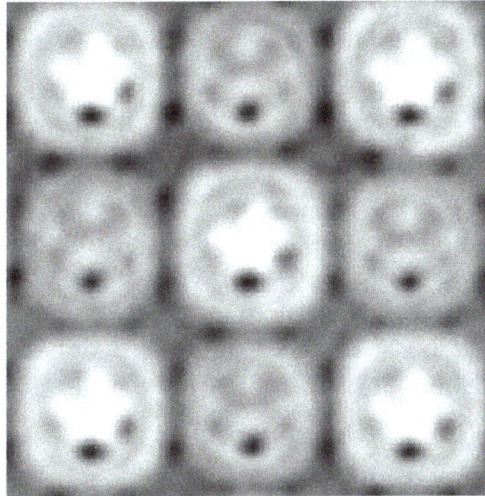

Electron micrograph of 2D crystals of the LH1-Reaction center photosynthetic unit.

Transforming Light Energy into Charge Separation

Reaction centers are present in all green plants, algae, and many bacteria. Although these species are separated by billions of years of evolution, the reaction centers are homologous for all photosynthetic species. In contrast, a large variety in light-harvesting complexes exist between the photosynthetic species. Green plants and algae have two different types of reaction centers that are part of larger supercomplexes known as photosystem I P700 and photosystem II P680. The structures of these supercomplexes are large, involving multiple light-harvesting complexes. The reaction center found in *Rhodopseudomonas* bacteria is currently best understood, since it was the first reaction center of known structure and has fewer polypeptide chains than the examples in green plants.

A reaction center is laid out in such a way that it captures the energy of a photon using pigment molecules and turns it into a usable form. Once the light energy has been absorbed directly by the pigment molecules, or passed to them by resonance transfer from a surrounding light-harvesting complex, they release two electrons into an electron transport chain.

Light is made up of small bundles of energy called photons. If a photon with the right amount of energy hits an electron, it will raise the electron to a higher energy level. Electrons are most stable at their lowest energy level, what is also called its ground state. In this state, the electron is in the orbit that has the least amount of energy. Electrons in higher energy levels can return to ground state in a manner analogous to a ball falling down a staircase. In doing so, the electrons release energy. This is the process that is exploited by a photosynthetic reaction center.

When an electron rises to a higher energy level, there is a corresponding decrease in the reduction potential of the molecule in which the electron resides occurs. This means that the molecule has a greater tendency to donate electrons, which is key to the conversion of light energy to chemical energy. In green plants, the electron transport chain has many electron acceptors including phaeophytin, quinone, plastoquinone, cytochrome bf, and ferredoxin, which result finally in the reduced molecule NADPH and the storage of energy. The passage of the electron through the electron transport chain also results in the pumping of protons (hydrogen ions) from the chloroplast's stroma and into the lumen, resulting in a proton gradient across the thylakoid membrane that can be used to synthesise ATP using the ATP synthase molecule. Both the ATP and NADPH are used in the Calvin cycle to fix carbon dioxide into triose sugars.

Bacteria

Structure

The bacterial photosynthetic reaction center has been an important model to understand the structure and chemistry of the biological process of capturing light energy. In the 1960s, Roderick Clayton was the first to purify the reaction center complex from purple bacteria. However, the first crystal structure was determined in 1984 by Hartmut Michel, Johann Deisenhofer and Robert Huber for which they shared the Nobel Prize in 1988. This was also significant, since it was the first structure for any membrane protein complex.

Four different subunits were found to be important for the function of the photosynthetic reaction center. The L and M subunits, shown in blue and purple in the image of the structure, both span the lipid bilayer of the plasma membrane. They are structurally similar to one another, both having 5 transmembrane alpha helices. Four bacteriochlorophyll b (BChl-b) molecules, two bacteriophaeophytin b molecules (BPh) molecules, two quinones (Q_A and Q_B), and a ferrous ion are associated with the L and M subunits. The H subunit, shown in gold, lies on the cytoplasmic side of the plasma membrane. A cytochrome subunit, here not shown, contains four c-type haems and is located on the periplasmic surface (outer) of the membrane. The latter sub-unit is not a general structural motif in photosynthetic bacteria. The L and M subunits bind the functional and light-interacting cofactors, shown here in green.

Reaction centers from different bacterial species may contain slightly altered bacterio-chlorophyll and bacterio-phaeophytin chromophores as functional co-factors. These alterations cause shifts in the colour of light that can be absorbed, thus creating specific niches for photosynthesis. The reaction center contains two pigments that serve to collect and transfer the energy from photon absorption: BChl and Bph. BChl roughly resembles the chlorophyll molecule found in green plants, but, due to minor structural differences, its peak absorption wavelength is shifted into the infrared, with wavelengths as long as 1000 nm. Bph has the same structure as BChl, but the central magnesium ion is replaced by two protons. This alteration causes both an absorbance maximum shift and a lowered redox-potential.

Mechanism

The process starts when light is absorbed by two BChl molecules (a dimer) that lie near the periplasmic side of the membrane. This pair of chlorophyll molecules, often called the "special pair", absorbs photons between 870 nm and 960 nm, depending on the species and, thus, is called P870 (for the species rhodobacter sphaeroides) or P960 (for rhodopseudomonas viridis), with *P* standing for "pigment"). Once P absorbs a photon, it ejects an electron, which is transferred through another molecule of Bchl to the BPh in the L subunit. This initial charge separation yields a positive charge on P and a negative charge on the BPh. This process takes place in 10 picoseconds (10^{-11} seconds).

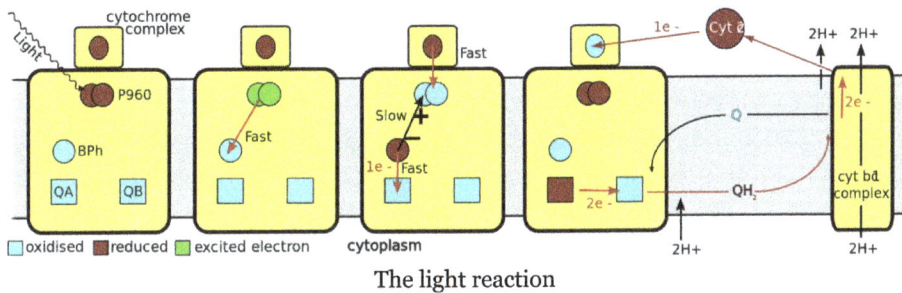

The light reaction

The charges on the specialpair $^+$ and the BPh$^-$ could undergo charge recombination in this state. This would waste the high-energy electron and convert the absorbed light energy into heat. Several factors of the reaction center structure serve to prevent this. First, the transfer of an electron from BPh$^-$ to P960$^+$ is relatively slow compared to two other redox reactions in the reaction center. The faster reactions involve the transfer of an electron from BPh$^-$ (BPh$^-$ is oxidised to BPh) to the electron acceptor quinone (Q_A), and the transfer of an electron to P960$^+$ (P960$^+$ is reduced to P960) from a heme in the cytochrome subunit above the reaction center.

The high-energy electron that resides on the tightly bound quinone molecule Q_A is transferred to an exchangeable quinone molecule Q_B. This molecule is loosely associated with the protein and is fairly easy to detach. Two of the high-energy electrons are required to fully reduce Q_B to QH_2, taking up two protons from the cytoplasm in the process. The reduced quinone QH_2 diffuses through the membrane to another protein complex (cytochrome bc_1-complex) where it is oxidised. In the process the reducing power of the QH_2 is used to pump protons across the membrane to the periplasmic space. The electrons from the cytochrome bc_1-complex are then transferred through a soluble cytochrome c intermediate, called cytochrome c_2, in the periplasm to the cytochrome subunit. Thus, the flow of electrons in this system is cyclical.

Green Plants

Oxygenic Photosynthesis

In 1772, the chemist Joseph Priestley carried out a series of experiments relating to the gases involved in respiration and combustion. In his first experiment, he lit a candle and placed it under an upturned jar. After a short period of time, the candle burned out. He carried out a similar experiment with a mouse in the confined space of the burning candle. He found that the mouse died a short time after the candle had been extinguished. However, he could revivify the foul air by placing green plants in the area and exposing them to light. Priestley's observations were some of the first experiments that demonstrated the activity of a photosynthetic reaction center.

In 1779, Jan Ingenhousz carried out more than 500 experiments spread out over 4 months in an attempt to understand what was really going on. He wrote up his discoveries in a book entitled *Experiments upon Vegetables*. Ingenhousz took green plants and immersed them in water inside a transparent tank. He observed many bubbles rising from the surface of the leaves whenever the plants were exposed to light. Ingenhousz collected the gas that was given off by the plants and performed several different tests in attempt to determine what the gas was. The test that finally revealed the identity of the gas was placing a smouldering taper into the gas sample and having it relight. This test proved it was oxygen, or, as Joseph Priestley had called it, 'de-phlogisticated air'.

In 1932, Professor Robert Emerson and an undergraduate student, William Arnold, used a repetitive flash technique to precisely measure small quantities of oxygen evolved by chlorophyll in the algae *Chlorella*. Their experiment proved the existence of a photosynthetic unit. Gaffron and Wohl later interpreted the experiment and realized that the light absorbed by the photosynthetic unit was transferred. This reaction occurs at the reaction center of photosystem II and takes place in cyanobacteria, algae and green plants.

Photosystem II

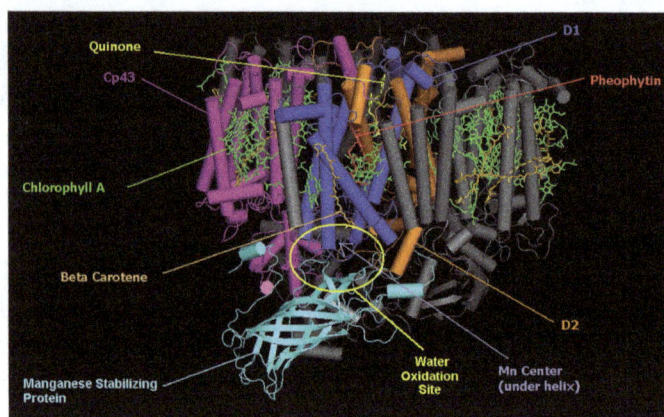

Cyanobacteria photosystem II, Monomer, PDB 2AXT.

Photosystem II is the photosystem that generates the two electrons that will eventually reduce $NADP^+$ in Ferredoxin-NADP-reductase. Photosystem II is present on the thylakoid membranes inside chloroplasts, the site of photosynthesis in green plants. The structure of Photosystem II is remarkably similar to the bacterial reaction center, and it is theorized that they share a common ancestor.

The core of photosystem II consists of two subunits referred to as D1 and D2. These two subunits are similar to the L and M subunits present in the bacterial reaction center. Photosystem II differs from the bacterial reaction center in that it has many additional subunits that bind additional chlorophylls to increase efficiency. The overall reaction catalysed by photosystem II is:

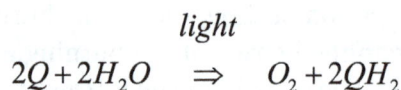

$$2Q + 2H_2O \quad \overset{light}{\Rightarrow} \quad O_2 + 2QH_2$$

Q represents plastoquinone, the oxidized form of Q. QH_2 represents plastoquinol, the reduced form of Q. This process of reducing quinone is comparable to that which takes place in the bacterial reaction center. Photosystem II obtains electrons by oxidizing water in a process called photolysis. Molecular oxygen is a byproduct of this process, and it is this reaction that supplies the atmosphere with oxygen. The fact that the oxygen from green plants originated from water was first deduced by the Canadian-born American biochemist Martin David Kamen. He used a natural, stable isotope of oxygen, O_{18} to trace the path of the oxygen, from water to gaseous molecular oxygen. This reaction is catalysed by a reactive center in photosystem II containing four manganese ions.

The reaction begins with the excitation of a pair of chlorophyll molecules similar to those in the bacterial reaction center. Due to the presence of chlorophyll a, as opposed to bacteriochlorophyll, photosystem II absorbs light at a shorter wavelength. The pair of chlorophyll molecules at the reaction center are often referred to as P680. When the photon has been absorbed, the resulting high-energy electron is transferred to a nearby phaeophytin molecule. This is above and to the right of the pair on the diagram and is coloured grey. The electron travels from the phaeophytin molecule through two plastoquinone molecules, the first tightly bound, the second loosely bound. The tightly bound molecule is shown above the phaeophytin molecule and is coloured red. The loosely bound molecule is to the left of this and is also coloured red. This flow of electrons is similar to that of the bacterial reaction center. Two electrons are required to fully reduce the loosely bound plastoquinone molecule to QH_2 as well as the uptake of two protons.

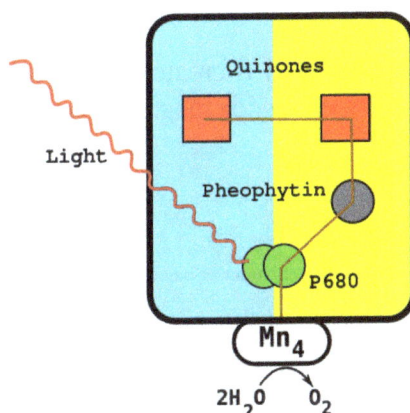

The difference between photosystem II and the bacterial reaction center is the source of the electron that neutralizes the pair of chlorophyll a molecules. In the bacterial reaction center, the electron is obtained from a reduced compound haem group in a cytochrome subunit or from a water-soluble cytochrome-c protein.

Once photoinduced charge separation has taken place, the P680 molecule carries a positive charge. P680 is a very strong oxidant and extracts electrons from two water molecules that are bound at the manganese center directly below the pair. This center, below and to the left of the pair in the diagram, contains four manganese ions, a calcium ion, a chloride ion, and a tyrosine residue. Manganese is efficient because it is capable of existing in four oxidation states: Mn^{2+}, Mn^{3+}, Mn^{4+} and Mn^{5+}. Manganese also forms strong bonds with oxygen-containing molecules such as water.

Every time the P680 absorbs a photon, it emits an electron, gaining a positive charge. This charge is neutralized by the extraction of an electron from the manganese center, which sits directly below it. The process of oxidizing two molecules of water requires four electrons. The water molecules that are oxidized in the manganese center are the source of the electrons that reduce the two molecules of Q to QH_2. To date, this water-splitting catalytic center cannot be reproduced by any man-made catalyst.

Photosystem I

After the electron has left photosystem II it is transferred to a cytochrome b6f complex and then to plastocyanin, a blue copper protein and electron carrier. The plastocyanin complex carries the electron that will neutralize the pair in the next reaction center, photosystem I.

As with photosystem II and the bacterial reaction center, a pair of chlorophyll *a* molecules initiates photoinduced charge separation. This pair is referred to as P700. 700 Is a reference to the wavelength at which the chlorophyll molecules absorb light maximally. The P700 lies in the center of the protein. Once photoinduced charge separation has been initiated, the electron travels down a pathway through a chlorophyll α molecule situated directly above the P700, through a quinone molecule situated directly above that, through three 4Fe-4S clusters, and finally to an interchangeable ferredoxin complex. Ferredoxin is a soluble protein containing a 2Fe-2S cluster coordinated by four cysteine residues. The positive charge left on the P700 is neutralized by the transfer of an electron from plastocyanin. Thus the overall reaction catalysed by photosystem I is:

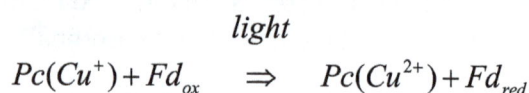

$$Pc(Cu^+) + Fd_{ox} \quad \overset{light}{\Rightarrow} \quad Pc(Cu^{2+}) + Fd_{red}$$

The cooperation between photosystems I and II creates an electron flow from H_2O to $NADP^+$. This pathway is called the 'Z-scheme' because the redox diagram from P680 to P700 resembles the letter z.

Photosynthetic Efficiency

The photosynthetic efficiency is the fraction of light energy converted into chemical energy during photosynthesis in plants and algae. Photosynthesis can be described by the simplified chemical reaction

$$6H_2O + 6CO_2 + energy \rightarrow C_6H_{12}O_6 + 6O_2$$

where $C_6H_{12}O_6$ is glucose (which is subsequently transformed into other sugars, cellulose, lignin, and so forth). The value of the photosynthetic efficiency is dependent on how light energy is defined – it depends on whether we count only the light that is absorbed, and on what kind of light is used. It takes eight (or perhaps 10 or more) photons to utilize one molecule of CO_2. The Gibbs free energy for converting a mole of CO_2 to glucose is 114 kcal, whereas eight moles of photons of wavelength 600 nm contains 381 kcal, giving a nominal efficiency of 30%. However, photosynthesis can occur with light up to wave-length 720 nm so long as there is also light at wavelengths below 680 nm to keep Photosystem II operating. Using longer wavelengths means less light energy is needed for the same number of photons and therefore for the same amount of photosynthesis. For actual sunlight, where only 45% of the light is in the photosynthetically active wavelength range, the theoretical maximum efficiency of solar energy conversion is approximately 11%. In actuality, however, plants do not absorb all incoming sunlight (due to reflection, respiration requirements of photosynthesis and the need for optimal solar radiation levels) and do not convert all harvested energy into biomass, which results in an overall photosynthetic efficiency of 3 to 6% of total solar radiation. If photosynthesis is inefficient, excess light energy must be dissipated to avoid damaging the photosynthetic apparatus. Energy can be dissipated as heat (non-photochemical quenching), or emitted as chlorophyll fluorescence.

Typical Efficiencies

Plants

Quoted values sunlight-to-biomass efficiency

Plant	Efficiency
Plants, typical	0.1% 0.2–2%
Typical crop plants	1–2%
Sugarcane	7–8% peak

The following is a breakdown of the energetics of the photosynthesis process from *Photosynthesis* by Hall and Rao:

Starting with the solar spectrum falling on a leaf, 47% lost due to photons outside the 400–700 nm active range (chlorophyll utilizes photons between 400 and 700 nm, extracting the energy of one 700 nm photon from each one) 30% of the in-band photons are lost due to incomplete absorption or photons hitting components other than chloroplasts 24% of the absorbed photon energy is lost due to degrading short wavelength photons to the 700 nm energy level 68% of the utilized energy is lost in conversion into d-glucose 35–45% of the glucose is consumed by the leaf in the processes of dark and photo respiration

Stated another way: 100% sunlight → non-bioavailable photons waste is 47%, leaving 53% (in the 400–700 nm range) → 30% of photons are lost due to incomplete absorption, leaving 37% (absorbed photon energy) → 24% is lost due to wavelength-mismatch degradation to 700 nm energy, leaving 28.2% (sunlight energy collected by chlorophyl) → 32% efficient conversion of ATP and NADPH to d-glucose, leaving 9% (collected as sugar) → 35–40% of sugar is recycled/consumed by the leaf in dark and photo-respiration, leaving 5.4% net leaf efficiency.

Many plants lose much of the remaining energy on growing roots. Most crop plants store ~0.25% to 0.5% of the sunlight in the product (corn kernels, potato starch, etc.). Sugar cane is exceptional in several ways, yielding peak storage efficiencies of ~8%.

Measuring the photosynthetic efficiency of wheat in the field using an LCpro-SD

Photosynthesis increases linearly with light intensity at low intensity, but at higher intensity this is no longer the case. Above about 10,000 lux or ~100 watts/square meter the rate no longer increases. Thus, most plants can only utilize ~10% of full mid-day sunlight intensity. This dramatically reduces average achieved photosynthetic efficiency in fields compared to peak laboratory results. However, real plants (as opposed to laboratory test samples) have lots of redundant, randomly oriented leaves. This helps to keep the average illumination of each leaf well below the mid-day peak enabling the plant to achieve a result closer to the expected laboratory test results using limited illumination.

Only if the light intensity is above a plant specific value, called the compensation point the plant assimilates more carbon and releases more oxygen by photosynthesis than it consumes by cellular respiration for its own current energy demand. Photosynthesis measurement systems are not designed to directly measure the amount of light absorb

From a 2010 study by the University of Maryland, photosynthesizing Cyanobacteria have been shown to be a significant species in the global carbon cycle, accounting for 20–30% of Earth's photosynthetic productivity and convert solar energy into biomass-stored chemical energy at the rate of ~450 TW.

Worldwide Figures

According to the cyanobacteria study above, this means the total photosynthetic productivity of earth is between ~1500–2250 TW, or 47,300–71,000 exajoules per year. Using this source's figure of 178,000 TW of solar energy hitting the Earth's surface, the total photosynthetic efficiency of the planet is 0.84% to 1.26%.

Efficiencies of Various Biofuel Crops

Popular choices for plant biofuels include: oil palm, soybean, castor oil, sunflower oil, safflower oil, corn ethanol, and sugar cane ethanol.

An analysis of a proposed Hawaiian oil palm plantation claimed to yield 600 gallons of biodiesel per acre per year. That comes to 2835 watts per acre or 0.7 W/m². Typical insolation in Hawaii is around 5.5 kWh/(m²day) or 230 W/m². For this particular oil palm plantation, if it delivered the claimed 600 gallons of biodiesel per acre per year, would be converting 0.3% of the incident solar energy to chemical fuel. Total photosynthetic efficiency would include more than just the biodiesel oil, so this 0.3% number is something of a lower bound.

Contrast this with a typical photovoltaic installation, which would produce an average of roughly 22 W/m² (roughly 10% of the average insolation), throughout the year. Furthermore, the photovoltaic panels would produce electricity, which is a high-quality form of energy, whereas converting the biodiesel into mechanical energy entails the loss of a large portion of the energy. On the other hand, a liquid fuel is much more convenient for a vehicle than electricity, which has to be stored in heavy, expensive batteries.

Most crop plants store ~0.25% to 0.5% of the sunlight in the product (corn kernels, potato starch, etc.), sugar cane is exceptional in several ways to yield peak storage efficiencies of ~8%.

Ethanol fuel in Brazil has a calculation that results in: "Per hectare per year, the biomass produced corresponds to 0.27 TJ. This is equivalent to 0.86 W/m². Assuming an average insolation of 225 W/

m², the photosynthetic efficiency of sugar cane is 0.38%." Sucrose accounts for little more than 30% of the chemical energy stored in the mature plant; 35% is in the leaves and stem tips, which are left in the fields during harvest, and 35% are in the fibrous material (bagasse) left over from pressing.

C3 vs. C4 and CAM Plants

C3 plants use the Calvin cycle to fix carbon. C4 plants use a modified Calvin cycle in which they separate Ribulose-1,5-bisphosphate carboxylase oxygenase (RuBisCO) from atmospheric oxygen, fixing carbon in their mesophyll cells and using oxaloacetate and malate to ferry the fixed carbon to RuBisCO and the rest of the Calvin cycle enzymes isolated in the bundle-sheath cells. The intermediate compounds both contain four carbon atoms, which gives C4. In Crassulacean acid metabolism (CAM), time isolates functioning RuBisCo (and the other Calvin cycle enzymes) from high oxygen concentrations produced by photosynthesis, in that O_2 is evolved during the day, and allowed to dissipate then, while at night atmospheric CO_2 is taken up and stored as malic or other acids. During the day, CAM plants close stomata and use stored acids as carbon sources for sugar, etc. production.

The C3 pathway requires 18 ATP for the synthesis of one molecule of glucose while the C4 pathway requires 20 ATP. Despite this reduced ATP efficiency, C4 is an evolutionary advancement, adapted to areas of high levels of light, where the reduced ATP efficiency is more than offset by the use of increased light. The ability to thrive despite restricted water availability maximizes the ability to use available light. The simpler C3 cycle which operates in most plants is adapted to wetter darker environments, such as many northern latitudes.. Corn, sugar cane, and sorghum are C4 plants. These plants are economically important in part because of their relatively high photosynthetic efficiencies compared to many other crops . Pineapple is a CAM plant.

PI Curve

The PI (photosynthesis-irradiance) curve is a graphical representation of the empirical relationship between solar irradiance and photosynthesis. A derivation of the Michaelis–Menten curve, it shows the generally positive correlation between light intensity and photosynthetic rate. Plotted along the x-axis is the independent variable, light intensity (irradiance), while the y-axis is reserved for the dependent variable, photosynthetic rate.

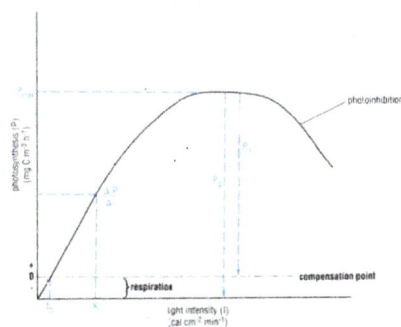

P v I curve

Introduction

The PI curve can be applied to terrestrial and marine reactions but is most commonly used to explain ocean-dwelling phytoplankton's photosynthetic response to changes in light intensity. Using this tool to approximate biological productivity is important because phytoplankton contribute ~50% of total global carbon fixation and are important suppliers to the marine food web.

Within the scientific community, the curve can be referred to as the PI, PE or Light Response Curve. While individual researchers may have their own preferences, all are readily acceptable for use in the literature. Regardless of nomenclature, the photosynthetic rate in question can be described in terms of carbon (C) fixed per unit per time. Since individuals vary in size, it is also useful to normalise C concentration to Chlorophyll a (an important photosynthetic pigment) to account for specific biomass.

History

As far back as 1905, marine researchers attempted to develop an equation to be used as the standard in establishing the relationship between solar irradiance and photosynthetic production. Several groups had relative success, but in 1976 a comparison study conducted by Alan Jassby and Trevor Platt, researchers at the Bedford Institute of Oceanography in Dartmouth, Nova Scotia, reached a conclusion that solidified the way in which a PI curve is developed. After evaluating the eight most-used equations, Jassby and Platt revealed that an adaptation of the Michaelis-Menten equation, previously used in enzyme kinetics, best served the relationship demonstration. Their findings were so conclusive that the Michaelis–Menten equation remains the standard for PI curve generation.

Equations

There are two simple derivations of the equation that are commonly used to generate the hyperbolic curve. The first assumes photosynthetic rate increases with increasing light intensity until Pmax is reached and continues to photosynthesise at the maximum rate thereafter.

$$P = P_{max}[I] / (KI + [I])$$

- P = photosynthetic rate at a given light intensity
  - o Commonly denoted in units such as (mg C m-3 h-1) or (μg C μg Chl-a-1 h-1)
- Pmax = the maximum potential photosynthetic rate per individual
- [I] = a given light intensity
  - o Commonly denoted in units such as (μMol photons m-2 s-1 or (Watts m-2 h-1)
- KI = half-saturation constant; the light intensity at which the photosynthetic rate proceeds at ½ Pmax

o   Units reflect those used for [I]

Both Pmax and the initial slope of the curve, ΔP/ΔI, are species-specific, and are influenced by a variety of factors, such as nutrient concentration, temperature and the physiological capabilities of the individual. Light intensity is influenced by latitudinal position and undergo daily and seasonal fluxes which will also affect the overall photosynthetic capacity of the individual. These three parameters are predictable and can be used to predetermine the general PI curve a population should follow.

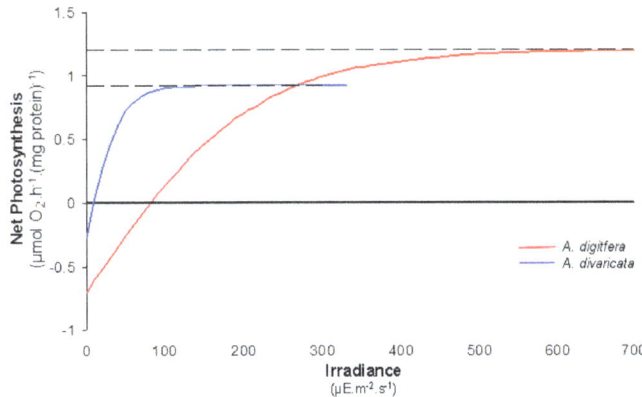

As can be seen in the graph, two species can have different responses to the same incremental changes in light intensity. Population A (in blue) has an initial rate higher than that of Population B (in red) and also exhibits a stronger rate change to increased light intensities at lower irradiance. Therefore, Population A will dominate in an environment with lower light availability. Although Population B has a slower photosynthetic response to increases in light intensity its Pmax is higher than that of Population A. This allows for eventual population dominance at greater light intensities. There are many determining factors influencing population success; using the PI curve to elicit predictions of rate flux to environmental changes is useful for monitoring phytoplankton bloom dynamics and ecosystem stability.

The second equation accounts for the phenomenon of photoinhibition. In the upper few meters of the ocean, phytoplankton may be subjected to irradiance levels that damage the chlorophyll-a pigment inside the cell, subsequently decreasing photosynthetic rate. The response curve depicts photoinhibition as a decrease in photosynthetic rate at light intensities stronger than those necessary for achievement of Pmax.

$$P = P_{max}(1 - e^{-\alpha I / P_{max}})e^{-\beta I / P_{max}}$$

Terms not included in the above equation are:

- βI = light intensity at the start of photoinhibition

- αI = a given light intensity

Examples

Data sets showing interspecific differences and population dynamics.

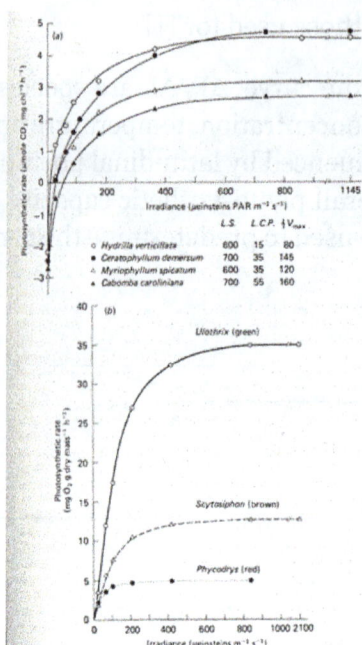

The hyperbolic response between photosynthesis and irradiance, depicted by the PI curve, is important for assessing phytoplankton population dynamics, which influence many aspects of the marine environment.

Artificial Photosynthesis

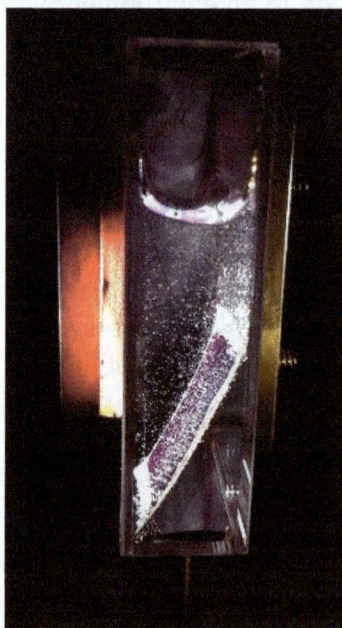

A sample of a photoelectric cell in a lab environment. Catalysts are added to the cell, which is submerged in water and illuminated by simulated sunlight. The bubbles seen are oxygen (forming on the front of the cell) and hydrogen (forming on the back of the cell).

Artificial photosynthesis is a chemical process that replicates the natural process of photosynthesis, a process that converts sunlight, water, and carbon dioxide into carbohydrates and oxygen.

The term, artificial photosynthesis, is commonly used to refer to any scheme for capturing and storing the energy from sunlight in the chemical bonds of a fuel (a solar fuel). Photocatalytic water splitting converts water into hydrogen ions and oxygen, and is a main research area in artificial photosynthesis. Light-driven carbon dioxide reduction is another studied process, that replicates natural carbon fixation.

Research developed in this field encompasses the design and assembly of devices for the direct production of solar fuels, photoelectrochemistry and its application in fuel cells, and the engineering of enzymes and photoautotrophic microorganisms for microbial biofuel and biohydrogen production from sunlight. Many, if not most, of the artificial approaches to artificial photosynthesis are bio-inspired, i.e., they rely on biomimetics.

Overview

The photosynthetic reaction can be divided into two half-reactions of oxidation and reduction, both of which are essential to producing fuel. In plant photosynthesis, water molecules are photo-oxidized to release oxygen and protons. The second stage of plant photosynthesis (also known as the Calvin-Benson cycle) is a light-independent reaction that converts carbon dioxide into glucose (fuel). Researchers of artificial photosynthesis are developing photocatalysts that are able to perform both of these reactions. Furthermore, the protons resulting from water splitting can be used for hydrogen production. These catalysts must be able to react quickly and absorb a large percentage of the incident solar photons.

Natural (left) versus artificial photosynthesis (right)

Whereas photovoltaics can provide energy directly from sunlight, the inefficiency of fuel production from photovoltaic electricity (indirect process) and the fact that sunshine is not constant throughout the day sets a limit to its use. One way of using natural photosynthesis is via the production of a biofuel, which is an indirect process that suffers from low energy conversion efficiency (due to photosynthesis' own low efficiency in converting sunlight to biomass), the cost of harvesting and transporting the fuel, and clashes with the increasing need of land mass for food production. Artificial photosynthesis aims then to produce a fuel from sunlight that can be conveniently stored and used when sunlight is not available, by using direct processes, that is, to produce a solar fuel. With the development of catalysts able to reproduce the key steps of photosynthesis, water and sunlight would ultimately be the only needed sources for clean energy production. The only by-product would be oxygen, and production of a solar fuel has the potential to be cheaper than gasoline.

One process for the creation of a clean and affordable energy supply is the development of photocatalytic water splitting under solar light. This method of sustainable hydrogen production is a key objective in the development of alternative energy systems. It is also predicted to be one of the more, if not the most, efficient ways of obtaining hydrogen from water. The conversion of solar energy into hydrogen via a water-splitting process assisted by photosemiconductor catalysts is one of the most promising technologies in development. This process has the potential for large quantities of hydrogen to be generated in an ecologically sound manner. The conversion of solar energy into a clean fuel (H_2) under ambient conditions is one of the greatest challenges facing scientists in the twenty-first century.

Two approaches are generally recognized in the construction of solar fuel cells for hydrogen production:

- A homogeneous system is one where catalysts are not compartmentalized, that is, components are present in the same compartment. This means that hydrogen and oxygen are produced in the same location. This can be a drawback, since they compose an explosive mixture, demanding gas product separation. Also, all components must be active in approximately the same conditions (e.g., pH).

- A heterogeneous system has two separate electrodes, an anode and a cathode, making possible the separation of oxygen and hydrogen production. Furthermore, different components do not necessarily need to work in the same conditions. However, the increased complexity of these systems makes them harder to develop and more expensive.

Another area of research within artificial photosynthesis is the selection and manipulation of photosynthetic microorganisms, namely green microalgae and cyanobacteria, for the production of solar fuels. Many strains are able to produce hydrogen naturally, and scientists are working to improve them. Algae biofuels such as butanol and methanol are produced both at laboratory and commercial scales. This approach has benefited from the development of synthetic biology, which is also being explored by the J. Craig Venter Institute to produce a synthetic organism capable of biofuel production.

History

The artificial photosynthesis was first anticipated by the Italian chemist Giacomo Ciamician in 1912. In a lecture that was later published in Science he proposed a switch from the use of fossil fuels to radiant energy provided by the sun and captured by technical photochemistry devices. In this switch he saw a possibility to close the gap between the rich north and poor south and ventured a guess that this switch from coal to solar energy would "not be harmful to the progress and to human happiness."

In the late 60s, Akira Fujishima discovered the photocatalytic properties of titanium dioxide, the so-called Honda-Fujishima effect, which could be used for hydrolysis.

The Swedish Consortium for Artificial Photosynthesis, the first of its kind, was established in 1994 as a collaboration between groups of three different universities, Lund, Uppsala and Stockholm, being presently active around Lund and the Ångström Laboratories in Uppsala. The consortium was built with a multidisciplinary approach to focus on learning from natural photosynthesis and applying this knowledge in biomimetic systems.

Research into artificial photosynthesis is undergoing a boom at the beginning of the 21st century. In 2000, Commonwealth Scientific and Industrial Research Organisation (CSIRO) researchers publicized their intent to focus on carbon dioxide capture and its conversion to hydrocarbons. In 2003, the Brookhaven National Laboratory announced the discovery of an important intermediate step in the reduction of CO_2 to CO (the simplest possible carbon dioxide reduction reaction), which could lead to better catalysts.

One of the drawbacks of artificial systems for water-splitting catalysts is their general reliance on scarce, expensive elements, such as ruthenium or rhenium. In 2008, with the funding of the United States Air Force Office of Scientific Research, MIT chemist and head of the Solar Revolution Project Daniel G. Nocera and postdoctoral fellow Matthew Kanan attempted to circumvent this issue by using a catalyst containing the cheaper and more abundant elements cobalt and phosphate. The catalyst was able to split water into oxygen and protons using sunlight, and could potentially be coupled to a hydrogen gas producing catalyst such as platinum. Furthermore, while the catalyst broke down during catalysis, it could self-repair. This experimental catalyst design was considered a major breakthrough in the field by many researchers.

Whereas CO is the prime reduction product of CO_2, more complex carbon compounds are usually desired. In 2008, Andrew B. Bocarsly reported the direct conversion of carbon dioxide and water to methanol using solar energy in a highly efficient photochemical cell.

While Nocera and coworkers had accomplished water splitting to oxygen and protons, a light-driven process to produce hydrogen is desirable. In 2009, the Leibniz Institute for Catalysis reported inexpensive iron carbonyl complexes able to do just that. In the same year, researchers at the University of East Anglia also used iron carbonyl compounds to achieve photoelectrochemical hydrogen production with 60% efficiency, this time using a gold electrode covered with layers of indium phosphide to which the iron complexes were linked. Both of these processes used a molecular approach, where discrete nanoparticles are responsible for catalysis.

Visible light water splitting with a one piece multijunction cell was first demonstrated and patented by William Ayers at Energy Conversion Devices in 1983. This group demonstrated water photolysis into hydrogen and oxygen, now referred to as an "artificial leaf" or "wireless solar water splitting" with a low cost, thin film amorphous silicon multijunction cell directly immersed in water. Hydrogen evolved on the front amorphous silicon surface decorated with various catalysts while oxygen evolved from the back metal substrate which also eliminated the hazard of mixed hydrogen/oxygen gas evolution. A Nafion membrane above the immersed cell provided a path for proton transport. The higher photovoltage available from the multijuction thin film cell with visible light was a major advance over previous photolysis attempts with UV sensitive single junction cells. The group's patent also lists several other semiconductor multijunction compositions in addition to amorphous silicon.

In 2009, F. del Valle and K. Domen showed the impact of the thermal treatment in a closed atmosphere using Cd1-xZnxS photocatalysts. Cd1-xZnxS solid solution reports high activity in hydrogen production from water splitting under sunlight irradiation. A mixed heterogeneous/molecular approach by researchers at the University of California, Santa Cruz, in 2010, using both nitrogen-doped and cadmium selenide quantum dots-sensitized titanium dioxide nanoparticles and nanowires, also yielded photoproduced hydrogen.

Artificial photosynthesis remained an academic field for many years. However, in the beginning of 2009, Mitsubishi Chemical Holdings was reported to be developing its own artificial photosynthesis research by using sunlight, water and carbon dioxide to "create the carbon building blocks from which resins, plastics and fibers can be synthesized." This was confirmed with the establishment of the KAITEKI Institute later that year, with carbon dioxide reduction through artificial photosynthesis as one of the main goals.

In 2010, the DOE established, as one of its Energy Innovation Hubs, the Joint Center for Artificial

Photosynthesis. The mission of JCAP is to find a cost-effective method to produce fuels using only sunlight, water, and carbon-dioxide as inputs. JCAP is led by a team from Caltech, led by Professor Nathan Lewis and brings together more than 120 scientists and engineers from Caltech and its lead partner, Lawrence Berkeley National Laboratory. JCAP also draws on the expertise and capabilities of key partners from Stanford University, the University of California at Berkeley, UCSB, UCI, and UCSD, and the Stanford Linear Accelerator. In addition, JCAP serves as a central hub for other solar fuels research teams across the United States, including 20 DOE Energy Frontier Research Center. The program has a budget of $122M over five years, subject to Congressional appropriation

Also in 2010, a team led by professor David Wendell at the University of Cincinnati successfully demonstrated photosynthesis in an artificial construct consisting of enzymes suspended in a foam housing.

In 2011, Daniel Nocera and his research team announced the creation of the first practical artificial leaf. In a speech at the 241st National Meeting of the American Chemical Society, Nocera described an advanced solar cell the size of a poker card capable of splitting water into oxygen and hydrogen, approximately ten times more efficient than natural photosynthesis. The cell is mostly made of inexpensive materials that are widely available, works under simple conditions, and shows increased stability over previous catalysts: in laboratory studies, the authors demonstrated that an artificial leaf prototype could operate continuously for at least forty-five hours without a drop in activity. In May 2012, Sun Catalytix, the startup based on Nocera's research, stated that it will not be scaling up the prototype as the device offers few savings over other ways to make hydrogen from sunlight. Leading experts in the field have supported a proposal for a Global Project on Artificial Photosynthesis as a combined energy security and climate change solution. Conferences on this theme have been held at Lord Howe Island in 2011, at Chicheley Hall in the UK in 2014 and at Canberra and Lord Howe island in 2016.

Current Research

In energy terms, natural photosynthesis can be divided in three steps:

- Light-harvesting complexes in bacteria and plants capture photons and transduce them into electrons, injecting them into the photosynthetic chain.

- Proton-coupled electron transfer along several cofactors of the photosynthetic chain, causing local, spatial charge separation.

- Redox catalysis, which uses the aforementioned transferred electrons to oxidize water to dioxygen and protons; these protons can in some species be utilized for dihydrogen production.

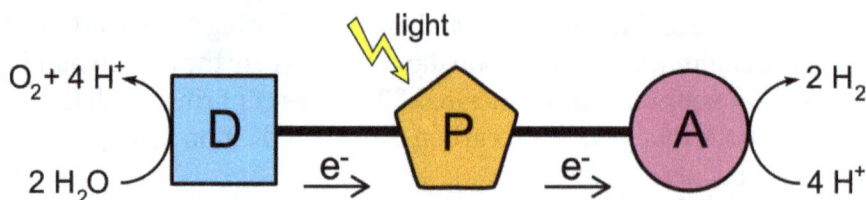

A triad assembly, with a photosensitizer (P) linked in tandem to a water oxidation catalyst (D) and a hydrogen evolving catalyst (A). Electrons flow from D to A when catalysis occurs.

Using biomimetic approaches, artificial photosynthesis tries to construct systems doing the same type of processes. Ideally, a triad assembly could oxidize water with one catalyst, reduce protons with another and have a photosensitizer molecule to power the whole system. One of the simplest designs is where the photosensitizer is linked in tandem between a water oxidation catalyst and a hydrogen evolving catalyst:

- The photosensitizer transfers electrons to the hydrogen catalyst when hit by light, becoming oxidized in the process.

- This drives the water splitting catalyst to donate electrons to the photosensitizer. In a triad assembly, such a catalyst is often referred to as a donor. The oxidized donor is able to perform water oxidation.

The state of the triad with one catalyst oxidized on one end and the second one reduced on the other end of the triad is referred to as a charge separation, and is a driving force for further electron transfer, and consequently catalysis, to occur. The different components may be assembled in diverse ways, such as supramolecular complexes, compartmentalized cells, or linearly, covalently linked molecules.

Research into finding catalysts that can convert water, carbon dioxide, and sunlight to carbohydrates or hydrogen is a current, active field. By studying the natural oxygen-evolving complex (OEC), researchers have developed catalysts such as the "blue dimer" to mimic its function or inorganic-based materials such as Birnessite with the similar building block as the OEC. Photoelectrochemical cells that reduce carbon dioxide into carbon monoxide (CO), formic acid (HCOOH) and methanol (CH_3OH) are under development. However, these catalysts are still very inefficient.

Hydrogen Catalysts

Hydrogen is the simplest solar fuel to synthesize, since it involves only the transference of two electrons to two protons. It must, however, be done stepwise, with formation of an intermediate hydride anion:

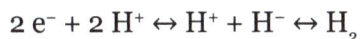

$$2\,e^- + 2\,H^+ \leftrightarrow H^+ + H^- \leftrightarrow H_2$$

The proton-to-hydrogen converting catalysts present in nature are hydrogenases. These are enzymes that can either reduce protons to molecular hydrogen or oxidize hydrogen to protons and electrons. Spectroscopic and crystallographic studies spanning several decades have resulted in a good understanding of both the structure and mechanism of hydrogenase catalysis. Using this information, several molecules mimicking the structure of the active site of both nickel-iron and iron-iron hydrogenases have been synthesized. Other catalysts are not structural mimics of hydrogenase but rather functional ones. Synthesized catalysts include structural H-cluster models, a dirhodium photocatalyst, and cobalt catalysts.

Water-oxidizing Catalysts

Water oxidation is a more complex chemical reaction than proton reduction. In nature, the oxygen-evolving complex performs this reaction by accumulating reducing equivalents (electrons) in a manganese-calcium cluster within photosystem II (PS II), then delivering them to water molecules, with the resulting production of molecular oxygen and protons:

$$2\ H_2O \rightarrow O_2 + 4\ H^+ + 4e^-$$

Without a catalyst (natural or artificial), this reaction is very endothermic, requiring high temperatures (at least 2500 K).

The exact structure of the oxygen-evolving complex has been hard to determine experimentally. As of 2011, the most detailed model was from a 1.9 Å resolution crystal structure of photosystem II. The complex is a cluster containing four manganese and one calcium ions, but the exact location and mechanism of water oxidation within the cluster is unknown. Nevertheless, bio-inspired manganese and manganese-calcium complexes have been synthesized, such as $[Mn_4O_4]$ cubane-type clusters, some with catalytic activity.

Some ruthenium complexes, such as the dinuclear μ-oxo-bridged "blue dimer" (the first of its kind to be synthesized), are capable of light-driven water oxidation, thanks to being able to form high valence states. In this case, the ruthenium complex acts as both photosensitizer and catalyst.

Many metal oxides have been found to have water oxidation catalytic activity, including ruthenium(IV) oxide (RuO_2), iridium(IV) oxide (IrO_2), cobalt oxides (including nickel-doped Co_3O_4), manganese oxide (including layered MnO_2 (birnessite), Mn_2O_3), and a mix of Mn_2O_3 with $CaMn_2O_4$. Oxides are easier to obtain than molecular catalysts, especially those from relatively abundant transition metals (cobalt and manganese), but suffer from low turnover frequency and slow electron transfer properties, and their mechanism of action is hard to decipher and, therefore, to adjust.

Recently Metal-Organic Framework (MOF)-based materials have been shown to be a highly promising candidate for water oxidation with first row transition metals. The stability and tunability of this system is projected to be highly beneficial for future development.

Photosensitizers

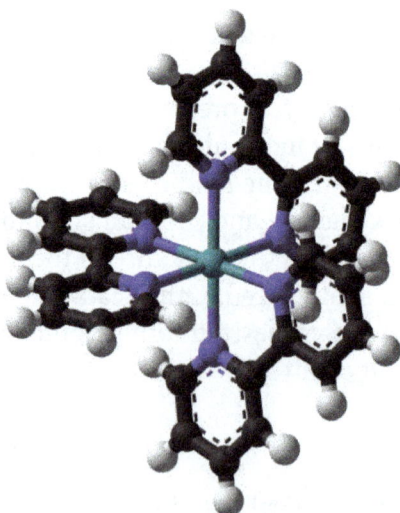

Structure of $[Ru(bipy)_3]^{2+}$, a broadly used photosensitizer.

Nature uses pigments, mainly chlorophylls, to absorb a broad part of the visible spectrum. Artificial systems can use either one type of pigment with a broad absorption range or combine several pigments for the same purpose.

Ruthenium polypyridine complexes, in particular tris(bipyridine)ruthenium(II) and its derivatives, have been extensively used in hydrogen photoproduction due to their efficient visible light absorption and long-lived consequent metal-to-ligand charge transfer excited state, which makes the complexes strong reducing agents. Other noble metal-containing complexes used include ones with platinum, rhodium and iridium.

Metal-free organic complexes have also been successfully employed as photosensitizers. Examples include eosin Y and rose bengal. Pyrrole rings such as porphyrins have also been used in coating nanomaterials or semiconductors for both homogeneous and heterogeneous catalysis.

As part of current research efforts artificial photonic antenna systems are being studied to determine efficient and sustainable ways to collect light for artificial photosynthesis. Gion Calzaferri (2009) describes one such antenna that uses zeolite L as a host for organic dyes, to mimic plant's light collecting systems. The material may be interfaced to an external device via a stopcock intermediate.

Carbon Dioxide Reduction Catalysts

In nature, carbon fixation is done by green plants using the enzyme RuBisCO as a part of the Calvin cycle. RuBisCO is a rather slow catalyst compared to the vast majority of other enzymes, incorporating only a few molecules of carbon dioxide into ribulose-1,5-bisphosphate per minute, but does so at atmospheric pressure and in mild, biological conditions. The resulting product is further reduced and eventually used in the synthesis of glucose, which in turn is a precursor to more complex carbohydrates, such as cellulose and starch. The process consumes energy in the form of ATP and NADPH.

Artificial CO_2 reduction for fuel production aims mostly at producing reduced carbon compounds from atmospheric CO_2. Some transition metal polyphosphine complexes have been developed for this end; however, they usually require previous concentration of CO_2 before use, and carriers (molecules that would fixate CO_2) that are both stable in aerobic conditions and able to concentrate CO_2 at atmospheric concentrations haven't been yet developed. The simplest product from CO_2 reduction is carbon monoxide (CO), but for fuel development, further reduction is needed, and a key step also needing further development is the transfer of hydride anions to CO.

Other Materials and Components

Charge separation is a key property of dyad and triad assemblies. Some nanomaterials employed are fullerenes (such as carbon nanotubes), a strategy that explores the pi-bonding properties of these materials. Diverse modifications (covalent and non-covalent) of carbon nanotubes have been attempted to increase the efficiency of charge separation, including the addition of ferrocene and pyrrole-like molecules such as porphyrins and phthalocyanines.

Since photodamage is usually a consequence in many of the tested systems after a period of exposure to light, bio-inspired photoprotectants have been tested, such as carotenoids (which are used in photosynthesis as natural protectants).

Light-driven Methodologies Under Development

Photoelectrochemical Cells

Photoelectrochemical cells are a heterogeneous system that use light to produce either electricity or hydrogen. The vast majority of photoelectrochemical cells use semiconductors as catalysts. There have been attempts to use synthetic manganese complex-impregnated Nafion as a working electrode, but it has been since shown that the catalytically active species is actually the broken-down complex.

A promising, emerging type of solar cell is the dye-sensitized solar cell. This type of cell still depends on a semiconductor (such as TiO_2) for current conduction on one electrode, but with a coating of an organic or inorganic dye that acts as a photosensitizer; the counter electrode is a platinum catalyst for H_2 production. These cells have a self-repair mechanism and solar-to-electricity conversion efficiencies rivaling those of solid-state semiconductor ones.

Photocatalytic Water Splitting in Homogeneous Systems

Direct water oxidation by photocatalysts is a more efficient usage of solar energy than photoelectrochemical water splitting because it avoids an intermediate thermal or electrical energy conversion step.

Bio-inspired manganese clusters have been shown to possess water oxidation activity when adsorbed on clays together with ruthenium photosensitizers, although with low turnover numbers.

As mentioned above, some ruthenium complexes are able to oxidize water under solar light irradiation. Although their photostability is still an issue, many can be reactivated by a simple adjustment of the conditions they work in. Improvement of catalyst stability has been tried resorting to polyoxometalates, in particular ruthenium-based ones.

Whereas a fully functional artificial system is usually envisioned when constructing a water splitting device, some mixed approaches have been tried. One of these involve the use of a gold electrode to which photosystem II is linked; an electric current is detected upon illumination.

Hydrogen-producing Artificial Systems

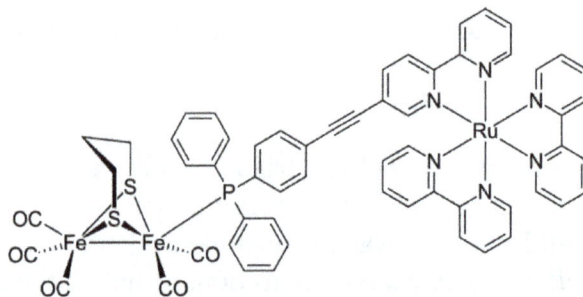

A H-cluster FeFe hydrogenase model compound covalently linked to a ruthenium photosensitizer. The ruthenium complex absorbs light and transduces its energy to the iron compound, which can then reduce protons to H_2.

The simplest photocatalytic hydrogen production unit consists of a hydrogen-evolving catalyst linked to a photosensitizer. In this dyad assembly, a so-called sacrificial donor for the photosensitizer is needed, that is, one that is externally supplied and replenished; the photosensitizer donates the necessary reducing equivalents to the hydrogen-evolving catalyst, which uses protons from a solution where it is immersed or dissolved in. Cobalt compounds such as cobaloximes are some of the best hydrogen catalysts, having been coupled to both metal-containing and metal-free photosensitizers. The first H-cluster models linked to photosensitizers (mostly ruthenium photosensitizers, but also porphyrin-derived ones) were prepared in the early 2000s. Both types of assembly are under development to improve their stability and increase their turnover numbers, both necessary for constructing a sturdy, long-lived solar fuel cell.

As with water oxidation catalysis, not only fully artificial systems have been idealized: hydrogenase enzymes themselves have been engineered for photoproduction of hydrogen, by coupling the enzyme to an artificial photosensitizer, such as $[Ru(bipy)_3]^{2+}$ or even photosystem I.

NADP⁺/NADPH Coenzyme-inspired Catalyst

In natural photosynthesis, the $NADP^+$ coenzyme is reducible to NADPH through binding of a proton and two electrons. This reduced form can then deliver the proton and electrons, potentially as a hydride, to reactions that culminate in the production of carbohydrates (the Calvin cycle). The coenzyme is recyclable in a natural photosynthetic cycle, but this process is yet to be artificially replicated.

A current goal is to obtain an NADPH-inspired catalyst capable of recreating the natural cyclic process. Utilizing light, hydride donors would be regenerated and produced where the molecules are continuously used in a closed cycle. Brookhaven chemists are now using a ruthenium-based complex to serve as the acting model. The complex is proven to perform correspondingly with NADP+/NADPH, behaving as the foundation for the proton and two electrons needed to convert acetone to isopropanol.

Currently, Brookhaven researchers are aiming to find ways for light to generate the hydride donors. The general idea is to use this process to produce fuels from carbon dioxide.

Photobiological Production of Fuels

Some photoautotrophic microorganisms can, under certain conditions, produce hydrogen. Nitrogen-fixing microorganisms, such as filamentous cyanobacteria, possess the enzyme nitrogenase, responsible for conversion of atmospheric N_2 into ammonia; molecular hydrogen is a byproduct of this reaction, and is many times not released by the microorganism, but rather taken up by a hydrogen-oxidizing (uptake) hydrogenase. One way of forcing these organisms to produce hydrogen is then to annihilate uptake hydrogenase activity. This has been done on a strain of *Nostoc punctiforme*: one of the structural genes of the NiFe uptake hydrogenase was inactivated by insertional mutagenesis, and the mutant strain showed hydrogen evolution under illumination.

Many of these photoautotrophs also have bidirectional hydrogenases, which can produce hydrogen under certain conditions. However, other energy-demanding metabolic pathways can compete with the necessary electrons for proton reduction, decreasing the efficiency of the overall process; also, these hydrogenases are very sensitive to oxygen.

Several carbon-based biofuels have also been produced using cyanobacteria, such as 1-butanol.

Synthetic biology techniques are predicted to be useful in this field. Microbiological and enzymatic engineering have the potential of improving enzyme efficiency and robustness, as well as constructing new biofuel-producing metabolic pathways in photoautotrophs that previously lack them, or improving on the existing ones. Another field under development is the optimization of photobioreactors for commercial application.

Employed Research Techniques

Research in artificial photosynthesis is necessarily a multidisciplinary field, requiring a multitude of different expertise. Some techniques employed in making and investigating catalysts and solar cells include:

- Organic and inorganic chemical synthesis.

- Electrochemistry methods, such as photoelectrochemistry, cyclic voltammetry, electrochemical impedance spectroscopy Dielectric spectroscopy, and bulk electrolysis.

- Spectroscopic methods:

     o   fast techniques, such as time-resolved spectroscopy and ultrafast laser spectroscopy;

     o   magnetic resonance spectroscopies, such as nuclear magnetic resonance, electron paramagnetic resonance;

     o   X-ray spectroscopy methods, including x-ray absorption such as XANES and EXAFS, but also x-ray emission.

- Crystallography.

- Molecular biology, microbiology and synthetic biology methodologies.

Advantages, Disadvantages, and Efficiency

Advantages of solar fuel production through artificial photosynthesis include:

- The solar energy can be immediately converted and stored. In photovoltaic cells, sunlight is converted into electricity and then converted again into chemical energy for storage, with some necessary loss of energy associated with the second conversion.

- The byproducts of these reactions are environmentally friendly. Artificially photosynthesized fuel would be a carbon-neutral source of energy, which could be used for transportation or homes.

Disadvantages include:

- Materials used for artificial photosynthesis often corrode in water, so they may be less stable than photovoltaics over long periods of time. Most hydrogen catalysts are very sensitive to oxygen, being inactivated or degraded in its presence; also, photodamage may occur over time.

- The overall cost is not yet advantageous enough to compete with fossil fuels as a commercially viable source of energy.

A concern usually addressed in catalyst design is efficiency, in particular how much of the incident light can be used in a system in practice. This is comparable with photosynthetic efficiency, where light-to-chemical-energy conversion is measured. Photosynthetic organisms are able to collect about 50% of incident solar radiation, however the theoretical limit of photosynthetic efficiency is 4.6 and 6.0% for C3 and C4 plants respectively. In reality, the efficiency of photosynthesis is much lower and is usually below 1%, with some exceptions such as sugarcane in tropical climate. In contrast, the highest reported efficiency for artificial photosynthesis lab prototypes is 22.4%. However, plants are efficient in using CO_2 at atmospheric concentrations, something that artificial catalysts still cannot perform.

Photocatalytic Water Splitting

Photocatalytic water splitting is an artificial photosynthesis process with photocatalysis in a photoelectrochemical cell used for the dissociation of water into its constituent parts, hydrogen (H2) and oxygen (O2), using either artificial or natural light. Theoretically, only solar energy (photons), water, and a catalyst are needed. This topic is the focus of much research, but thus far no technology has been commercialized

Hydrogen fuel production has gained increased attention as oil and other nonrenewable fuels become increasingly depleted and expensive. Methods such as photocatalytic water splitting are being investigated to produce hydrogen, a clean-burning fuel. Water splitting holds particular promise since it utilizes water, an inexpensive renewable resource. Photocatalytic water splitting has the simplicity of using a powder in solution and sunlight to produce H2 and O2 from water and can provide a clean, renewable energy, without producing greenhouse gases or having many adverse effects on the atmosphere.

Concepts

When H2O is split into O2 and H2, the stoichiometric ratio of its products is 2:1:

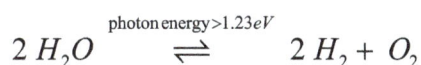

$$2\,H_2O \overset{\text{photon energy} > 1.23eV}{\rightleftharpoons} 2\,H_2 + O_2$$

The process of water-splitting is a highly endothermic process ($\Delta H > 0$). Water splitting occurs naturally in photosynthesis when photon energy is absorbed and converted into the chemical energy through a complex biological pathway. However, production of hydrogen from water requires large amounts of input energy, making it incompatible with existing energy generation. For this reason, most commercially produced hydrogen gas is produced from natural gas.

There are several strict requirements for a photocatalyst to be useful for water splitting. The minimum potential difference (voltage) needed to split water is 1.23V at 0 pH. Since the minimum band gap for successful water splitting at pH=0 is 1.23 eV, corresponding to light of 1008 nm, the electrochemical requirements can theoretically reach down into infrared light, albeit with negligible catalytic activity. These values are true only for a completely reversible reaction at standard temperature and pressure (1 bar and 25 °C).

Theoretically, infrared light has enough energy to split water into hydrogen and oxygen; however, this reaction is kinetically very slow because the wavelength is greater than 380 nm. The potential must be less than 3.0V to make efficient use of the energy present across the full spectrum of sunlight. Water splitting can transfer charges, but not be able to avoid corrosion for long term stability. Defects within crystalline photocatalysts can act as recombination sites, ultimately lowering efficiency.

Under normal conditions due to the transparency of water to visible light photolysis can only occur with a radiation wavelength of 180 nm or shorter. We see then that, assuming a perfect system, the minimum energy input is 6.893 eV.

Materials used in photocatalytic water splitting fulfill the band requirements outlined previously and typically have dopants and/or co-catalysts added to optimize their performance. A sample semiconductor with the proper band structure is titanium dioxide (TiO_2). However, due to the relatively positive conduction band of TiO_2, there is little driving force for H_2 production, so TiO_2 is typically used with a co-catalyst such as platinum (Pt) to increase the rate of H_2 production. It is routine to add co-catalysts to spur H_2 evolution in most photocatalysts due to the conduction band placement. Most semiconductors with suitable band structures to split water absorb mostly UV light; in order to absorb visible light, it is necessary to narrow the band gap. Since the conduction band is fairly close to the reference potential for H_2 formation, it is preferable to alter the valence band to move it closer to the potential for O_2 formation, since there is a greater natural overpotential.

Photocatalysts can suffer from catalyst decay and recombination under operating conditions. Catalyst decay becomes a problem when using a sulfide-based photocatalyst such as cadmium sulfide (CdS), as the sulfide in the catalyst is oxidized to elemental sulfur at the same potentials used to split water. Thus, sulfide-based photocatalysts are not viable without sacrificial reagents such as sodium sulfide to replenish any sulfur lost, which effectively changes the main reaction to one of hydrogen evolution as opposed to water splitting. Recombination of the electron-hole pairs needed for photocatalysis can occur with any catalyst and is dependent on the defects and surface area of the catalyst; thus, a high degree of crystallinity is required to avoid recombination at the defects.

The conversion of solar energy to hydrogen by means of photocatalysis is one of the most interesting ways to achieve clean and renewable energy systems. However, if this process is assisted by photocatalysts suspended directly in water instead of using a photovoltaic and electrolytic system the reaction is in just one step, and can therefore be more efficient.

Method of Evaluation

Photocatalysts must conform to several key principles in order to be considered effective at water splitting. A key principle is that H_2 and O_2 evolution should occur in a stoichiometric 2:1 ratio; significant deviation could be due to a flaw in the experimental setup and/or a side reaction, both of which do not indicate a reliable photocatalyst for water splitting. The prime measure of photocatalyst effectiveness is quantum yield (QY), which is:

$$QY\ (\%) = (\text{Photochemical reaction rate}) / (\text{Photon absorption rate}) \times 100\%$$

This quantity is a reliable determination of how effective a photocatalyst is; however, it can be misleading due to varying experimental conditions. To assist in comparison, the rate of gas evolution can also be used; this method is more problematic on its own because it is not normalized, but it can be useful for a rough comparison and is consistently reported in the literature. Overall, the best photocatalyst has a high quantum yield and gives a high rate of gas evolution.

The other important factor for a photocatalyst is the range of light absorbed; though UV-based photocatalysts will perform better per photon than visible light-based photocatalysts due to the higher photon energy, far more visible light reaches the Earth's surface than UV light. Thus, a less efficient photocatalyst that absorbs visible light may ultimately be more useful than a more efficient photocatalyst absorbing solely light with smaller wavelengths.

Photocatalyst Systems

$Cd_{1-x}Zn_xS$

Solid solutions $Cd_{1-x}Zn_xS$ with different Zn concentration ($0.2 < x < 0.35$) has been investigated in the production of hydrogen from aqueous solutions containing SO_3^{2-}/S^{2-} as sacrificial reagents under visible light. Textural, structural and surface catalyst properties were determined by N_2 adsorption isotherms, UV–vis spectroscopy, SEM and XRD and related to the activity results in hydrogen production from water splitting under visible light irradiation. It was found that the crystallinity and energy band structure of the $Cd_{1-x}Zn_xS$ solid solutions depend on their Zn atomic concentration. The hydrogen production rate was found to increase gradually when the Zn concentration on photocatalysts increases from 0.2 to 0.3. Subsequent increase in the Zn fraction up to 0.35 leads to lower hydrogen production. Variation in photoactivity is analyzed in terms of changes in crystallinity, level of conduction band and light absorption ability of $Cd_{1-x}Zn_xS$ solid solutions derived from their Zn atomic concentration.

$NaTaO_3$:La

$NaTaO_3$:La yields the highest water splitting rate of photocatalysts without using sacrificial reagents. This UV-based photocatalyst was shown to be highly effective with water splitting rates of 9.7 mmol/h and a quantum yield of 56%. The nanostep structure of the material promotes water splitting as edges functioned as H_2 production sites and the grooves functioned as O_2 production sites. Addition of NiO particles as cocatalysts assisted in H_2 production; this step was done by using an impregnation method with an aqueous solution of $Ni(NO_3)_2 \cdot 6H_2O$ and evaporating the solution in the presence of the photocatalyst. $NaTaO_3$ has a conduction band higher than that of NiO, so photogenerated electrons are more easily transferred to the conduction band of NiO for H_2 evolution.

$K_3Ta_3B_2O_{12}$

$K_3Ta_3B_2O_{12}$, another catalyst activated by solely UV light and above, does not have the performance or quantum yield of $NaTaO_3$:La. However, it does have the ability to split water without the assistance of cocatalysts and gives a quantum yield of 6.5% along with a water splitting rate of 1.21 mmol/h. This ability is due to the pillared structure of the photocatalyst, which involves TaO 6 pillars connected by BO3 triangle units. Loading with NiO did not assist the photocatalyst due to the highly active H2 evolution sites.

(Ga.82Zn.18)(N.82O.18)

(Ga.82Zn.18)(N.82O.18) has the highest quantum yield in visible light for visible light-based photocatalysts that do not utilize sacrificial reagents as of October 2008. The photocatalyst gives a quantum yield of 5.9% along with a water splitting rate of 0.4 mmol/h. Tuning the catalyst was done by increasing calcination temperatures for the final step in synthesizing the catalyst. Temperatures up to 600 °C helped to reduce the number of defects, though temperatures above 700 °C destroyed the local structure around zinc atoms and was thus undesirable. The treatment ultimately reduced the amount of surface Zn and O defects, which normally function as recombination sites, thus limiting photocatalytic activity. The catalyst was then loaded with Rh2-yCryO3 at a rate of 2.5 wt % Rh and 2 wt% Cr to yield the best performance.

Cobalt Based Systems

Photocatalysts based on cobalt have been reported. Members are tris(bipyridine) cobalt(II), compounds of cobalt ligated to certain cyclic polyamines, and certain cobaloximes.

In 2014 researchers announced an approach that connected a chromophore to part of a larger organic ring that surrounded a cobalt atom. The process is less efficient than using a platinum catalyst, cobalt is less expensive, potentially reducing total costs. The process uses one of two supramolecular assemblies based on Co(II)-templated coordination of Ru(bpy)32+ (bpy = 2,2′-bipyridyl) analogues as photosensitizers and electron donors to a cobaloxime macrocycle. The Co(II) centres of both assemblies are high spin, in contrast to most previously described cobaloximes. Transient absorption optical spectroscopies include that charge recombination occurs through multiple ligand states present within the photosensitizer modules.

Bismuth Vanadate

Bismuth based systems have been demonstrated to have an efficiency of 5% with the advantage of a very simple and cheap catalyst.

Tungsten Diselenide (WSe$_2$)

Tungsten diselenide may have a role in future hydrogen fuel production, as a recent discovery in 2015 by scientists in Switzerland revealed that the compound's own photocatalytic properties might be a key to significantly more efficient electrolysis of water to produce hydrogen fuel.

III-V Semiconductor Systems

Systems based on the material class of III-V semiconductors, such as InGaP, enable currently the highest solar-to-hydrogen efficiencies of up to 14%. Long-term stability of these high-cost high-efficiency systems does, however, remain an issue.

References

- Robert Bud; Deborah Jean Warner (1998). Instruments of Science: An Historical Encyclopedia. Science Museum, London, and National Museum of American History, Smithsonian Institution. ISBN 978-0-8153-1561-2.

- Vesselinka Petrova-Koch; Rudolf Hezel; Adolf Goetzberger (2009). High-Efficient Low-Cost Photovoltaics: Recent Developments. Springer. pp. 1–. ISBN 978-3-540-79358-8.

- Mee, C.; Crundell, M.; Arnold, B.; Brown, W. (2011). International A/AS Level Physics. Hodder Education. p. 241. ISBN 978-0-340-94564-3.

- Fromhold, A. T. (1991). Quantum Mechanics for Applied Physics and Engineering. Courier Dover Publications. pp. 5–6. ISBN 978-0-486-66741-6.

- Sears, F. W.; Zemansky, M. W.; Young, H. D. (1983). University Physics (6th ed.). Addison-Wesley. pp. 843–844. ISBN 0-201-07195-9.

- Holler, F. James; Skoog, Douglas A. and Crouch, Stanley R. (2006) Principles Of Instrumental Analysis. Cengage Learning. ISBN 0495012017

- Valeur, Bernard, Berberan-Santos, Mario (2012). Molecular Fluorescence: Principles and Applications. Wiley-VCH. ISBN 978-3-527-32837-6.

- Symmetry and Spectroscopy: An introduction to vibrational and electronic spectroscopy (Reprint ed.). Dover Publications. ISBN 0-486-66144-X.

- Gerischer, Heinz (1985). "Semiconductor electrodes and their interaction with light". In Schiavello, Mario. Photoelectrochemistry, Photocatalysis and Photoreactors Fundamentals and Developments. Springer. p. 39. ISBN 978-90-277-1946-1.

- Thomson, J. J. (2005). Conduction of Electricity Through Gases. Watchmaker Publishing. ISBN 978-1-929148-49-3. Retrieved 9 July 2011.

- Buchwald, Jed; Warwick, Andrew, eds. (2004). Histories of the Electron: The Birth of Microphysics (PDF) (illustrated, reprint ed.). MIT Press. pp. 21–23. ISBN 978-0-262-52424-7.

6

A Comprehensive Study of Bioluminescence

Bioluminescence is the light emitted by a living organism. It occurs in a number of marine animals and as well as fungi. The light emitted by living organisms is emitted because of biophoton. The topics discussed in the chapter are of great importance to broaden the existing knowledge on bioluminescence.

Bioluminescence

Bioluminescence is the production and emission of light by a living organism. It is a form of chemiluminescence. Bioluminescence occurs widely in marine vertebrates and invertebrates, as well as in some fungi, microorganisms including some bioluminescent bacteria and terrestrial invertebrates such as fireflies. In some animals, the light is produced by symbiotic organisms such as *Vibrio* bacteria.

Flying and glowing firefly, *Photinus pyralis*

The principal chemical reaction in bioluminescence involves the light-emitting pigment luciferin and the enzyme luciferase, assisted by other proteins such as aequorin in some species. The enzyme catalyzes the oxidation of luciferin. In some species, the type of luciferin requires cofactors such as calcium or magnesium ions, and sometimes also the energy-carrying molecule adenosine triphosphate (ATP). In evolution, luciferins vary little: one in particular, coelenterazine, is found in nine different animal (phyla), though in some of these, the animals obtain it through their diet. Conversely, luciferases vary widely in different species. Bioluminescence has arisen over forty times in evolutionary history.

Female Glowworm, *Lampyris noctiluca*

Both Aristotle and Pliny the Elder mentioned that damp wood sometimes gives off a glow and many centuries later Robert Boyle showed that oxygen was involved in the process, both in wood and in glow-worms. It was not until the late nineteenth century that bioluminescence was properly investigated. The phenomenon is widely distributed among animal groups, especially in marine environments where dinoflagellates cause phosphorescence in the surface layers of water. On land it occurs in fungi, bacteria and some groups of invertebrates, including insects.

The uses of bioluminescence by animals include counter-illumination camouflage, mimicry of other animals, for example to lure prey, and signalling to other individuals of the same species, such as to attract mates. In the laboratory, luciferase-based systems are used in genetic engineering and for biomedical research. Other researchers are investigating the possibility of using bioluminescent systems for street and decorative lighting, and a bioluminescent plant has been created.

History

Before the development of the safety lamp for use in coal mines, dried fish skins were used in Britain and Europe as a weak source of light. This experimental form of illumination avoided the necessity of using candles which risked sparking explosions of firedamp. Another safe source of illumination in mines was bottles containing fireflies. In 1920, the American zoologist E. Newton Harvey published a monograph, *The Nature of Animal Light*, summarizing early work on bioluminescence. Harvey notes that Aristotle mentions light produced by dead fish and flesh, and that both Aristotle and Pliny the Elder (in his *Natural History*) mention light from damp wood. He also records that Robert Boyle experimented on these light sources, and showed that both they and the glow-worm require air for light to be produced. Harvey notes that in 1753, J. Baker identified the flagellate *Noctiluca* "as a luminous animal" "just visible to the naked eye", and in 1854 Johann Florian Heller (1813-1871) identified strands (hyphae) of fungi as the source of light in dead wood.

Tuckey, in his posthumous 1818 *Narrative of the Expedition to the Zaire*, described catching the animals responsible for luminescence. He mentions pellucids, crustaceans (to which he ascribes the milky whiteness of the water), and cancers (shrimps and crabs). Under the microscope he described the "luminous property" to be in the brain, resembling "a most brilliant amethyst about the size of a large pin's head".

Charles Darwin noticed bioluminescence in the sea, describing it in his *Journal*:

While sailing in these latitudes on one very dark night, the sea presented a wonderful and most beautiful spectacle. There was a fresh breeze, and every part of the surface, which during the day is seen as foam, now glowed with a pale light. The vessel drove before her bows two billows of liquid phosphorus, and in her wake she was followed by a milky train. As far as the eye reached, the crest of every wave was bright, and the sky above the horizon, from the reflected glare of these livid flames, was not so utterly obscure, as over the rest of the heavens.

Darwin also observed a luminous "jelly-fish of the genus Dianaea" and noted that "When the waves scintillate with bright green sparks, I believe it is generally owing to minute crustacea. But there can be no doubt that very many other pelagic animals, when alive, are phosphorescent." He guessed that "a disturbed electrical condition of the atmosphere" was probably responsible. Daniel Pauly comments that Darwin "was lucky with most of his guesses, but not here", noting that biochemistry was too little known, and that the complex evolution of the marine animals involved "would have been too much for comfort".

Osamu Shimomura isolated the photoprotein aequorin and its cofactor coelenterazine from the crystal jelly *Aequorea victoria* in 1961.

Bioluminescence attracted the attention of the United States Navy in the Cold War, since submarines in some waters can create a bright enough wake to be detected; a German submarine was sunk in the First World War, having been detected in this way. The navy was interested in predicting when such detection would be possible, and hence guiding their own submarines to avoid detection.

Among the anecdotes of navigation by bioluminescence, the Apollo 13 astronaut Jim Lovell recounted how as a navy pilot he had found his way back to his aircraft carrier USS *Shangri-La* when his navigation systems failed. Turning off his cabin lights, he saw the glowing wake of the ship, and was able to fly to it and land safely.

The French pharmacologist Raphaël Dubois carried out work on bioluminescence in the late nineteenth century. He studied click beetles (*Pyrophorus*) and the marine bivalve mollusc *Pholas dactylus*. He refuted the old idea that bioluminescence came from phosphorus,[a] and demonstrated that the process was related to the oxidation of a specific compound, which he named luciferin, by an enzyme. He sent Harvey siphons from the mollusc preserved in sugar. Harvey had become interested in bioluminescence as a result of visiting the South Pacific and Japan and observing phosphorescent organisms there. He studied the phenomenon for many years. His research aimed to demonstrate that luciferin, and the enzymes that act on it is to produce light, were interchangeable between species, showing that all bioluminescent organisms had a common ancestor. However, he found this hypothesis to be false, with different organisms having major differences in the composition of their light-producing proteins. He spent the next thirty years purifying and studying the components, but it fell to the young Japanese chemist Osamu Shimomura to be the first to obtain crystalline luciferin. He used the sea firefly *Vargula hilgendorfii*, but it was another ten years before he discovered the chemical's structure and was able to publish his 1957 paper *Crystalline Cypridina Luciferin*. More recently, Martin Chalfie, Osamu Shimomura and Roger Y. Tsien won the 2008 Nobel Prize in Chemistry for their 1961 discovery and development of green fluorescent protein as a tool for biological research.

Evolution

Bioluminescence in fish began at least by the Cretaceous period. About 1,500 fish species are known to be bioluminescent, and this feature evolved independently at a minimum of 27 times. Of these 27 occasions, 17 involved the taking up of bioluminous bacteria from the surrounding water while in the others, the intrinsic light evolved through chemical synthesis. These fish have become surprisingly diverse in the deep ocean and control their light with the help of their nervous system, using it not just to lure prey or hide from predators, but also for communication.

Chemical Mechanism

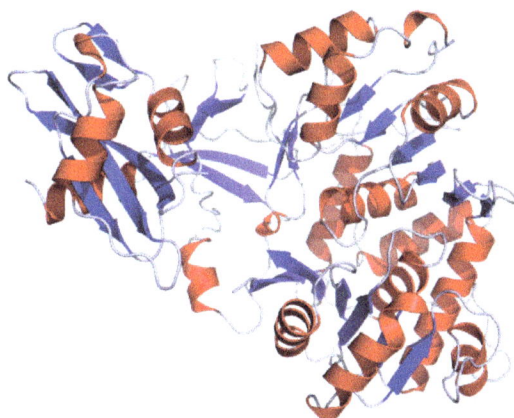

Protein folding structure of the luciferase of the firefly *Photinus pyralis*.
The enzyme is a much larger molecule than luciferin.

Bioluminescence is a form of chemiluminescence where light energy is released by a chemical reaction. Fireflies, anglerfish, and other organisms produce the light-emitting pigment luciferin and the enzyme luciferase. Luciferin reacts with oxygen to create light:

$$L + O2 + ATP-> [Luciferase]$$
$$[Mg^{2+}]oxy - L + CO2 + AMP + PP + light$$

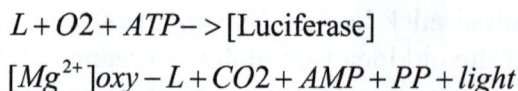

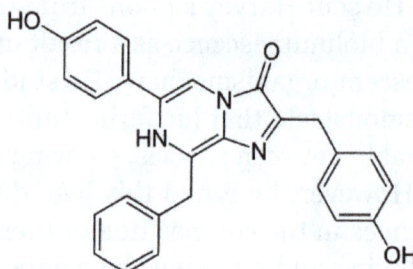

Coelenterazine is a luciferin found in many different marine phyla
from comb jellies to vertebrates. Like all luciferins, it is oxidised to produce light.

Carbon dioxide (CO_2), adenosine monophosphate (AMP) and phosphate groups (PP) are released as waste products. Luciferase catalyzes the reaction, which may be mediated by cofactors such as calcium (Ca^{2+}) or magnesium (Mg^{2+}) ions, and for some types of luciferin (L) also the energy-carrying molecule adenosine triphosphate (ATP). The reaction can occur either inside or outside the cell. In bacteria such as *Aliivibrio*, the expression of genes related to bioluminescence is controlled by the lux operon.

In evolution, luciferins generally vary little: one in particular, coelenterazine, is the light emitting pigment for nine ancient phyla (groups of very different organisms), including polycystine radiolaria, Cercozoa (Phaeodaria), protozoa, comb jellies, cnidaria including jellyfish and corals, crustaceans, molluscs, arrow worms and vertebrates (ray-finned fish). Not all these organisms synthesize coelenterazine: some of them obtain it through their diet. Conversely, luciferase enzymes vary widely and tend to be different in each species. Overall, bioluminescence has arisen over forty times in evolutionary history.

Luciferin-luciferase reactions are not the only way that organisms produce light. The parchment worm *Chaetopterus* (a marine Polychaete) makes use of another photoprotein, aequorin, instead of luciferase. When calcium ions are added, the aequorin's rapid catalysis creates a brief flash quite unlike the prolonged glow produced by luciferase. In a second, much slower, step luciferin is regenerated from the oxidised (oxyluciferin) form, allowing it to recombine with aequorin, in readiness for a subsequent flash. Photoproteins are thus enzymes, but with unusual reaction kinetics.

In the hydrozoan jellyfish *Aequorea victoria*, some of the blue light released by aequorin in contact with calcium ions is absorbed by green fluorescent protein; it in turn releases green light.

Distribution

Huge numbers of bioluminescent dinoflagellates creating phosphorescence in breaking waves

Bioluminescence occurs widely among animals, especially in the open sea, including fish, jellyfish, comb jellies, crustaceans, and cephalopod molluscs; in some fungi and bacteria; and in various terrestrial invertebrates including insects. Many, perhaps most deep-sea animals produce light. Most marine light-emission is in the blue and green light spectrum. However, some loose-jawed fish emit red and infrared light, and the genus *Tomopteris* emits yellow light.

The most frequently encountered bioluminescent organisms may be the dinoflagellates present in the surface layers of the sea, which are responsible for the sparkling phosphorescence sometimes seen at night in disturbed water. At least eighteen genera exhibit luminosity. A different effect is the thousands of square miles of the ocean which shine with the light produced by bioluminescent bacteria, known as mareel or the milky seas effect.

Non-marine bioluminescence is less widely distributed, the two best-known cases being in fireflies and glow worms. Other invertebrates including insect larvae, annelids and arachnids possess bio-luminescent abilities. Some forms of bioluminescence are brighter (or exist only) at night, following a circadian rhythm.

Uses in Nature

Bioluminescence has several functions in different taxa. Haddock et al. (2010) list as more or less definite functions in marine organisms the following: defensive functions of startle, counterillumi-nation (camouflage), misdirection (smoke screen), distractive body parts, burglar alarm (making predators easier for higher predators to see), and warning to deter settlers; offensive functions of lure, stun or confuse prey, illuminate prey, and mate attraction/recognition. It is much easier for researchers to detect that a species is able to produce light than to analyse the chemical mecha-nisms or to prove what function the light serves. In some cases the function is unknown, as with species in three families of earthworm (Oligochaeta), such as *Diplocardia longa* where the coe-lomic fluid produces light when the animal moves. The following functions are reasonably well established in the named organisms.

Counterillumination Camouflage

Principle of counterillumination camouflage in firefly squid, *Watasenia scintillans*. When seen from below by a predator, the bioluminescence helps to match the squid's brightness and colour to the sea surface above.

In many animals of the deep sea, including several squid species, bacterial bioluminescence is used for camouflage by counterillumination, in which the animal matches the overhead environmental light as seen from below. In these animals, photoreceptors control the illumination to match the brightness of the background. These light organs are usually separate from the tissue containing the bioluminescent bacteria. However, in one species, *Euprymna scolopes*, the bacteria are an integral component of the animal's light organ.

Attraction

A fungus gnat from New Zealand, *Arachnocampa luminosa*, lives in the predator-free environment of caves and its larvae emit bluish-green light. They dangle silken threads that glow and attract flying insects, and wind in their fishing-lines when prey becomes entangled. The bioluminescence of the larvae of another fungus gnat from North America which lives on streambanks and under overhangs has a similar function. *Orfelia fultoni* builds sticky little webs and emits light of a deep blue colour. It has an inbuilt biological clock and, even when kept in total darkness, turns its light on and off in a circadian rhythm.

Fireflies use light to attract mates. Two systems are involved according to species; in one, females emit light from their abdomens to attract males; in the other, flying males emit signals to which the sometimes sedentary females respond. Click beetles emit an orange light from the abdomen when flying and a green light from the thorax when they are disturbed or moving about on the ground. The former is probably a sexual attractant but the latter may be defensive. Larvae of the click beetle *Pyrophorus nyctophanus* live in the surface layers of termite mounds in Brazil. They light up the mounds by emitting a bright greenish glow which attracts the flying insects on which they feed.

In the marine environment, use of luminescence for mate attraction is chiefly known among ostracods, small shrimplike crustaceans, especially in the Cyprididae family. Pheromones may be used for long-distance communication, with bioluminescence used at close range to enable mates to "home in". A polychaete worm, the Bermuda fireworm creates a brief display, a few nights after the full moon, when the female lights up to attract males.

Defense

Many cephalopods, including at least 70 genera of squid, are bioluminescent. Some squid and small crustaceans use bioluminescent chemical mixtures or bacterial slurries in the same way as many squid use ink. A cloud of luminescent material is expelled, distracting or repelling a potential predator, while the animal escapes to safety. The deep sea squid *Octopoteuthis deletron* may autotomise portions of its arms which are luminous and continue to twitch and flash, thus distracting a predator while the animal flees.

Dinoflagellates may use bioluminescence for defence against predators. They shine when they detect a predator, possibly making the predator itself more vulnerable by attracting the attention of predators from higher trophic levels. Grazing copepods release any phytoplankton cells that flash, unharmed; if they were eaten they would make the copepods glow, attracting predators, so the phytoplankton's bioluminescence is defensive. The problem of shining stomach contents is solved (and the explanation corroborated) in predatory deep-sea fishes: their stomachs have a black lining able to keep the light from any bioluminescent fish prey which they have swallowed from attracting larger predators.

The sea-firefly is a small crustacean living in sediment. At rest it emits a dull glow but when disturbed it darts away leaving a cloud of shimmering blue light to confuse the predator. During World War II it was gathered and dried for use by the Japanese military as a source of light during clandestine operations.

The larvae of railroad worms (*Phrixothrix*) have paired photic organs on each body segment, able to glow with green light; these are thought to have a defensive purpose. They also have organs on the head which produce red light; they are the only terrestrial organisms to emit light of this colour.

Warning

Aposematism is a widely used function of bioluminescence, providing a warning that the creature concerned is unpalatable. It is suggested that many firefly larvae glow to repel predators; millipedes glow for the same purpose. Some marine organisms are believed to emit light for a similar reason. These include scale worms, jellyfish and brittle stars but further research is needed to fully establish the function of the luminescence. Such a mechanism would be of particular advantage to soft-bodied cnidarians if they were able to deter predation in this way. The limpet *Latia neritoides* is the only known freshwater gastropod that emits light. It produces greenish luminescent mucus which may have an anti-predator function. The marine snail *Hinea brasiliana* uses flashes of light, probably to deter predators. The blue-green light is emitted through the translucent shell, which functions as an efficient diffuser of light.

Communication

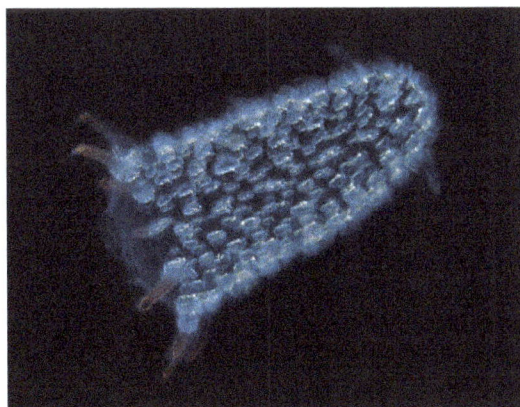

Pyrosoma, a colonial tunicate; each individual zooid in the colony flashes a blue-green light.

Communication in the form of quorum sensing plays a role in the regulation of luminescence in many species of bacteria. Small extracellularly secreted molecules stimulate the bacteria to turn on genes for light production when cell density, measured by concentration of the secreted molecules, is high.

Pyrosomes are colonial tunicates and each zooid has a pair of luminescent organs on either side of the inlet siphon. When stimulated by light, these turn on and off, causing rhythmic flashing. No neural pathway runs between the zooids, but each responds to the light produced by other individuals, and even to light from other nearby colonies. Communication by light emission between the zooids enables coordination of colony effort, for example in swimming where each zooid provides part of the propulsive force.

Some bioluminous bacteria infect nematodes that parasitize Lepidoptera larvae. When these caterpillars die, their luminosity may attract predators to the dead insect thus assisting in the dispersal of both bacteria and nematodes. A similar reason may account for the many species of fungi that emit light. Species in the genera *Armillaria*, *Mycena*, *Omphalotus*, *Panellus*, *Pleurotus* and others do this, emitting usually greenish light from the mycelium, cap and gills. This may attract night-flying insects and aid in spore dispersal, but other functions may also be involved.

Quantula striata is the only known bioluminescent terrestrial mollusc. Pulses of light are emitted from a gland near the front of the foot and may have a communicative function, although the adaptive significance is not fully understood.

Mimicry

A deep sea anglerfish, *Bufoceratias wedli*, showing the esca (lure)

Bioluminescence is used by a variety of animals to mimic other species. Many species of deep sea fish such as the anglerfish and dragonfish make use of aggressive mimicry to attract prey. They have an appendage on their heads called an esca that contains bioluminescent bacteria able to produce a long-lasting glow which the fish can control. The glowing esca is dangled or waved about to lure small animals to within striking distance of the fish.

The cookiecutter shark uses bioluminescence to camouflage its underside by counterillumination, but a small patch near its pectoral fins remains dark, appearing as a small fish to large predatory fish like tuna and mackerel swimming beneath it. When such fish approach the lure, they are bitten by the shark.

Female *Photuris* fireflies sometimes mimic the light pattern of another firefly, *Photinus*, to attract its males as prey. In this way they obtain both food and the defensive chemicals named lucibufagins, which *Photuris* cannot synthesize.

South American giant cockroaches of the genus *Lucihormetica* were believed to be the first known example of defensive mimicry, emitting light in imitation of bioluminescent, poisonous click beetles. However, doubt has been cast on this assertion, and there is no conclusive evidence that the cockroaches are bioluminescent.

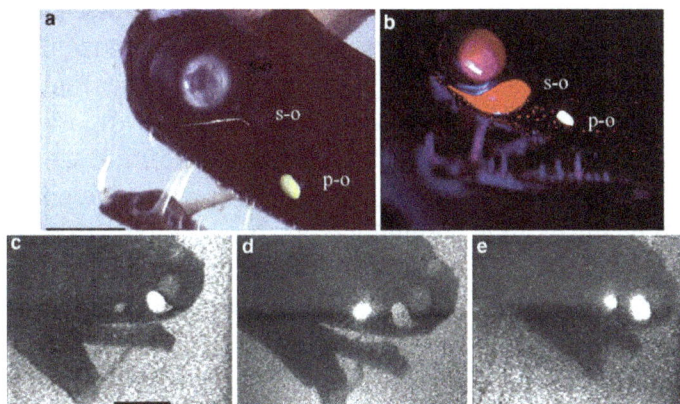

Flashing of photophores of black dragonfish, *Malacosteus niger*, showing red fluorescence

Illumination

While most marine bioluminescence is green to blue, some deep sea barbeled dragonfishes in the genera *Aristostomias*, *Pachystomias* and *Malacosteus* emit a red glow. This adaptation allows the fish to see red-pigmented prey, which are normally invisible in the deep ocean environment where red light has been filtered out by the water column.

The black dragonfish (also called the northern stoplight loosejaw) *Malacosteus niger* is believed to be one of the only fish to produce a red glow. Its eyes, however, are insensitive to this wavelength; it has an additional retinal pigment which fluoresces blue-green when illuminated. This alerts the fish to the presence of its prey. The additional pigment is thought to be assimilated from chlorophyll derivatives found in the copepods which form part of its diet.

Biotechnology

Biology and Medicine

Bioluminescent organisms are a target for many areas of research. Luciferase systems are widely used in genetic engineering as reporter genes, each producing a different colour by fluorescence, and for biomedical research using bioluminescence imaging. For example, the firefly luciferase gene was used as early as 1986 for research using transgenic tobacco plants. *Vibrio* bacteria symbiose with marine invertebrates such as the Hawaiian bobtail squid (*Euprymna scolopes*), are key experimental models for bioluminescence. Bioluminescent activated destruction is an experimental cancer treatment. See also optogenetics which involves the use of light to control cells in living tissue, typically neurons, that have been genetically modified to express light-sensitive ion channels, and also see biophoton, a photon of non-thermal origin in the visible and ultraviolet spectrum emitted from a biological system.

Light Production

The structures of photophores, the light producing organs in bioluminescent organisms, are being investigated by industrial designers. Engineered bioluminescence could perhaps one day be used to reduce the need for street lighting, or for decorative purposes if it becomes possible to produce light that is both bright enough and can be sustained for long periods at a workable price. The gene that makes the tails of fireflies glow has been added to mustard plants. The plants glow faintly for an hour when touched, but a sensitive camera is needed to see the glow. University of Wisconsin–Madison is researching the use of genetically engineered bioluminescent E. coli bacteria, for use as bioluminescent bacteria in a light bulb. In June 2013 the Glowing Plant project raised nearly $500,000 on the crowdfunding site Kickstarter to create a bioluminescent plant. An iGEM team from Cambridge (England) has started to address the problem that luciferin is consumed in the light-producing reaction by developing a genetic biotechnology part that codes for a luciferin re-generating enzyme from the North American firefly; this enzyme "helps to strengthen and sustain light output". In 2016, Glowee, a French company started selling bioluminescent lights, tageting shop fronts and municipal street signs as their main markets. France has a law that forbids re-tailers and offices from illumunating their windows between 1 and 7 in the morning in order to minimise energy consumption and pollution. Glowee hoped their product would get round this ban. They used bacteria called Aliivibrio fischeri which glow in the dark but the maximum lifetime of their product was three days.

Biophoton

Biophotons are photons of light in the ultraviolet and low visible light range that are produced by a biological system. They are non-thermal in origin, and the emission of biophotons is technically a type of bioluminescence, though bioluminescence is generally reserved for higher luminance lu-ciferin/luciferase systems. The term *biophoton* used in this narrow sense should not be confused with the broader field of biophotonics, which studies the general interaction of light with biological systems.

Biological tissues typically produce an observed radiant emittance in the visible and ultraviolet frequencies ranging from 10^{-17} to 10^{-23} W/cm² (approx 1-1000 photons/cm²/second). This low lev-el of light has a much weaker intensity than the visible light produced by bioluminescence, but biophotons are detectable above the background of thermal radiation that is emitted by tissues at their normal temperature.

While detection of biophotons has been reported by several groups, hypotheses that such biopho-tons indicate the state of biological tissues and facilitate a form of cellular communication are still under investigation, and claims that biophotons are responsible for physical healing are unsup-ported. Alexander Gurwitsch, who discovered the existence of biophotons, was awarded the Stalin Prize in 1941 for his mitogenic radiation work.

Detection and Measurement

Biophotons may be detected with photomultipliers or by means of an ultra low noise CCD camera to produce an image, using an exposure time of typically 15 minutes for plant materials. Photo-multiplier tubes have also been used to measure biophoton emissions from fish eggs, and some applications have measured biophotons from animals and humans.

The typical observed radiant emittance of biological tissues in the visible and ultraviolet frequencies ranges from 10^{-17} to 10^{-23} W/cm² with a photon count from a few to nearly 1000 photons per cm² in the range of 200nm to 800nm.

Proposed Physical Mechanisms

Chemi-excitation via oxidative stress by reactive oxygen species and/or catalysis by enzymes (i.e., peroxidase, lipoxygenase) is a common event in the biomolecular milieu. Such reactions can lead to the formation of triplet excited species, which release photons upon returning to a lower energy level in a process analogous to phosphorescence. That this process is a contributing factor to spontaneous biophoton emission has been indicated by studies demonstrating that biophoton emission can be increased by depleting assayed tissue of antioxidants or by addition of carbonyl derivatizing agents. Further support is provided by studies indicating that emission can be increased by addition of reactive oxygen species.

Plants

Imaging of biophotons from leaves has been used as a method for Assaying R Gene Responses. These genes and their associated proteins are responsible for pathogen recognition and activation of defense signaling networks leading to the hypersensitive response, which is one of the mechanisms of the resistance of plants to pathogen infection. It involves the generation of reactive oxygen species (ROS), which have crucial roles in signal transduction or as toxic agents leading to cell death.

Biophoton have been observed in stressed plant's roots, too. In healthy cells, the concentration of ROS is minimized by a system of biological antioxidants. However, heat shock and other stresses changes the equilibrium between oxidative stress and antioxidant activity, for example, the rapid rise in temperature induces biophoton emission by ROS.

Theoretical Biophysics

Hypothesized Involvement in Cellular Communication

In the 1920s, the Russian embryologist Alexander Gurwitsch reported "ultraweak" photon emissions from living tissues in the UV-range of the spectrum. He named them "mitogenetic rays" because his experiments convinced him that they had a stimulating effect on cell division.

Biophotons were claimed to have been employed by the Stalin regime to diagnose cancer. The method has not been tested in the West. However, failure to replicate his findings and the fact that, though cell growth can be stimulated and directed by radiation this is possible only at much higher amplitudes, evoked a general skepticism about Gurwitsch's work. In 1953 Irving Langmuir dubbed Gurwitsch's ideas pathological science. Commercial products, therapeutic claims and services supposedly based on his work appear at present to be best regarded as such.

But in the later 20th century Gurwitsch's daughter Anna, Colli, Quickenden and Inaba separately returned to the subject, referring to the phenomenon more neutrally as "dark luminescence", "low level luminescence", "ultraweak bioluminescence", or "ultraweak chemiluminescence". Their common basic hypothesis was that the phenomenon was induced from rare oxidation processes and radical reactions. In the 1970s Fritz-Albert Popp and his research group at the University of Marburg (Germany) showed that the spectral distribution of the emission fell over a wide range of wavelengths, from 200 to 750 nm. Popp proposed that the radiation might be both semi-periodic and coherent.

One biophoton mechanism focuses on injured cells that are under higher levels of oxidative stress, which is one source of light, and can be deemed to constitute a "distress signal" or background chemical process is yet to be demonstrated. The difficulty of teasing out the effects of any supposed biophotons amid the other numerous chemical interactions between cells makes it difficult to devise a testable hypothesis. A 2010 review article discusses various published theories on this kind of signaling.

Pseudoscience

Many claims with no scientific proof have been made for cures and diagnosis using biophotons. An appraisal of "biophoton therapy" by the IOCOB notes that biophoton therapy claims to treat a wide variety of diseases, such as malaria, Lyme disease, multiple sclerosis, schizophrenia, and depression, but that all these claims remain unproven. Dr. F. Popp, a researcher who investigates biophoton emission, concludes that the complexity of cellular chemical reactions in living systems is such that it excludes the possibility to create a machine to selectively heal systems using biophotons, and that "there are always charlatans who believe in these miracles."

Quantum Medicine

This claims:

"The quantum level possesses the highest level of coherence within the human organism. Sick individuals with weak immune systems or cancer have poor and chaotic coherence with disturbed biophoton cellular communication. Therefore, disease can be seen as the result of disturbances on the cellular level that act to distort the cell's quantum perspective. This causes electrons to become misplaced in protein molecules and metabolic processes become derailed as a result. Once cellular metabolism is compromised the cell becomes isolated from the regulated process of natural growth control."

A review of the *American Academy of Quantum Medicine* concludes that many quantum medicine practitioners are not licensed as health care professionals, that quantum medicine uses scientific terminology but is nonsense, and that the practitioners have created "a nonexistent 'energy system' to help peddle products and procedures to their clients."

Luciferase

Luciferase is a generic term for the class of oxidative enzymes that produce bioluminescence, and is distinct from a photoprotein. The name is derived from Lucifer, the root of which means 'light-bearer' (*lucem ferre*). One example is the firefly luciferase (EC 1.13.12.7) from the firefly *Photinus pyralis*. "Firefly luciferase" as a laboratory reagent often refers to *P. pyralis* luciferase although recombinant luciferases from several other species of fireflies are also commercially available.

Examples

A variety of organisms regulate their light production using different luciferases in a variety of light-emitting reactions. The most famous are the fireflies, although the enzyme exists in organisms as different as the Jack-O-Lantern mushroom *(Omphalotus olearius)* and many marine creatures.

Firefly and Click Beetle

The luciferases of fireflies - of which there are over 2000 species - and of the other Elateroidea (click beetles and relatives in general) are diverse enough to be useful in molecular phylogeny. In fireflies, the oxygen required is supplied through a tube in the abdomen called the abdominal trachea. One well-studied luciferase is that of the Photinini firefly *Photinus pyralis*, which has an optimum pH of 7.8.

Sea Pansy

Also well studied is the fancy sea pansy, *Renilla reniformis*. In this organism, the luciferase (Renilla-luciferin 2-monooxygenase) is closely associated with a luciferin-binding protein as well as a green fluorescent protein (GFP). Calcium triggers release of the luciferin (coelenterazine) from the luciferin binding protein. The substrate is then available for oxidation by the luciferase, where it is degraded to coelenteramide with a resultant release of energy. In the absence of GFP, this energy would be released as a photon of blue light (peak emission wavelength 482 nm). However, due to the closely associated GFP, the energy released by the luciferase is instead coupled through resonance energy transfer to the fluorophore of the GFP, and is subsequently released as a photon of green light (peak emission wavelength 510 nm). The catalyzed reaction is:

- coelenterazine + O_2 → coelenteramide + CO_2 + photon of light

Bacterial

Bacterial bioluminescence is seen in Photobacterium species, *Vibrio fischeri, Vibrio haweyi, and Vibrio harveyi*. Light emission in some bioluminescent bacteria utilizes 'antenna' such as 'lumazine protein' to accept the energy from the primary excited state on the luciferase, resulting in an excited lulnazine chromophore which emits light that is of a shorter wavelength (more blue), while in others use a yellow fluorescent protein (YFP) with FMN as the chromophore and emits light that is red-shifted relative to that from luciferase.

Dinoflagellate

Dinoflagellate luciferase is a multi-domain protein, consisting of an N-terminal domain, and three catalytic domains, each of which preceded by a helical bundle domain. The structure of the dinoflagellate luciferase catalytic domain has been solved. The core part of the domain is a 10 stranded beta barrel that is structurally similar to lipocalins and FABP. The N-terminal domain is conserved between dinoflagellate luciferase and luciferin binding proteins (LBPs). It has been suggested that this region may mediate an interaction between LBP and luciferase or their association with the vacuolar membrane. The helical bundle domain has a three helix bundle structure that holds four important histidines that are thought to play a role in the pH regulation of the enzyme. There is a large pocket in the β-barrel of the dinoflagellate luciferase at pH 8 to accommodate the tetrapyrrole substrate but there is no opening to allow the substrate to enter. Therefore, a significant conformational change must occur to provide access and space for a ligand in the active site and the source for this change is through the four N-terminal histidine residues. At pH 8, it can be seen that the unprotonated histidine residues are involved in a network of hydrogen bonds at the interface of the helices in the bundle that block substrate access to the active site and disruption of this interaction by protonation (at pH 6.3) or by replacement of the histidine residues by alanine causes a large molecular motion of the bundle, separating the helices by 11Å and opening the catalytic site. Logically, the histidine residues cannot be replaced by alanine in nature but this experimental replacement further confirms that the larger histidine residues block the active site. Additionally, three Gly-Gly sequences, one in the N-terminal helix and two in the helix-loop-helix motif, could serve as hinges about which the chains rotate in order to further open the pathway to the catalytic site and enlarge the active site.

A dinoflagellate luciferase is capable of emitting light due to its interaction with its substrate (luciferin) and the luciferin-binding protein (LBP) in the scintillon organelle found in dinoflagellates. The luciferase acts in accordance with luciferin and LBP in order to emit light but each component functions at a different pH. Luciferase and its domains are not active at pH 8 but they are extremely active at the optimum pH of 6.3 whereas LBP binds luciferin at pH 8 and releases it at pH 6.3. Consequently, luciferin is only released to react with an active luciferase when the scintillon is acidified to pH 6.3. Therefore, in order to lower the pH, voltage-gated channels in the scintillon membrane are opened to allow the entry of protons from a vacuole possessing an action potential produced from a mechanical stimulation. Hence, it can be seen that the action potential in the vacuolar membrane leads to acidification and this in turn allows the luciferin to be released to react with luciferase in the scintillon, producing a flash of blue light.

Copepod

Newer luciferases have recently been identified that, unlike other luciferases above, are naturally secreted molecules. One such example is the Metridia luciferase (MetLuc) that is derived from the marine copepod *Metridia longa*. The *Metridia longa* secreted luciferase gene encodes a 24 kDa protein containing an N-terminal secretory signal peptide of 17 amino acid residues. The sensitivity and high signal intensity of this luciferase molecule proves advantageous in many reporter studies. Some of the benefits of using a secreted reporter molecule like MetLuc is its no-lysis protocol that allows one to be able to conduct live cell assays and multiple assays on the same cell.

Mechanism of Reaction

The chemical reaction catalyzed by firefly luciferase takes place in two steps:

- luciferin + ATP → luciferyl adenylate + PP_i

- luciferyl adenylate + O_2 → oxyluciferin + AMP + light

Light is produced because the reaction forms oxyluciferin in an electronically excited state. The reaction releases a photon of light as oxyluciferin goes back to the ground state.

Luciferyl adenylate can additionally participate in a side reaction with O_2 to form hydrogen peroxide and dehydroluciferyl-AMP. About 20% of the luciferyl adenylate intermediate is oxidized in this pathway.

The reaction catalyzed by bacterial luciferase is also an oxidative process:

- $FMNH_2 + O_2 + RCHO \rightarrow FMN + RCOOH + H_2O + light$

In the reaction, a reduced flavin mononucleotide oxidizes a long-chain aliphatic aldehyde to an aliphatic carboxylic acid. The reaction forms an excited hydroxyflavin intermediate, which is dehydrated to the product FMN to emit blue-green light.

Nearly all of the energy input into the reaction is transformed into light. The reaction is 80% to 90% efficient. As a comparison, the incandescent light bulb only converts about 10% of its energy into light. and a 150 lumen per Watt (lm/W) LED converts 20% of input energy to visible light.

Firefly luciferase generates light from luciferin in a multistep process. First, D-luciferin is adenylated by MgATP to form luciferyl adenylate and pyrophosphate. After activation by ATP, luciferyl adenylate is oxidized by molecular oxygen to form a dioxetanone ring. A decarboxylation reaction forms an excited state of oxyluciferin, which tautomerizes between the keto-enol form. The reaction finally emits light as oxyluciferin returns to the ground state.

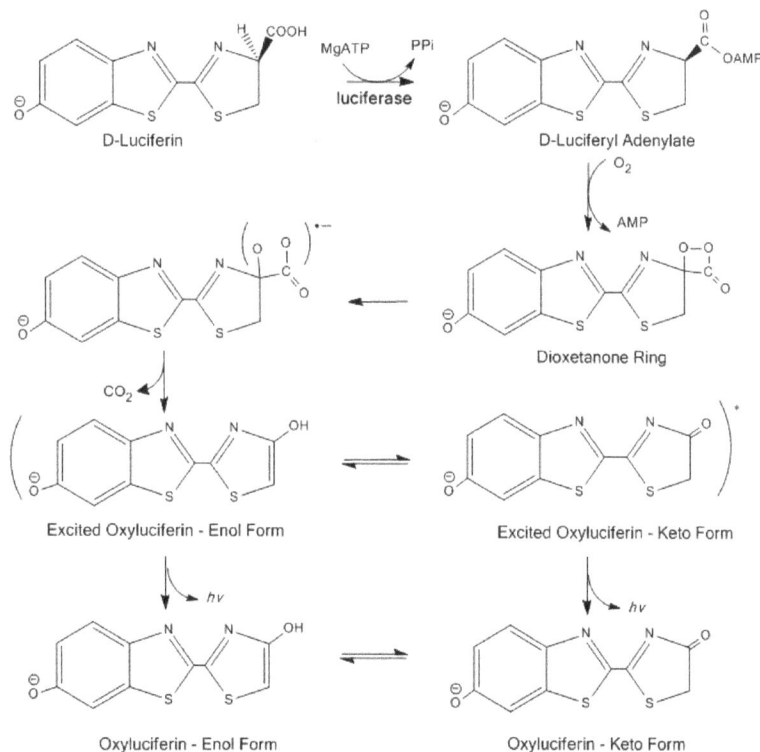

Mechanism for luciferase.

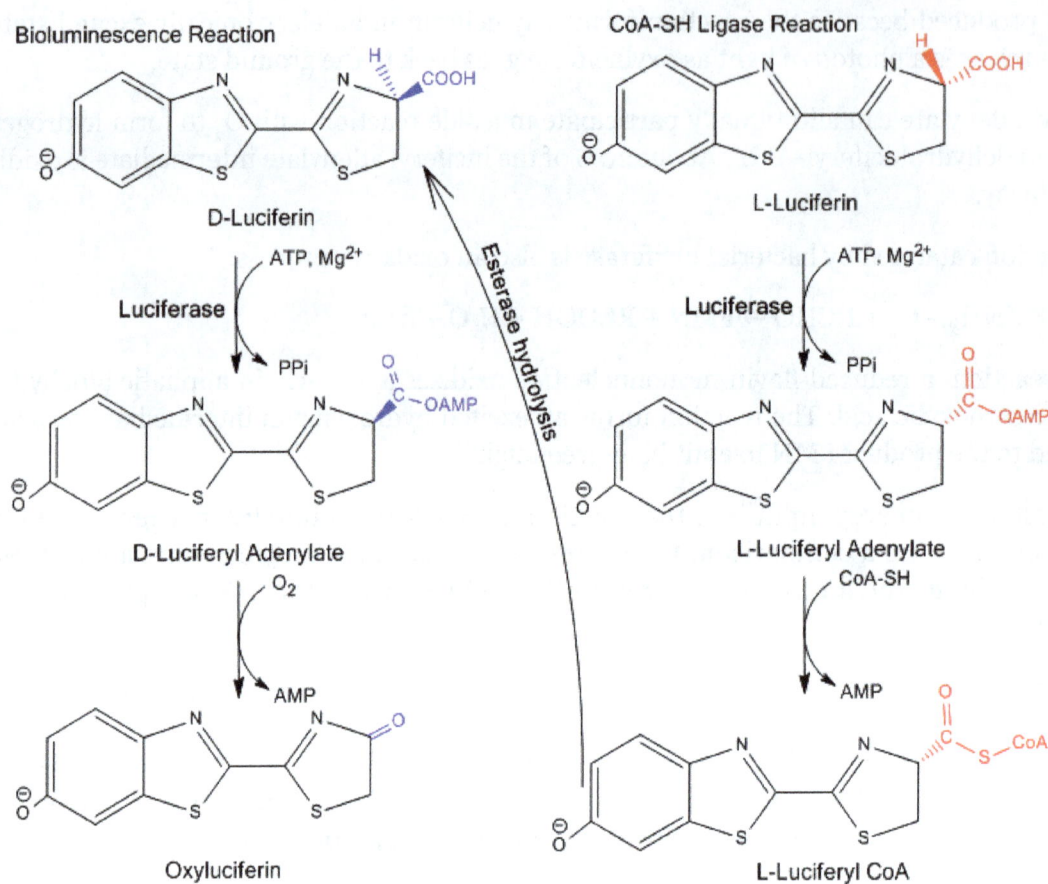

Luciferase has two modes of enzyme activity: bioluminescence activity and CoA synthetase activity.

Bifunctionality

Luciferase can function in two different pathways: a bioluminescence pathway and a CoA-ligase pathway. In both pathways, luciferase initially catalyzes an adenylation reaction with MgATP. However, in the CoA-ligase pathway, CoA can displace AMP to form luciferyl CoA.

Fatty acyl-CoA synthetase similarly activates fatty acids with ATP, followed by displacement of AMP with CoA. Because of their similar activities, luciferase is able to replace fatty acyl-CoA synthetase and convert long-chain fatty acids into fatty-acyl CoA for beta oxidation.

Structure

The protein structure of firefly luciferase consists of two compact domains: the N-terminal domain and the C-terminal domain. The N-terminal domain is composed of two β-sheets in an αβαβα structure and a β barrel. The two β-sheets stack on top of each other, with the β-barrel covering the end of the sheets.

The C-terminal domain is connected to the N-terminal domain by a flexible hinge, which can separate the two domains. The amino acid sequences on the surface of the two domains facing each other are conserved in bacterial and firefly luciferase, thereby strongly suggesting that the active site is located in the cleft between the domains.

During a reaction, luciferase has a conformational change and goes into a "closed" form with the two domains coming together to enclose the substrate. This ensures that water is excluded from the reaction and does not hydrolyze ATP or the electronically excited product.

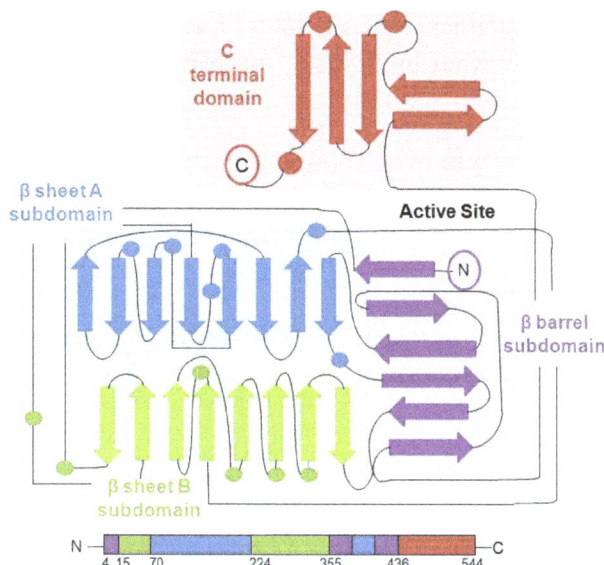

Diagram of the secondary structure of firefly luciferase. Arrows represent β-strands and circles represent α-helices. The locations of each of the subdomains in the sequence of luciferase is shown in the bottom diagram.

Spectral Differences in Bioluminescence

Firefly luciferase bioluminescence color can vary between yellow-green (λ_{max} = 550 nm) to red (λ_{max} = 620). There are currently several different mechanisms describing how the structure of luciferase affects the emission spectrum of the photon and effectively the color of light emitted.

One mechanism proposes that the color of the emitted light depends on whether the product is in the keto or enol form. The mechanism suggests that red light is emitted from the keto form of oxyluciferin, while green light is emitted from the enol form of oxyluciferin. However, 5,5-dimethyloxyluciferin emits green light even though it is constricted to the keto form because it cannot tautomerize.

Another mechanism proposes that twisting the angle between benzothiazole and thiazole rings in oxyluciferin determines the color of bioluminescence. This explanation proposes that a planar form with an angle of 0° between the two rings corresponds to a higher energy state and emits a higher-energy green light, whereas an angle of 90° puts the structure in a lower energy state and emits a lower-energy red light.

The most recent explanation for the bioluminescence color examines the microenvironment of the excited oxyluciferin. Studies suggest that the interactions between the excited state product and nearby residues can force the oxyluciferin into an even higher energy form, which results in the emission of green light. For example, Arg 218 has electrostatic interactions with other nearby residues, restricting oxyluciferin from tautomerizing to the enol form. Similarly, other results have indicated that the microenvironment of luciferase can force oxyluciferin into a more rigid, high-energy structure, forcing it to emit a high-energy green light.

Regulation

D-luciferin is the substrate for firefly luciferase's bioluminescence reaction, while L-luciferin is the substrate for luciferyl-CoA synthetase activity. Both reactions are inhibited by the substrate's enantiomer: L-luciferin and D-luciferin inhibit the bioluminescence pathway and the CoA-ligase pathway, respectively. This shows that luciferase can differentiate between the isomers of the luciferin structure.

L-luciferin is able to emit a weak light even though it is a competitive inhibitor of D-luciferin and the bioluminescence pathway. Light is emitted because the CoA synthesis pathway can be converted to the bioluminescence reaction by hydrolyzing the final product via an esterase back to D-luciferin.

Luciferase activity is additionally inhibited by oxyluciferin and allosterically activated by ATP. When ATP binds to the enzyme's two allosteric sites, luciferase's affinity to bind ATP in its active site increases.

Applications

Luciferase can be produced in the lab through genetic engineering for a number of purposes. Luciferase genes can be synthesized and inserted into organisms or transfected into cells. Mice, silkworms, and potatoes are just a few of the organisms that have already been engineered to produce the protein.

In the luciferase reaction, light is emitted when luciferase acts on the appropriate luciferin substrate. Photon emission can be detected by light sensitive apparatus such as a luminometer or modified optical microscopes. This allows observation of biological processes. Since light excitation is not needed for luciferase bioluminescence, there is minimal autofluorescence and therefore virtually background-free fluorescence. Therefore, as little as 0.02pg can still be accurately measured using a standard scintillation counter.

In biological research, luciferase is commonly used as a reporter to assess the transcriptional activity in cells that are transfected with a genetic construct containing the luciferase gene under the control of a promoter of interest. Additionally proluminescent molecules that are converted to luciferin upon activity of a particular enzyme can be used to detect enzyme activity in coupled or two-step luciferase assays. Such substrates have been used to detect caspase activity and cytochrome P450 activity, among others.

Luciferase can also be used to detect the level of cellular ATP in cell viability assays or for kinase activity assays. Luciferase can act as an ATP sensor protein through biotinylation. Biotinylation will immobilize luciferase on the cell-surface by binding to a streptavidin-biotin complex. This allows luciferase to detect the efflux of ATP from the cell and will effectively display the real-time release of ATP through bioluminescence. Luciferase can additionally be made more sensitive for ATP detection by increasing the luminescence intensity by changing certain amino acid residues in the sequence of the protein.

Whole animal imaging (referred to as *in vivo* or, occasionally, *ex vivo* imaging) is a powerful technique for studying cell populations in live animals, such as mice. Different types of cells (e.g. bone marrow stem cells, T-cells) can be engineered to express a luciferase allowing their non-invasive visualization inside a live animal using a sensitive charge-couple device camera (CCD camera).This technique has been used to follow tumorigenesis and response of tumors to treatment in animal models. However, environmental factors and therapeutic interferences may cause some discrepancies between tumor burden and bioluminescence intensity in relation to changes in proliferative activity. The intensity of the signal measured by in vivo imaging may depend on various factors, such as D-luciferin absorption through the peritoneum, blood flow, cell membrane permeability, availability of co-factors, intracellular pH and transparency of overlying tissue, in addition to the amount of luciferase.

The Glowing Plant project plans to use bacterial bio-luminescent systems to engineer novelty glowing *Arabidopsis thaliana* plants. Longer term they hypothesize that maybe such systems could be used to create eco-friendly sustainable light sources.

Luciferase is a heat-sensitive protein that is used in studies on protein denaturation, testing the protective capacities of heat shock proteins. The opportunities for using luciferase continue to expand.

Luciferin

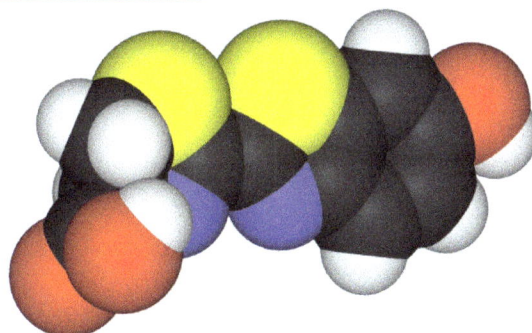

This is a space-filling model of firefly luciferin. Color coding:
yellow=sulfur; blue=nitrogen; black=carbon; red=oxygen; white=hydrogen.

Luciferin is a generic term for the light-emitting compound found in organisms that generate bioluminescence. Luciferins typically undergo an enzyme-catalysed oxidation and the resulting excited state intermediate emits light upon decaying to its ground state. This may refer to molecules that are substrates for both luciferases and photoproteins.

Types

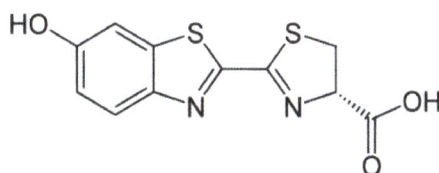

This structure of firefly luciferin is reversed (left to right) from the space-filling model shown above

Luciferins are a class of small-molecule substrates that are oxidized in the presence of the enzyme luciferase to produce oxyluciferin and energy in the form of light. It is not known just how many types of luciferins there are, but some of the better-studied compounds are listed below. There are many types of luciferins, yet all share the use of reactive oxygen species to emit light.

Firefly

Firefly luciferin is the luciferin found in many Lampyridae species. It is the substrate of luciferase (EC 1.13.12.7) responsible for the characteristic yellow light emission from fireflies. The chemistry is unusual, as adenosine triphosphate (ATP) is required for light emission.

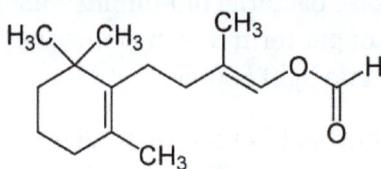

Latia luciferin

Snail

Latia luciferin is, in terms of chemistry, (E)-2-methyl-4-(2,6,6-trimethyl-1-cyclohex-1-yl)-1-buten-1-ol formate and is from the freshwater snail *Latia neritoides*.

Bacterial

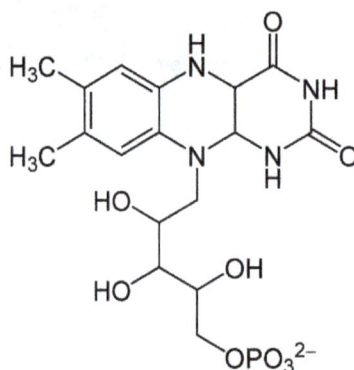

Bacterial luciferin

Bacterial luciferin is a type of luciferin found in bacteria, some of which live within the specialized tissues of some squid and fish. The molecule contains a reduced riboflavin phosphate.

Coelenterazine

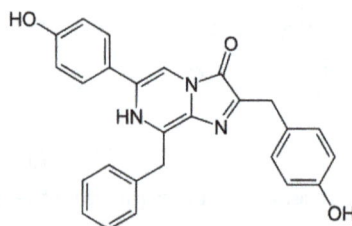

Coelenterazine

Coelenterazine is found in radiolarians, ctenophores, cnidarians, squid, brittle stars, copepods, chaetognaths, fish, and shrimp. It is the prosthetic group in the protein aequorin responsible for the blue light emission.

Dinoflagellate

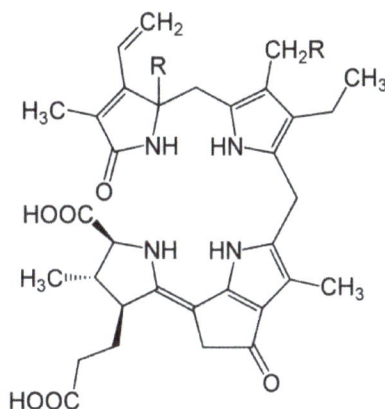

Luciferin of dinoflagellates (R = H) resp. of euphausiid shrimps (R = OH). The latter is also called *Component F*.

Dinoflagellate luciferin is a chlorophyll derivative (i. e. a Tetrapyrrole) and is found in some dino-flagellates, which are often responsible for the phenomenon of nighttime glowing waves (historically this was called phosphorescence, but is a misleading term). A very similar type of luciferin is found in some types of euphausiid shrimp.

Vargulin

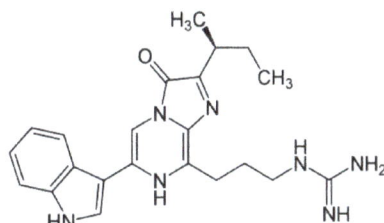

Vargulin (cypridinluciferin)

Vargulin is found in certain ostracods and deep-sea fish, to be specific, Porichys. Like the compound coelenterazine, it is an imidazopyrazinone and emits primarily blue light in the animals.

References

- Smiles, Samuel (1862). Lives of the Engineers. Volume III (George and Robert Stephenson). London: John Murray. p. 107. ISBN 0-7153-4281-9.

- Pauly, Daniel (13 May 2004). Darwin's Fishes: An Encyclopedia of Ichthyology, Ecology, and Evolution. Cambridge University Press. pp. 15–16. ISBN 978-1-139-45181-9.

- Huth, John Edward (15 May 2013). The Lost Art of Finding Our Way. Harvard University Press. p. 423. ISBN 978-0-674-07282-4.

- Pieribone, Vincent; Gruber, David F. (2005). Aglow in the Dark: The Revolutionary Science of Biofluorescence. Harvard University Press. pp. 35–41. ISBN 978-0-674-01921-8.

- Shimomura, Osamu (2012). Bioluminescence: Chemical Principles and Methods. World Scientific. p. 234. ISBN 978-981-4366-08-3.

- Marcellin, Frances (26 February 2016). "Glow-in-the-dark bacterial lights could illuminate shop windows". The New Scientist. Retrieved 4 March 2016.

- Yong, Ed (8 June 2016). "Surprising History of Glowing Fish". Phenomena. National Geographic. Retrieved 11 June 2016.

- Davis, Matthew P.; Sparks, John S.; Smith, W. Leo (2016-06-08). "Repeated and Widespread Evolution of Bioluminescence in Marine Fishes". PLOS ONE. 11 (6): e0155154. doi:10.1371/journal.pone.0155154. ISSN 1932-6203.

- Tuckey, James Hingston (May 1818). Thomson, Thomas, ed. Narrative of the Expedition to the Zaire. Annals of Philosophy. volume XI. p. 392. Retrieved 22 April 2015.

- Reshetiloff, Kathy (1 July 2001). "Chesapeake Bay night-lights add sparkle to woods, water". Bay Journal. Retrieved 16 December 2014.

- Branham, Marc. "Glow-worms, railroad-worms (Insecta: Coleoptera: Phengodidae)". Featured Creatures. University of Florida. Retrieved 29 November 2014.

- Nordgren, I. K.; Tavassoli, A. (2014). "A bidirectional fluorescent two-hybrid system for monitoring protein-protein interactions". Molecular BioSystems. 10 (3): 485–490. doi:10.1039/c3mb70438f. PMID 24382456.

- Ludwig Institute for Cancer Research (21 April 2003). "Firefly Light Helps Destroy Cancer Cells; Researchers Find That The Bioluminescence Effects Of Fireflies May Kill Cancer Cells From Within". Science Daily. Retrieved 4 December 2014.

7

Applications of Photochemistry

The applications of photochemistry that have been discussed in the following section are ultraviolet germicidal irradiation, UV curing, light therapy, laser hair removal etc. The section serves as a source to understand the major applications related to photochemistry.

Ultraviolet Germicidal Irradiation

Ultraviolet germicidal irradiation (UVGI) is a disinfection method that uses short-wavelength ultraviolet (UV-C) light to kill or inactivate microorganisms by destroying nucleic acids and disrupting their DNA, leaving them unable to perform vital cellular functions. UVGI is used in a variety of applications, such as food, air, and water purification.

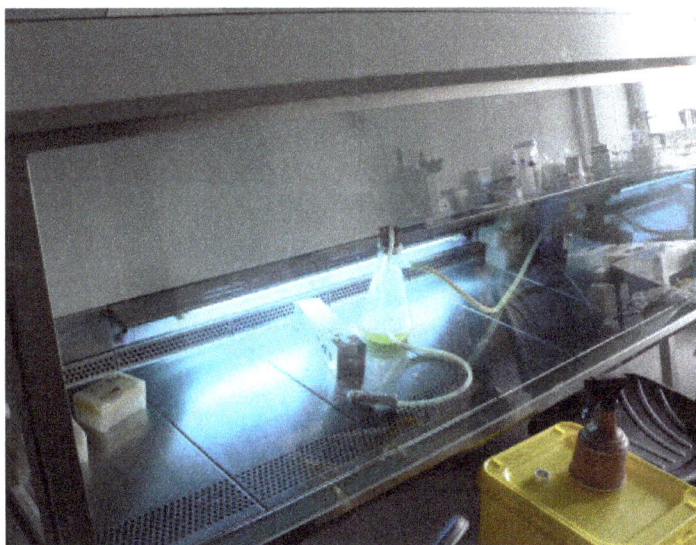

A low-pressure mercury-vapor discharge tube floods the inside of a biosafety cabinet with shortwave UV light when not in use, sterilizing microbiological contaminants from irradiated surfaces.

UV-C light is weak at the Earth's surface as the ozone layer of the atmosphere blocks it. UVGI devices can produce strong enough UV-C light in circulating air or water systems to make them inhospitable environments to microorganisms such as bacteria, viruses, molds and other pathogens. UVGI can be coupled with a filtration system to sanitize air and water.

The application of UVGI to disinfection has been an accepted practice since the mid-20th century. It has been used primarily in medical sanitation and sterile work facilities. Increasingly it has been employed to sterilize drinking and wastewater, as the holding facilities are enclosed and can be circulated to ensure a higher exposure to the UV. In recent years UVGI has found renewed application in air purifiers.

History

In 1878, Arthur Downes and Thomas P. Blunt published a paper describing the sterilization of bacteria exposed to short-wavelength light. UV has been a known mutagen at the cellular level for more than one hundred years. The 1903 Nobel Prize for Medicine was awarded to Niels Finsen for his use of UV against lupus vulgaris, tuberculosis of the skin.

Using UV light for disinfection of drinking water dates back to the year 1910 in Marseille, France. The prototype plant was taken out of service after only a short time, due to reliability problems. In 1955, UV water treatment systems were applied in Austria and Switzerland; by 1985 about 1,500 plants were in use in Europe. In 1998 it was discovered that protozoa such as cryptosporidium and giardia were more vulnerable to UV light than previously thought; this opened the way to wide-scale use of UV water treatment in North America. By 2001, over 6,000 UV water treatment plants were operating in Europe.

Over the years, UV costs have declined as researchers develop and use new UV methods to disinfect water and wastewater. Currently, several countries have developed regulations that allow systems to disinfect their drinking water supplies with UV light.

Method of Operation

UV light is electromagnetic radiation with wavelengths shorter than visible light. UV can be separated into various ranges, with short-wavelength UV (UVC) considered "germicidal UV". At certain wavelengths, UV is mutagenic to bacteria, viruses and other microorganisms. Particularly at wavelengths around 260 nm–270 nm, UV breaks molecular bonds within microorganismal DNA, producing thymine dimers that can kill or disable the organisms. Mercury-based lamps emit UV light at the 253.7 nm line Ultraviolet Light Emitting Diodes (UV-C LED) lamps emit UV light at selectable wavelengths between 255 and 280 nm. It is a process similar to the effect of longer wavelengths (UVB) producing sunburn in humans. Microorganisms have less protection from UV and cannot survive prolonged exposure to it.

A UVGI system is designed to expose environments such as water tanks, sealed rooms and forced air systems to germicidal UV. Exposure comes from germicidal lamps that emit germicidal UV electromagnetic radiation at the correct wavelength, thus irradiating the environment. The forced flow of air or water through this environment ensures the exposure.

Effectiveness

The effectiveness of germicidal UV depends on the length of time a microorganism is exposed to UV, the intensity and wavelength of the UV radiation, the presence of particles that can protect the microorganisms from UV, and a microorganism's ability to withstand UV during its exposure.

In many systems, redundancy in exposing microorganisms to UV is achieved by circulating the air or water repeatedly. This ensures multiple passes so that the UV is effective against the highest number of microorganisms and will irradiate resistant microorganisms more than once to break them down.

"Sterilization" is often misquoted as being achievable. While it is theoretically possible in a controlled environment, it is very difficult to prove and the term "disinfection" is generally used by companies offering this service as to avoid legal reprimand. Specialist companies will often advertise a certain log reduction e.g., 99.9999% effective, instead of sterilization. This takes into consideration a phenomenon known as light and dark repair (photoreactivation and base excision repair, respectively), in which a cell can repair DNA that has been damaged by UV light.

The effectiveness of this form of disinfection depends on line-of-sight exposure of the microorganisms to the UV light. Environments where design creates obstacles that block the UV light are not as effective. In such an environment, the effectiveness is then reliant on the placement of the UVGI system so that line of sight is optimum for disinfection.

Dust and films coating the bulb lower UV output. Therefore, bulbs require periodic cleaning and replacement to ensure effectiveness. The lifetime of germicidal UV bulbs varies depending on design. Also, the material that the bulb is made of can absorb some of the germicidal rays.

Lamp cooling under airflow can also lower UV output; thus, care should be taken to shield lamps from direct airflow, or to add additional lamps to compensate for the cooling effect.

Increases in effectiveness and UV intensity can be achieved by using reflection. Aluminum has the highest reflectivity rate versus other metals and is recommended when using UV.

One method for gauging UV effectiveness in water disinfection applications is to compute UV dose. The U.S. EPA publishes UV dosage guidelines for water treatment applications. UV dose cannot be measured directly but can be inferred based on the known or estimated inputs to the process:

- Flow rate (contact time)

- Transmittance (light reaching the target)

- Turbidity (cloudiness)

- Lamp age or fouling or outages (reduction in UV intensity)

In air and surface disinfection applications the UV effectiveness is estimated by calculating the UV dose which will be delivered to the microbial population. The UV dose is calculated as follows:

UV dose $\mu Ws/cm^2$ = UV intensity $\mu W/cm^2$ x Exposure time (seconds)

The UV intensity is specified for each lamp at a distance of 1 meter. UV intensity is inversely proportional to the square of the distance so it decreases at longer distances. Alternatively, it rapidly increases at distances shorter than 1m. In the above formula the UV intensity must always be adjusted for distance unless the UV dose is calculated at exactly 1m from the lamp. Also, to ensure effectiveness the UV dose must be calculated at the end of lamp life (EOL is specified in number of hours when the lamp is expected to reach 80% of its initial UV output) and at the furthest distance from the lamp on the periphery of the target area. In some applications coating is applied to the lamp to make it shatter-proof. The coating is Fluoro Ethylene Polymer which completely encapsulates the lamp and contains the shards and mercury in case of accidental breakage. The coating decreases the UV intensity up to 20%.

To accurately predict what UV dose will be delivered to the target the UV intensity, adjusted for distance, coating and end of lamp life, will be multiplied by the exposure time. In static applications the exposure time can be as long as needed for an effective UV dose to be reached. In case of rapidly moving air, in AC air ducts for example, the exposure time is short so the UV intensity must be increased by introducing multiple UV lamps or even banks of lamps. Also, the UV installation must be located in a long straight duct section with the lamps perpendicular to the air flow to maximize the exposure time.

These calculations actually predict the UV fluence and it is assumed that the UV fluence will be equal to the UV dose. The UV dose is the amount of germicidal UV energy absorbed by a microbial population over a period of time. If the microorganisms are planktonic (free floating) the UV fluence will be equal the UV dose. However, if the microorganisms are protected by mechanical particles, such as dust and dirt, or have formed biofilm a much higher UV fluence will be needed for an effective UV dose to be introduced to the microbial population.

Inactivation of Microorganisms

The degree of inactivation by ultraviolet radiation is directly related to the UV dose applied to the water. The dosage, a product of UV light intensity and exposure time, is usually measured in microjoules per square centimeter, or equivalently as microwatt seconds per square centimeter ($\mu W \cdot s/cm^2$). Dosages for a 90% kill of most bacteria and viruses range from 2,000 to 8,000 $\mu W \cdot s/cm^2$. Larger parasites such as cryptosporidium require a lower dose for inactivation. As a result, the U.S. Environmental Protection Agency has accepted UV disinfection as a method for drinking water plants to obtain cryptosporidium, giardia or virus inactivation credits. For example, for one-decimal-logarithm reduction of cryptosporidium, a minimum dose of 2,500 $\mu W \cdot s/cm^2$ is required based on the U.S. EPA UV Guidance Manual published in 2006.

Strengths and Weaknesses

Advantages

UV water treatment devices can be used for well water and surface water disinfection. UV treatment compares favorably with other water disinfection systems in terms of cost, labor, and the need for technically trained personnel for operation. Water chlorination treats larger organisms and offers residual disinfection, but these systems are expensive because they need special operator training and a steady supply of a potentially hazardous material. Finally, boiling of water is the most reliable treatment method but it demands labor, and imposes a high economic cost. UV treatment is rapid and, in terms of primary energy use, approximately 20,000 times more efficient than boiling.

Disadvantages

UV disinfection is most effective for treating high-clarity, purified reverse osmosis distilled water. Suspended particles are a problem because microorganisms buried within particles are shielded from the UV light and pass through the unit unaffected. However, UV systems can be coupled with a pre-filter to remove those larger organisms that would otherwise pass through the UV system unaffected. The pre-filter also clarifies the water to improve light transmittance and therefore UV dose throughout the entire water column. Another key factor of UV water treatment is the flow rate—if the flow is too high, water will pass through without sufficient UV exposure. If the flow is too low, heat may build up and damage the UV lamp.

A disadvantage of UVGI is that while water treated by chlorination is resistant to reinfection (until the chlorine off-gasses), UVGI water is not resistant to reinfection. UVGI water must be transported or delivered in such a way as to avoid reinfection.

Safety

In UVGI systems the lamps are shielded or are in environments that limit exposure, such as a closed water tank or closed air circulation system, often with interlocks that automatically shut off the UV lamps if the system is opened for access by human beings.

For human beings, skin exposure to germicidal wavelengths of UV light can produce rapid sunburn and skin cancer. Exposure of the eyes to this UV radiation can produce extremely painful inflammation of the cornea and temporary or permanent vision impairment, up to and including blindness in some cases. UV can damage the retina of the eye.

Another potential danger is the UV production of ozone, which can be harmful to health. The U.S. Environmental Protection Agency designated 0.05 parts per million (ppm) of ozone to be a safe level. Lamps designed to release UVC and higher frequencies are doped so that any UV light below 254 nm wavelengths will not be released, to minimize ozone production. A full-spectrum lamp will release all UV wavelengths, and will produce ozone when UVC hits oxygen (O_2) molecules.

UV-C radiation is able to break down chemical bonds. This leads to rapid aging of plastics, insulation, gaskets, and other materials. Note that plastics sold to be "UV-resistant" are tested only for UV-B, as UV-C doesn't normally reach the surface of the Earth. When UV is used near plastic, rubber, or insulation, care should be taken to shield said components; metal tape or aluminum foil will suffice.

The American Conference of Governmental Industrial Hygienists (ACGIH) Committee on Physical Agents has established a TLV for UV-C exposure to avoid such skin and eye injuries among those most susceptible. For 254 nm UV, this TLV is 6 mJ/cm² over an eight-hour period. The TLV function differs by wavelengths because of variable energy and potential for cell damage. This TLV is supported by the International Commission on Non-Ionizing Radiation Protection and is used in setting lamp safety standards by the Illuminating Engineering Society of North America. When TUSS was planned, and until quite recently, this TLV was interpreted as if eye exposure in rooms was continuous over eight hours and at the highest eye-level irradiance found in the room. In those highly unlikely conditions, a 6.0 mJ/cm² dose is reached under the ACGIH TLV after just eight hours of continuous exposure to an irradiance of 0.2 µW/cm². Thus, 0.2 µW/cm² was widely interpreted as the upper permissible limit of irradiance at eye height.

Uses

Air Disinfection

UVGI can be used to disinfect air with prolonged exposure. Disinfection is a function of UV intensity and time. For this reason, it is not as effective on moving air, or when the lamp is perpendicular to the flow, as exposure times are dramatically reduced. Air purification UVGI systems can be free-standing units with shielded UV lamps that use a fan to force air past the UV light. Other systems are installed in forced air systems so that the circulation for the premises moves microorganisms past the lamps. Key to this form of sterilization is placement of the UV lamps and a good filtration system to remove the dead microorganisms. For example, forced air systems by design impede line-of-sight, thus creating areas of the environment that will be shaded from the UV light. However, a UV lamp placed at the coils and drain pans of cooling systems will keep microorganisms from forming in these naturally damp places.

ASHRAE covers UVGI and its applications in indoor air quality and building maintenance in "Ultraviolet Lamp Systems", Chapter 16 of its 2008 Handbook, *HVAC Systems and Equipment*. Its 2011 Handbook, *HVAC Applications*, covers "Ultraviolet air and surface treatment" in Chapter 60.

Water Disinfection

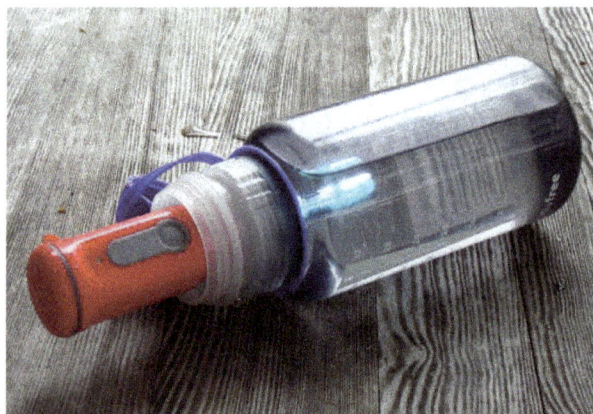

A portable, battery-powered, low-pressure mercury-vapor discharge lamp for water sterilization.

Ultraviolet disinfection of water is a purely physical, chemical-free process. Even parasites such as *cryptosporidia* or *giardia*, which are extremely resistant to chemical disinfectants, are efficiently reduced. UV can also be used to remove chlorine and chloramine species from water; this process is called photolysis, and requires a higher dose than normal disinfection. The sterilized microorganisms are not removed from the water. UV disinfection does not remove dissolved organics, inorganic compounds or particles in the water. However, UV-oxidation processes can be used to simultaneously destroy trace chemical contaminants and provide high-level disinfection, such as the world's largest indirect potable reuse plant in New York which opened the Catskill-Delaware Water Ultraviolet Disinfection Facility on 8 October 2013. A total of 56 energy-efficient UV reactors were installed to treat 2.2 billion US gallons (8,300,000 m^3) a day to serve New York City.

It used to be thought that UV disinfection was more effective for bacteria and viruses, which have more-exposed genetic material, than for larger pathogens that have outer coatings or that form cyst states (e.g., Giardia) that shield their DNA from UV light. However, it was recently discovered that ultraviolet radiation can be somewhat effective for treating the microorganism Cryptosporidium. The findings resulted in the use of UV radiation as a viable method to treat drinking water. Giardia in turn has been shown to be very susceptible to UV-C when the tests were based on infectivity rather than excystation. It has been found that protists are able to survive high UV-C doses but are sterilized at low doses.

Developing Countries

A 2006 project at University of California, Berkeley produced a design for inexpensive water disinfection in resource deprived settings. The project was designed to produce an open source design that could be adapted to meet local conditions. In a somewhat similar proposal in 2014, Australian students designed a system using chip packet foil to reflect solar UV radiation into a glass tube that should disinfect water without power.

Wastewater Treatment

Ultraviolet in sewage treatment is commonly replacing chlorination. This is in large part because of concerns that reaction of the chlorine with organic compounds in the waste water stream could synthesize potentially toxic and long lasting chlorinated organics and also because of the environmental risks of storing chlorine gas or chlorine containing chemicals. Individual wastestreams to be treated by UVGI must be tested to ensure that the method will be effective due to potential interferences such as suspended solids, dyes, or other substances that may block or absorb the UV radiation. According to the World Health Organization, "UV units to treat small batches (1 to several liters) or low flows (1 to several liters per minute) of water at the community level are estimated to have costs of US$20 per megaliter, including the cost of electricity and consumables and the annualized capital cost of the unit."

Large-scale urban UV wastewater treatment is performed in cities such as Edmonton, Alberta. The use of ultraviolet light has now become standard practice in most municipal wastewater treatment processes. Effluent is now starting to be recognized as a valuable resource, not a problem that needs to be dumped. Many wastewater facilities are being renamed as water reclamation facilities, whether the wastewater is discharged into a river, used to irrigate crops, or injected into an aquifer for later recovery. Ultraviolet light is now being used to ensure water is free from harmful organisms.

Aquarium and Pond

Ultraviolet sterilizers are often used to help control unwanted microorganisms in aquaria and ponds. UV irradiation ensures that pathogens cannot reproduce, thus decreasing the likelihood of a disease outbreak in an aquarium.

Aquarium and pond sterilizers are typically small, with fittings for tubing that allows the water to flow through the sterilizer on its way from a separate external filter or water pump. Within the sterilizer, water flows as close as possible to the ultraviolet light source. Water pre-filtration is critical as water turbidity lowers UVC penetration. Many of the better UV sterilizers have long dwell times and limit the space between the UVC source and the inside wall of the UV sterilizer device.

Laboratory Hygiene

UVGI is often used to disinfect equipment such as safety goggles, instruments, pipettors, and other devices. Lab personnel also disinfect glassware and plasticware this way. Microbiology laboratories use UVGI to disinfect surfaces inside biological safety cabinets ("hoods") between uses.

Food and Beverage Protection

Since the U.S. Food and Drug Administration issued a rule in 2001 requiring that virtually all fruit and vegetable juice producers follow HACCP controls, and mandating a 5-log reduction in pathogens, UVGI has seen some use in sterilization of juices such as fresh-pressed apple cider.

Technology

Lamps

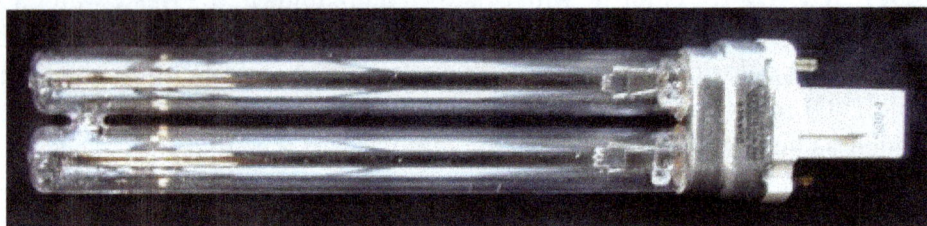

A 9 W germicidal lamp in a compact fluorescent lamp form factor

Germicidal UV for disinfection is most typically generated by a mercury-vapor lamp. Low-pressure mercury vapor has a strong emission line at 254 nm, which is within the range of wavelengths that demonstrate strong disinfection effect. The optimal wavelengths for disinfection are close to 270 nm.

Lamps are either amalgam or medium-pressure lamps. Low-pressure UV lamps offer high efficiencies (approx 35% UVC) but lower power, typically 1 W/cm power density (power per unit of arc length). Amalgam UV lamps are a higher-power version of low-pressure lamps. They operate at higher temperatures and have a lifetime of up to 16,000 hours. Their efficiency is slightly lower than that of traditional low-pressure lamps (approx 33% UVC output) and power density is approximately 2–3 W/cm. Medium-pressure UV lamps have a broad and pronounced peak-line spectrum and a high radiation output but lower UVC efficiency of 10% or less. Typical power density is 30 W/cm^3 or greater.

Depending on the quartz glass used for the lamp body, low-pressure and amalgam UV emit radiation at 254 nm and also at 185 nm, which has chemical effects. UV radiation at 185 nm is used to generate ozone.

The UV lamps for water treatment consist of specialized low-pressure mercury-vapor lamps that produce ultraviolet radiation at 254 nm, or medium-pressure UV lamps that produce a polychromatic output from 200 nm to visible and infrared energy. The UV lamp never contacts the water; it is either housed in a quartz glass sleeve inside the water chamber or mounted external to the water which flows through the transparent UV tube. Water passing through the flow chamber is exposed to UV rays which are absorbed by suspended solids, such as microorganisms and dirt, in the stream.

Compact and versatile options with UV-C LEDs

Light Emitting Diodes (LEDs)

Recent developments in LED technology have led to commercially available UV-C LEDs. UV-C LEDs use semiconductors to emit light between 255 nm-280 nm. The wavelength emission is tuneable by adjusting the material of the semiconductor. The reduced size of LEDs open up options for small reactor systems allowing for point-of-use applications and integration into medical devices. Low power consumption of semiconductors introduce UV disinfection systems that utilized small solar cells in remote or Third World applications.

Water Treatment Systems

Sizing of a UV system is affected by three variables: flow rate, lamp power, and UV transmittance in the water. Manufacturers typically developed sophisticated Computational Fluid Dynamics (CFD) models validated with bioassay testing. This involves testing the UV reactor's disinfection performance with either MS2 or T1 bacteriophages at various flow rates, UV transmittance, and power levels in order to develop a regression model for system sizing. For example, this is a requirement for all drinking water systems in the United States per the EPA UV Guidance Manual.

The flow profile is produced from the chamber geometry, flow rate, and particular turbulence model selected. The radiation profile is developed from inputs such as water quality, lamp type (power, germicidal efficiency, spectral output, arc length), and the transmittance and dimension of the quartz sleeve. Proprietary CFD software simulates both the flow and radiation profiles. Once the 3D model of the chamber is built, it is populated with a grid or mesh that comprises thousands of small cubes.

Points of interest—such as at a bend, on the quartz sleeve surface, or around the wiper mechanism—use a higher resolution mesh, whilst other areas within the reactor use a coarse mesh. Once the mesh is produced, hundreds of thousands of virtual particles are "fired" through the chamber. Each particle has several variables of interest associated with it, and the particles are "harvested" after the reactor. Discrete phase modeling produces delivered dose, head loss, and other chamber-specific parameters.

When the modeling phase is complete, selected systems are validated using a professional third party to provide oversight and to determine how closely the model is able to predict the reality of system performance. System validation uses non-pathogenic surrogates such as MS 2 phage or *Bacillus subtilis* to determine the Reduction Equivalent Dose (RED) ability of the reactors. Most systems are validated to deliver 40 mJ/cm² within an envelope of flow and transmittance.

To validate effectiveness in drinking-water systems, the method described in the EPA UV Guidance Manual is typically used by the U.S., whilst Europe has adopted Germany's DVGW 294 standard. For wastewater systems, the NWRI/AwwaRF Ultraviolet Disinfection Guidelines for Drinking Water and Water Reuse protocols are typically used, especially in wastewater reuse applications.

UV Curing

UV curing is a speed curing process in which high intensity ultraviolet light is used to create a photochemical reaction that instantly cures inks, adhesives and coatings. UV Curing is adaptable to printing, coating, decorating, stereolithography and assembling of a variety of products and materials owing to some of its key attributes, it is: a low temperature process, a high speed process, and a solventless process—cure is by polymerization rather than by evaporation. Originally introduced in the 1960s this technology has streamlined and increased automation in many industries in the manufacturing sector.

Applications

UV curing is used whenever there is a need for curing and drying of inks, adhesives and coatings. UV-cured adhesive has become a high-speed replacement for two-part adhesives, eliminating the need for solvent removal, ratio mixing and potential life concern. It is used in the screen printing process where the UV curing systems are used to cure screen-printed products, which range from T-shirts to 3d and cylindrical parts. It is used in fine instrument finishing (guitars, violins, ukuleles, etc.), pool cue manufacturing and other wood craft industries. Printing with UV curable inks provides the ability to print on a very wide variety of substrates such as plastics, paper, canvas, glass, metal, foam boards, tile, films, and many other materials.

Other industries that take advantage of UV curing include medicine, automobiles, cosmetics (for example artificial fingernails and gel nail polish, food, science, education and art. This curable ink has efficiently met the requirements of the publication sector on variety of paper and board.

Advantages of UV Curing

The primary advantage of curing finishes and inks with ultraviolet lie in the speed at which the final product can be readied for shipping. In addition to speeding up production, this also can reduce flaws and errors, as the amount of time that dust, flies or any airborne object has to settle upon the object is reduced. This can increase the quality of the finished item, and allow for greater consistency.

The other obvious benefit is that manufacturers can devote less space to finishing items, since they don't have to wait for them to dry. This creates an efficiency that ripples through the entire manufacturing process.

Types of UV Curing

Mercury vapor lamps are the industry standard for curing products with ultraviolet light. The bulbs work by high voltage passing through, vaporizing the mercury. An arc is created within the mercury which emits a spectral output in the UV region of the light spectrum. The light intensity occurs in the 240 nm-270 nm and 350-380-nm. This intense spectrum of light is what causes the rapid curing of the different applications being used.

In the last few years an emerging type of UV curing technology called UV LED curing has entered the marketplace. This technology is growing rapidly in popularity and has many advantages over mercury based lamps although is not the right fit for every application

Fluorescent lamps made specifically for UV curing are also available. These have the ability to dial in to specific frequencies at a lower price point as fluorescent lamps are an established technology and the spectrum is easily controlled by the type of phosphor used. They can produce frequencies that LEDs and mercury vapor lamps can not, including multiple frequencies. They are somewhat less efficient than LEDs or Mercury vapor but cost a fraction of the price of the other systems. The allow for curing all around an item by using multiple tubes and off the shelf ballast systems.

Types of Ultraviolet Lamps

Mercury Vapor Lamp (H type)

The mercury lamp has an output in the short wave UV range between 220 and 320 nm (nanometers) and a spike of energy in the longwave range at 365 nm. The H lamp is a good choice for clear coatings and thin ink layers and produces hard surface cures and high gloss finishes.

Mercury Vapor Lamp with Iron Additive (D type)

The addition of iron to the lamp yields a strong output in the longwave range between 350 and 400 nm while the mercury component maintains good output in the short wavelength range. The D lamp is a good choice for curing heavily pigmented inks, adhesives, and thick laydowns of clear materials.

Mercury Vapor Lamp with Gallium Additive (V type)

The addition of gallium to the lamp yields a strong output in the longwave range between 400 and 450 nm. This makes the V lamp a good choice for curing white pigmented inks and base coats containing titanium dioxide which blocks the most shortwave UV.

Fluorescent Lamps

Fluorescent lamps are used for UV curing in a number of applications. In particular, these are used where the excessive heat of mercury vapor is undesirable, or when an item needs more than a single source of light and instead the item needs to be surrounded by light, such as musical instruments. Fluorescent lamps can be created that produce ultraviolet anywhere within the UVA/UVB spectrum. Additionally, lamps that have multiple peaks are possible, allowing a wider variety of photoinitiators to be used. While fluorescent lamps are less efficient at producing UV than mercury vapor, newer initiators require less total energy, offsetting this disadvantage. Fluorescent lamps in a wide variety of sizes and wattages are available.

LEDs

UV LED devices are capable of emitting a narrow spectrum of radiation (+/- 10 nm), while mercury lamps have a broader spectral distribution. Fluorescent ultraviolet lamps can be fairly narrow, although not as narrow as LEDs.

LEDs are much more expensive but last up to 10 times longer, and like fluorescent tubes, can be cycled on and off frequently as they require no startup or cool down period. While they can't produce the same spectrum as mercury vapor or fluorescent tubes, photoinitiators can be formulated to work with them easily.

Light Therapy

Light therapy or phototherapy, classically referred to as heliotherapy, consists of exposure to daylight or to specific wavelengths of light using polychromatic polarised light, lasers, light-emitting diodes, fluorescent lamps, dichroic lamps or very bright, full-spectrum light. The light is administered for a prescribed amount of time and, in some cases, at a specific time of day.

One common use of the term is associated with the treatment of skin disorders, chiefly psoriasis, acne vulgaris, eczema and neonatal jaundice.

Light therapy which strikes the retina of the eyes is used to treat diabetic retinopathy and also circadian rhythm disorders such as delayed sleep phase disorder and can also be used to treat seasonal affective disorder, with some support for its use also with non-seasonal psychiatric disorders.

Medical Uses

Skin Conditions

The treatments involve exposing the skin to ultraviolet light. The exposures can be to small area of the skin or over the whole body surface, like in a tanning bed. The most common treatment is with narrowband UVB (NB-UVB) with a wavelength of 311-313 nanometer. It was found that this is the safest treatment Full body phototherapy can be delivered at doctor's office or at home using a large high power UVB booth.

Psoriasis

For psoriasis, UVB phototherapy has been shown to be effective. A feature of psoriasis is localized inflammation mediated by the immune system. Ultraviolet radiation is known to suppress the immune system and reduce inflammatory responses. Light therapy for skin conditions like psoriasis usually use NB-UVB (311 nm wavelength) though it may use UV-A (315–400 nm wavelength) or UV-B (280–315 nm wavelength) light waves. UV-A, combined with psoralen, a drug taken orally, is known as PUVA treatment. In UVB phototherapy the exposure time is very short, seconds to minutes depending on intensity of lamps and the person's skin pigment and sensitivity. The time is controlled with a timer that turns off the lamps after the treatment time ends.

Vitiligo

One percent of the population suffer from vitiligo, and narrowband UVB phototherapy is an effective treatment. "NB-UVB phototherapy results in satisfactory repigmentation in our vitiligo patients and should be offered as a treatment option."

Acne Vulgaris

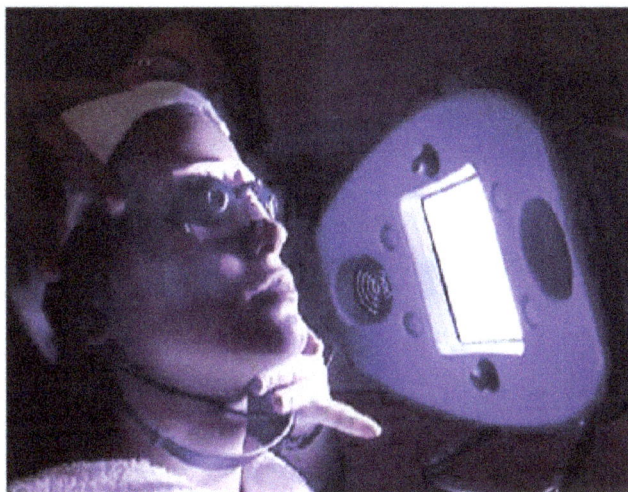

High intensity blue light (425nm) used for the treatment of acne.

Evidence for light therapy and lasers in acne vulgaris as of 2012 is not sufficient to recommend them. There is moderate evidence for the efficacy of blue and blue-red light therapies in treating mild acne, but most studies are of low quality. While light therapy appears to provide short term benefit, there is a lack of long term outcome data or data in those with severe acne.

Cancer

According to the American Cancer Society, there is some evidence that ultraviolet light therapy may be effective in helping treat certain kinds of skin cancer, and ultraviolet blood irradiation therapy is established for this application. However, alternative uses of light for cancer treatment – light box therapy and colored light therapy – are not supported by evidence.

Other skin Conditions

Phototherapy can be effective in the treatment of eczema, atopic dermatitis, polymorphous light eruption, cutaneous T-cell lymphoma and lichen planus. Narrowband UVB lamps, 311–313 nanometer is the most common treatment.

Wound Healing

Low level laser therapy has been studied as a potential treatment for chronic wounds. Reviews of the scientific literature do not support the widespread use of this technique due to inconsistent results and low research quality. Higher power lasers have also been used to close acute wounds as an alternative to stitching.

Retinal Conditions

There is preliminary evidence that light therapy is an effective treatment for diabetic retinopathy and diabetic macular oedema.

Mood and Sleep Related

Seasonal Affective Disorder

Full sunlight or exposure to bright light from a light box is used to treat seasonal affective disorder (SAD). Light boxes for SAD are designed to filter out most UV light, which can cause eye and skin damage. Mayo Clinic states that light therapy is a proven treatment for seasonal affective disorder. It is considered a first-line treatment. Controlled-trial comparisons with antidepressants show equal effectiveness, with less expense and more rapid onset of therapeutic benefit, though a minority of patients may not respond to it. Direct sunlight, reflected into the windows of a home or office by a computer-controlled mirror device called a heliostat, has also been used as a type of light therapy for the treatment of SAD.

The effectiveness of light therapy for treating SAD may be linked to the fact that light therapy makes up for lost sunlight exposure and resets the body's internal clock. Studies show that light therapy helps reduce the debilitating and depressive behaviors of SAD, such as excessive sleepiness and fatigue, with results lasting for at least 1 month. Light therapy is preferred over antidepressants in the treatment of SAD because it is a relatively safe and easy therapy.

It is possible that response to light therapy for SAD could be season dependent. Morning therapy has provided the best results because light in the early morning aids in regulating the circadian rhythm.

Non-seasonal Depression

Light therapy has also been suggested in the treatment of non-seasonal depression and other psychiatric disturbances, including major depressive disorder, bipolar disorder and postpartum depression. A meta-analysis by the Cochrane Collaboration concluded that "for patients suffering from non-seasonal depression, light therapy offers modest though promising antidepressive efficacy." A 2008 systematic review concluded that "overall, bright light therapy is an excellent candidate for inclusion into the therapeutic inventory available for the treatment of nonseasonal depression today, as adjuvant therapy to antidepressant medication, or eventually as stand-alone treatment for specific subgroups of depressed patients." A 2015 review found that supporting evidence for light therapy was weak due to serious methodological flaws. However, a 2016 randomized double blind clinical trial (which used a sham light treatment) found that light therapy in combination with Prozac for 8 weeks as well as light therapy alone resulted in a significantly superior decrease in depression scores than placebo.

Circadian Rhythm Sleep Disorders and Jet Lag

In the management of circadian rhythm disorders such as delayed sleep phase disorder (DSPD), the timing of light exposure is critical. For DSPD, the light must be provided to the retina as soon after spontaneous awakening as possible to achieve the desired effect, as shown by the phase response curve for light in humans. Use upon awakening may also be effective for non-24-hour sleep–wake disorder. Some users have reported success with lights that turn on shortly *before* awakening (dawn simulation). Evening use is recommended for people with advanced sleep phase disorder. Some, but not all, totally blind people whose retinae are intact, may benefit from light therapy.

Situational CRSD

Light therapy has been tested for individuals with shift work sleep disorder, and for jet lag.

Sleep disorder in Parkinson's disease

Light therapy has been trialed in treating sleep disorders experienced by patients with Parkinson's disease.

Neonatal Jaundice (Postnatal Jaundice)

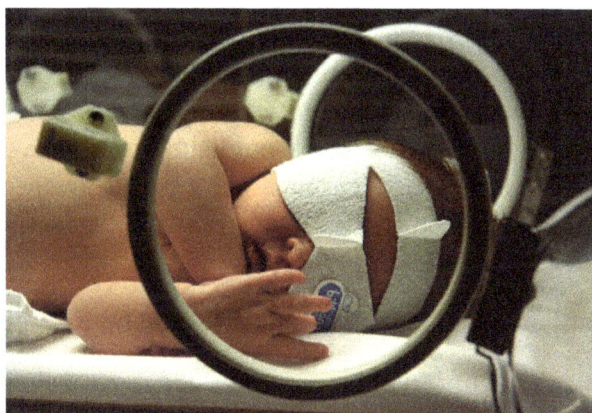

A newborn infant undergoing white-light phototherapy to treat neonatal jaundice.

Light therapy is used to treat cases of neonatal jaundice through the isomerization of the bilirubin and consequently transformation into compounds that the newborn can excrete via urine and stools. A common treatment of neonatal jaundice is the bili light or billiblanket.

Techniques

Photodynamic Therapy

Photodynamic therapy is a form of phototherapy using nontoxic light-sensitive compounds that are exposed selectively to light, whereupon they become toxic to targeted malignant and other diseased cells

One of the treatments is using blue light with aminolevulinic acid for the treatment of actinic keratosis. This is not a U.S. FDA-approved treatment for acne vulgaris.

Light Boxes

The brightness and color temperature of light from a light box are quite similar to daylight.

The production of the hormone melatonin, a sleep regulator, is inhibited by light and permitted by darkness as registered by photosensitive ganglion cells in the retina. To some degree, the reverse is true for serotonin, which has been linked to mood disorders. Hence, for the purpose of manipulating melatonin levels or timing, light boxes providing very specific types of artificial illumination to the retina of the eye are effective.

Light therapy uses either a light box which emits up to 10,000 lux of light at a specified distance, much brighter than a customary lamp, or a lower intensity of specific wavelengths of light from the blue (460 nm) to the green (525 nm) areas of the visible spectrum. A 1995 study showed that green light therapy at doses of 350 lux produces melatonin suppression and phase shifts equivalent to 10,000 lux white light therapy, but another study published in May 2010 suggests that the blue light often used for SAD treatment should perhaps be replaced by green or white illumination, because of a possible involvement of the cones in melatonin suppression.

In treatment, the patient's eyes are to be at a prescribed distance from the light source with the light striking the (lower) retina. This does not require looking directly into the light.

Considering three major factors – clinical efficacy, ocular and dermatologic safety, and visual comfort, the Center for Environmental Therapeutics (CET) recommends the following criteria for light box selection:

- Light boxes should have been tested successfully in peer-reviewed clinical trials.

- The box should provide 10,000 lux of illumination at a comfortable sitting distance. Product specifications are often missing or unverified; illuminance can be controlled using a light meter.

- Fluorescent lamps should have a smooth diffusing screen that filters out ultraviolet (UV) rays. UV rays are harmful to the eyes and skin.

- Blue light is known to be superior to red light in managing depressive symptoms which have a seasonal pattern.

- The light should be projected downward toward the eyes at an angle to minimize aversive visual glare.

- Smaller is not better; when using a compact light box, even small head movements will take the eyes out of the therapeutic range of the light.

Risks and Complications

Ultraviolet light causes progressive damage to human skin. This is mediated by genetic damage, collagen damage, as well as destruction of vitamin A and vitamin C in the skin and free radical generation. Ultraviolet light is also known to be a factor in formation of cataracts.

Researchers have questioned whether limiting blue light exposure could reduce the risk of age-related macular degeneration.

Modern phototherapy lamps used in the treatment of seasonal affective disorder and sleep disorders either filter out or do not emit ultraviolet light and are considered safe and effective for the intended purpose, as long as photosensitizing drugs are not being taken at the same time and in the absence of any existing eye conditions. Light therapy is a mood altering treatment, and just as with drug treatments, there is a possibility of triggering a manic state from a depressive state, causing anxiety and other side effects. While these side effects are usually controllable, it is recommended that patients undertake light therapy under the supervision of an experienced clinician, rather than attempting to self-medicate.

It is reported that bright light therapy may activate the production of reproductive hormones, such as testosterone, luteinizing hormone, follicle-stimulating hormone, and estradiol.

There are few absolute contraindications to light therapy, although there are some circumstances in which caution is required. These include when a patient has a condition that might render the eyes more vulnerable to phototoxicity, has a tendency toward mania, has a photosensitive skin condition, or is taking a photosensitizing herb (such as St. John's wort) or medication. Patients with porphyria should avoid most forms of light therapy. Patients on certain drugs such as methotrexate or chloroquine should use caution with light therapy as there is a chance that these drugs could cause porphyria.

Side Effects

Side effects of light therapy for sleep phase disorders include jumpiness or jitteriness, headache, eye irritation and nausea. Some non-depressive physical complaints, such as poor vision and skin rash or irritation, may improve with light therapy.

History

Many ancient cultures practiced various forms of heliotherapy, including people of Ancient Greece, Ancient Egypt, and Ancient Rome. The Inca, Assyrian and early German settlers also worshipped the sun as a health bringing deity. Indian medical literature dating to 1500 BCE describes a treatment combining herbs with natural sunlight to treat non-pigmented skin areas. Buddhist literature from about 200 CE and 10th-century Chinese documents make similar references.

The Faroese physician Niels Finsen is believed to be the father of modern phototherapy. He developed the first artificial light source for this purpose. Finsen used short wavelength light to treat lupus vulgaris, a skin infection caused by *Mycobacterium tuberculosis*. He thought that the beneficial effect was due to ultraviolet light killing the bacteria, but recent studies showed that his lens and filter system did not allow such short wavelengths to pass through, leading instead to the conclusion that light of approximately 400 nanometers generated reactive oxygen that would kill the bacteria. Finsen also used red light to treat smallpox lesions. He received the Nobel Prize in Physiology or Medicine in 1903. Scientific evidence for some of his treatments is lacking, and later eradication of smallpox and development of antibiotics for tuberculosis rendered light therapy obsolete for these diseases.

Since then a large array of treatments using controlled light have been developed. Though the popular consumer understanding of "light therapy" is associated with treating seasonal affective disorder, circadian rhythm disorders and skin conditions like psoriasis, other applications include the use of low level laser, red light, near-infrared and ultraviolet lights for pain management, hair growth, skin treatments, and accelerated wound healing.

Photodynamic Therapy

Photodynamic therapy (PDT), sometimes called photochemotherapy, is a form of phototherapy involving light and a photosensitizing chemical substance, used in conjunction with molecular oxygen to elicit cell death (phototoxicity). PDT has proven ability to kill microbial cells, including bacteria, fungi and viruses. PDT is popularly used in treating acne. It is used clinically to treat a wide range of medical conditions, including wet age-related macular degeneration, psoriasis, atherosclerosis and has shown some efficacy in anti-viral treatments, including herpes. It also treats malignant cancers including head and neck, lung, bladder and particular skin. The technology has also been tested for treatment of prostate cancer, both in a dog model and in prostate cancer patients.

It is recognised as a treatment strategy that is both minimally invasive and minimally toxic. Other light-based and laser therapies such as laser wound healing and rejuvenation, or intense pulsed light hair removal do not require a photosensitizer. Photosensitisers have been employed to sterilise blood plasma and water in order to remove blood-borne viruses and microbes and have been considered for agricultural uses, including herbicides and insecticides.

Photodynamic therapy's advantages lessen the need for delicate surgery and lengthy recuperation and minimal formation of scar tissue and disfigurement. A side effect is the associated photosensitisation of skin tissue.

Basics

PDT applications involve three components: a photosensitizer, a light source and tissue oxygen. The wavelength of the light source needs to be appropriate for exciting the photosensitizer to produce radicals and/or reactive oxygen species. These are free radicals (Type I) generated through electron abstraction or transfer from a substrate molecule and highly reactive state of oxygen known as singlet oxygen (Type II).

PDT is a multi-stage process. First a photosensitiser with negligible dark toxicity is administered, either systemically or topically, in the absence of light. When a sufficient amount of photosensitiser appears in diseased tissue, the photosensitiser is activated by exposure to light for a specified period. The light dose supplies sufficient energy to stimulate the photosensitiser, but not enough to damage neighbouring healthy tissue. The reactive oxygen kills the target cells.

Photochemical Processes

When a photosensitiser is in its excited state (3Psen*) it can interact with molecular triplet oxygen ($^3O_2$) and produce radicals and reactive oxygen species (ROS), crucial to the Type II mechanism. These species include singlet oxygen ($^1O_2$), hydroxyl radicals (•OH) and superoxide (O_2^-) ions. They can interact with cellular components including unsaturated lipids, amino acid residues and nucleic acids. If sufficient oxidative damage ensues, this will result in target-cell death (only within the illuminated area).

Photochemical Mechanisms

When a chromophore molecule, such as a cyclic tetrapyrrolic molecule, absorbs a photon, one of its electrons is promoted into a higher-energy orbital, elevating the chromophore from the ground state (*So*) into a short-lived, electronically excited state (*Sn*) composed of vibrational sub-levels (*Sn′*). The excited chromophore can lose energy by rapidly decaying through these sub-levels via internal conversion (IC) to populate the first excited singlet state (S1), before quickly relaxing back to the ground state.

The decay from the excited singlet state (S1) to the ground state (So) is via fluorescence (S1 → So). Singlet state lifetimes of excited fluorophores are very short (τfl. = 10−9–10−6 seconds) since transitions between the same spin states (S → S or T → T) conserve the spin multiplicity of the electron and, according to the Spin Selection Rules, are therefore considered "allowed" transitions. Alternatively, an excited singlet state electron (S1) can undergo spin inversion and populate the lower-energy first excited triplet state (T1) via intersystem crossing (ISC); a spin-forbidden process, since the spin of the electron is no longer conserved. The excited electron can then undergo a second spin-forbidden inversion and depopulate the excited triplet state (T1) by decaying to the ground state (So) via phosphorescence (T1→ So). Owing to the spin-forbidden triplet to singlet transition, the lifetime of phosphorescence (τP = 10−3 − 1 second) is considerably longer than that of fluorescence.

Photosensitisers and Photochemistry

Tetrapyrrolic photosensitisers in the excited singlet state (1Psen*, S>0) are relatively efficient at intersystem crossing and can consequently have a high triplet-state quantum yield. The longer lifetime of this species is sufficient to allow the excited triplet state photosensitiser to interact with surrounding bio-molecules, including cell membrane constituents.

Photochemical Reactions

Excited triplet-state photosensitisers can react via Type-I and Type-II processes. Type-I processes can involve the excited singlet or triplet photosensitiser (1Psen*, S1; 3Psen*, T1), however due to the short lifetime of the excited singlet state, the photosensitiser can only react if it is intimately associated with a substrate. In both cases the interaction is with readily oxidisable or reducible substrates. Type-II processes involve the direct interaction of the excited triplet photosensitiser ($3Psen^*$, T1) with molecular oxygen ($3O_2$, $3\Sigma g$).

Type-I Processes

Type-I processes can be divided into Type I(i) and Type I(ii). Type I (i) involves the transfer of an electron (oxidation) from a substrate molecule to the excited state photosensitiser ($Psen^*$), generating a photosensitiser radical anion ($Psen\bullet-$) and a substrate radical cation ($Subs\bullet+$). The majority of the radicals produced from Type-I(i) reactions react instantaneously with oxygen, generating a mixture of oxygen intermediates. For example, the photosensitiser radical anion can react instantaneously with molecular oxygen ($3O_2$) to generate a superoxide radical anion ($O_2\bullet-$), which can go on to produce the highly reactive hydroxyl radical ($OH\bullet$), initiating a cascade of cytotoxic free radicals; this process is common in the oxidative damage of fatty acids and other lipids.

The Type-I process (ii) involves the transfer of a hydrogen atom (reduction) to the excited state photosensitiser ($Psen^*$). This generates free radicals capable of rapidly reacting with molecular oxygen and creating a complex mixture of reactive oxygen intermediates, including reactive peroxides.

Type-II Processes

Type-II processes involve the direct interaction of the excited triplet state photosensitiser ($3Psen^*$) with ground state molecular oxygen ($3O_2$, $3\Sigma g$); a spin allowed transition—the excited state photosensitiser and ground state molecular oxygen are of the same spin state (T).

When the excited photosensitiser collides with molecular oxygen, a process of triplet-triplet annihilation takes place ($3Psen^* \rightarrow 1Psen$ and $3O_2 \rightarrow 1O_2$). This inverts the spin of one oxygen molecule's ($3O_2$) outermost antibonding electrons, generating two forms of singlet oxygen ($1\Delta g$ and $1\Sigma g$), while simultaneously depopulating the photosensitiser's excited triplet state (T1 $\rightarrow$ S0). The higher-energy singlet oxygen state ($1\Sigma g$, $157kJ\ mol-1 > 3\Sigma g$) is very short-lived ($1\Sigma g \leq 0.33$ milliseconds (methanol), undetectable in H_2O/D_2O) and rapidly relaxes to the lower-energy excited state ($1\Delta g$, $94kJ\ mol-1 > 3\Sigma g$). It is, therefore, this lower-energy form of singlet oxygen ($1\Delta g$) that is implicated in cell injury and cell death.

The highly-reactive oxygen species ($1O_2$) produced via the Type-II process act near to their site generation and within a radius of approximately 20 nm, with a typical lifetime of approximately 40 nanoseconds in biological systems.

It is possible that (over a 6 µs period) singlet oxygen can diffuse up to approximately 300 nm *in vivo*. Singlet oxygen can theoretically only interact with proximal molecules and structures within this radius. ROS initiate reactions with many biomolecules, including amino acid residues in proteins, such as tryptophan; unsaturated lipids like cholesterol and nucleic acid bases, particularly guanosine and guanine derivatives, with the latter base more susceptible to ROS. These interactions cause damage and potential destruction to cellular membranes and enzyme deactivation, culminating in cell death.

It is probable that in the presence of molecular oxygen and as a direct result of the photoirradiation of the photosensitiser molecule, both Type-I and II pathways play a pivotal role in disrupting cellular mechanisms and cellular structure. Nevertheless, considerable evidence suggests that the Type-II photo-oxygenation process predominates in the induction of cell damage, a consequence of the interaction between the irradiated photosensitiser and molecular oxygen. Cells *in vivo* may be partially protected against the effects of photodynamic therapy by the presence of singlet oxygen scavengers (such as histidine). Certain skin cells are somewhat resistant to PDT in the absence of molecular oxygen; further supporting the proposal that the Type-II process is at the heart of photoinitiated cell death.

The efficiency of Type-II processes is dependent upon the triplet state lifetime τT and the triplet quantum yield (ΦT) of the photosensitiser. Both of these parameters have been implicated in phototherapeutic effectiveness; further supporting the distinction between Type-I and Type-II mechanisms. However, the success of a photosensitiser is not exclusively dependent upon a Type-II process. Multiple photosensitisers display excited triplet lifetimes that are too short to permit a Type-II process to occur. For example, the copper metallated octaethylbenzochlorin photosensitiser has a triplet state lifetime of less than 20 nanoseconds and is still deemed to be an efficient photodynamic agent.

Reactive Oxygen Species

In air and tissue, molecular oxygen occurs in a triplet state, whereas almost all other molecules are in a singlet state. Reactions between triplet and singlet molecules are forbidden by quantum mechanics, making oxygen relatively non-reactive at physiological conditions. A photosensitizer is a chemical compound that can be promoted to an excited state upon absorption of light and undergo intersystem crossing (ISC) with oxygen to produce singlet oxygen. This species is highly cytotoxic, rapidly attacking any organic compounds it encounters. It is rapidly eliminated from cells, in an average of 3 μs.

Photosensitisers

Many photosensitizers for PDT exist. They divide into porphyrins, chlorophylls and dyes. Examples include aminolevulinic acid (ALA), Silicon Phthalocyanine Pc 4, m-tetrahydroxyphenylchlorin (mTHPC) and mono-L-aspartyl chlorin e6 (NPe6).

Photosensitizers commercially available for clinical use include Allumera, Photofrin, Visudyne, Levulan, Foscan, Metvix, Hexvix, Cysview and Laserphyrin, with others in development, e.g. Antrin, Photochlor, Photosens, Photrex, Lumacan, Cevira, Visonac, BF-200 ALA, Amphinex and Azadipyrromethenes.

The major difference between photosensitizers is the parts of the cell that they target. Unlike in radiation therapy, where damage is done by targeting cell DNA, most photosensitizers target other cell structures. For example, mTHPC localizes in the nuclear envelope. In contrast, ALA localizes in the mitochondria and methylene blue in the lysosomes.

Cyclic Tetrapyrrolic Chromophores

Cyclic tetrapyrrolic molecules are fluorophores and photosensitisers. Cyclic tetrapyrrolic derivatives have an inherent similarity to the naturally occurring porphyrins present in living matter—consequently they have little or no toxicity in the absence of light.

Porphyrins

Porphyrins are a group of naturally occurring and intensely coloured compounds, whose name is drawn from the Greek word *porphura*, or purple. These molecules perform biologically important roles, including oxygen transport and photosynthesis and have applications in fields ranging from fluorescent imaging to medicine. Porphyrins are tetrapyrrolic molecules, with the heart of the skeleton a heterocyclic macrocycle, known as a porphine. The fundamental porphine frame consists of four pyrrolic sub-units linked on opposing sides (*a*-positions, numbered 1, 4, 6, 9, 11, 14, 16 and 19) through four methine (CH) bridges (5, 10, 15 and 20), known as the *meso*-carbon atoms/positions. The resulting conjugated planar macrocycle may be substituted at the *meso*- and/or *β*-positions (2, 3, 7, 8, 12, 13, 17 and 18): if the *meso*- and *β*-hydrogens are substituted with non-hydrogen atoms or groups, the resulting compounds are known as porphyrins.

The inner two protons of a free-base porphyrin can be removed by strong bases such as alkoxides, forming a dianionic molecule; conversely, the inner two pyrrolenine nitrogens can be protonated with acids such as trifluoroacetic acid affording a dicationic intermediate. The tetradentate anionic species can readily form complexes with most metals.

Absorption Spectroscopy

Porphyrin's highly conjugated skeleton produces a characteristic ultra-violet visible (UV-VIS) spectrum. The spectrum typically consists of an intense, narrow absorption band ($\varepsilon > 200000$ l mol−1 cm−1) at around 400 nm, known as the Soret or B band, followed by four longer wavelength (450–700 nm), weaker absorptions ($\varepsilon > 20000$ l mol−1 cm−1 (free-base porphyrins)) referred to as the Q bands.

The Soret band arises from a strong electronic transition from the ground state to the second excited singlet state ($S_0 \rightarrow S_2$); whereas the Q band is a result of a weak transition to the first excited singlet state ($S_0 \rightarrow S_1$). The dissipation of energy via internal conversion (IC) is so rapid that fluorescence is only observed from depopulation of the first excited singlet state to the lower-energy ground state ($S_1 \rightarrow S_0$).

Ideal Photosensitisers

The key characteristic of a photosensitiser is the ability to preferentially accumulate in diseased tissue and induce a desired biological effect via the generation of cytotoxic species. Specific criteria:

- Strong absorption with a high extinction coefficient in the red/near infrared region of the electromagnetic spectrum (600–850 nm)—allows deeper tissue penetration. (Tissue is much more transparent at longer wavelengths (~700–850 nm). Longer wavelengths allow the light to penetrate deeper and treat larger structures.)

- Suitable photophysical characteristics: a high-quantum yield of triplet formation ($\Phi_T \geq$ 0.5); a high singlet oxygen quantum yield ($\Phi_\Delta \geq$ 0.5); a relatively long triplet state lifetime (τ_T, μs range); and a high triplet-state energy ($\geq$ 94 kJ mol−1). Values of Φ_T= 0.83 and Φ_Δ

= 0.65 (haematoporphyrin); ΦT = 0.83 and $\Phi\Delta$ = 0.72 (etiopurpurin); and ΦT = 0.96 and $\Phi\Delta$ = 0.82 (tin etiopurpurin) have been achieved

- Low dark toxicity and negligible cytotoxicity in the absence of light. (The photosensitizer should not be harmful to the target tissue until the treatment beam is applied.)

- Preferential accumulation in diseased/target tissue over healthy tissue

- Rapid clearance from the body post-procedure

- High chemical stability: single, well-characterised compounds, with a known and constant composition

- Short and high-yielding synthetic route (with easy translation into multi-gram scales/reactions)

- Simple and stable formulation

- Soluble in biological media, allowing intravenous administration. Otherwise, a hydrophilic delivery system must enable efficient and effective transportation of the photosensitiser to the target site via the bloodstream.

- Low photobleaching to prevent degradation of the photosensitizer so it can continue producing singlet oxygen

- Natural fluorescence (Many optical dosimetry techniques, such as fluorescence spectroscopy, depend on fluorescence.)

First Generation

While the disadvantages associated with first generation photosensitisers HpD and Photofrin (skin sensitivity and weak absorption at 630 nm) permitted some therapeutic use, they markedly reduced application to a wider field of disease. Second generation photosensitisers were key to the development of photodynamic therapy.

Second Generation

5-Aminolaevulinic Acid

5-Aminolaevulinic acid (ALA) is a prodrug used to treat and image multiple superficial cancers and tumours. ALA a key precursor in the biosynthesis of the naturally occurring porphyrin, haem.

Haem is synthesised in every energy-producing cell in the body and is a key structural component of haemoglobin, myoglobin and other haemproteins. The immediate precursor to haem is protoporphyrin IX (PPIX), an effective photosensitiser. Haem itself is not a photosensitiser, due to the coordination of a paramagnetic ion in the centre of the macrocycle, causing significant reduction in excited state lifetimes.

The haem molecule is synthesised from glycine and succinyl coenzyme A (succinyl CoA). The rate-limiting step in the biosynthesis pathway is controlled by a tight (negative) feedback mechanism in which the concentration of haem regulates the production of ALA. However, this controlled feedback can be by-passed by artificially adding excess exogenous ALA to cells. The cells respond by producing PPIX (photosensitiser) at a faster rate than the ferrochelatase enzyme can convert it to haem.

ALA, marketed as Levulan, has shown promise in photodynamic therapy (tumours) via both intravenous and oral administration, as well as through topical administration in the treatment of malignant and non-malignant dermatological conditions, including psoriasis, Bowen's disease and Hirsutism (Phase II/III clinical trials).

ALA accumulates more rapidly in comparison to other intravenously administered sensitisers. Typical peak tumour accumulation levels post-administration for PPIX are usually achieved within several hours; other (intravenous) photosensitisers may take up to 96 hours to reach peak levels. ALA is also excreted more rapidly from the body (~24 hours) than other photosensitisers, minimising photosensitivity side effects.

Esterified ALA derivatives with improved bioavailability have been examined. A methyl ALA ester (Metvix) is now available for basal cell carcinoma and other skin lesions. Benzyl (Benvix) and hexyl ester (Hexvix) derivatives are used for gastrointestinal cancers and for the diagnosis of bladder cancer.

Verteporfin

Benzoporphyrin derivative monoacid ring A (BPD-MA) marketed as Visudyne (Verteporfin, for injection) has been approved by health authorities in multiple jurisdictions, including US FDA, for the treatment of wet AMD beginning in 1999. It has also undergone Phase III clinical trials (USA) for the treatment of cutaneous non-melanoma skin cancer.

The chromophore of BPD-MA has a red-shifted and intensified long-wavelength absorption maxima at approximately 690 nm. Tissue penetration by light at this wavelength is 50% greater than that achieved for Photofrin (λmax. = 630 nm).

Verteporfin has further advantages over the first generation sensitiser Photofrin. It is rapidly absorbed by the tumour (optimal tumour-normal tissue ratio 30–150 minutes post-intravenous injection) and is rapidly cleared from the body, minimising patient photosensitivity (1–2 days).

Purlytin

Chlorin photosensitiser tin etiopurpurin is marketed as Purlytin. Purlytin has undergone Phase II clinical trials for cutaneous metastatic breast cancer and Kaposi's sarcoma in patients with AIDS (acquired immunodeficiency syndrome). Purlytin has been used successfully to treat the non-malignant conditions psoriasis and restenosis.

Chlorins are distinguished from the parent porphyrins by a reduced exocyclic double bond, decreasing the symmetry of the conjugated macrocycle. This leads to increased absorption in the long-wavelength portion of the visible region of the electromagnetic spectrum (650–680 nm). Purlytin is a purpurin; a degradation product of chlorophyll.

Purlytin has a tin atom chelated in its central cavity that causes a red-shift of approximately 20–30 nm (with respect to Photofrin and non-metallated etiopurpurin, λmax.SnEt2 = 650 nm). Purlytin has been reported to localise in skin and produce a photoreaction 7–14 days post-administration.

Foscan

Tetra(m-hydroxyphenyl)chlorin (mTHPC) is in clinical trials for head and neck cancers under the trade name Foscan. It has also been investigated in clinical trials for gastric and pancreatic cancers, hyperplasia, field sterilisation after cancer surgery and for the control of antibiotic-resistant bacteria.

Foscan has a singlet oxygen quantum yield comparable to other chlorin photosensitisers but lower drug and light doses (approximately 100 times more photoactive than Photofrin).

Foscan can render patients photosensitive for up to 20 days after initial illumination.

Lutex

Lutetium texaphyrin, marketed under the trade name Lutex and Lutrin, is a large porphyrin molecule. Texaphyrins are expanded porphyrins that have a penta-aza core. It offers strong absorption in the 730–770 nm region. Tissue transparency is optimal in this range. As a result, Lutex-based PDT can (potentially) be carried out more effectively at greater depths and on larger tumours.

Lutex has entered Phase II clinical trials for evaluation against breast cancer and malignant melanomas.

A Lutex derivative, Antrin, has undergone Phase I clinical trials for the prevention of restenosis of vessels after cardiac angioplasty by photoinactivating foam cells that accumulate within arteriolar plaques. A second Lutex derivative, Optrin, is in Phase I trials for AMD.

Texaphyrins also have potential as radiosensitisers (Xcytrin) and chemosensitisers. Xcytrin, a gadolinium texaphyrin (motexafin gadolinium), has been evaluated in Phase III clinical trials against brain metastases and Phase I clinical trials for primary brain tumours.

ATMPn

9-Acetoxy-2,7,12,17-tetrakis-(β-methoxyethyl)-porphycene has been evaluated as an agent for dermatological applications against psoriasis vulgaris and superficial non-melanoma skin cancer.

Zinc Phthalocyanine

A liposomal formulation of zinc phthalocyanine (CGP55847) has undergone clinical trials (Phase I/II, Switzerland) against squamous cell carcinomas of the upper aerodigestive tract. Phthalocyanines (PCs) are related to tetra-aza porphyrins. Instead of four bridging carbon atoms at the meso-positions, as for the porphyrins, PCs have four nitrogen atoms linking the pyrrolic sub-units. PCs also have an extended conjugate pathway: a benzene ring is fused to the β-positions of each of the four-pyrrolic sub-units. These rings strengthen the absorption of the chromophore at longer wavelengths (with respect to porphyrins). The absorption band of PCs is almost two orders of magnitude stronger than the highest Q band of haematoporphyrin. These favourable characteristics, along with the ability to selectively functionalise their peripheral structure, make PCs favourable photosensitiser candidates.

A sulphonated aluminium PC derivative (Photosense) has entered clinical trials (Russia) against skin, breast and lung malignancies and cancer of the gastrointestinal tract. Sulphonation significantly increases PC solubility in polar solvents including water, circumventing the need for alternative delivery vehicles.

PC4 is a silicon complex under investigation for the sterilisation of blood components against human colon, breast and ovarian cancers and against glioma.

A shortcoming of many of the metallo-PCs is their tendency to aggregate in aqueous buffer (pH 7.4), resulting in a decrease, or total loss, of their photochemical activity. This behaviour can be minimised in the presence of detergents.

Metallated cationic porphyrazines (PZ), including PdPZ+, CuPZ+, CdPZ+, MgPZ+, AlPZ+ and GaPZ+, have been tested *in vitro* on V-79 (Chinese hamster lung fibroblast) cells. These photosensitisers display substantial dark toxicity.

Naphthalocyanines

Naphthalocyanines (NCs) are an extended PC derivative. They have an additional benzene ring attached to each isoindole sub-unit on the periphery of the PC structure. Subsequently, NCs absorb strongly at even longer wavelengths (approximately 740–780 nm) than PCs (670–780 nm). This absorption in the near infrared region makes NCs candidates for highly pigmented tumours, including melanomas, which present significant absorption problems for visible light.

However, problems associated with NC photosensitisers include lower stability, as they decompose in the presence of light and oxygen. Metallo-NCs, which lack axial ligands, have a tendency to form H-aggregates in solution. These aggregates are photoinactive, thus compromising the photodynamic efficacy of NCs.

Silicon naphthalocyanine attached to copolymer PEG-PCL (poly(ethyleneglycol)-*block*-poly(ε-caprolactone)) accumulates selectively in cancer cells and reaches a maximum concentration after about one day. The compound provides real time near-infrared (NIR) fluorescence imaging with an extinction coefficient of 2.8×10^5 M^{-1} cm^{-1} and combinatorial phototherapy with dual photothermal and photodynamic therapeutic mechanisms that may be appropriate for adriamycin-resistant tumors. The particles had a hydrodynamic size of 37.66 ± 0.26 nm (polydispersity index = 0.06) and surface charge of -2.76 ± 1.83 mV.

Functional Groups

Altering the peripheral functionality of porphyrin-type chromophores can affect photodynamic activity.

Diamino platinum porphyrins show high anti-tumour activity, demonstrating the combined effect of the cytotoxicity of the platinum complex and the photodynamic activity of the porphyrin species.

Positively charged PC derivatives have been investigated. Cationic species are believed to selectively localise in the mitochondria.

Zinc and copper cationic derivatives have been investigated. The positively charged zinc complexed PC is less photodynamically active than its neutral counterpart *in vitro* against V-79 cells.

Water-soluble cationic porphyrins bearing nitrophenyl, aminophenyl, hydroxyphenyl and/or pyridiniumyl functional groups exhibit varying cytotoxicity to cancer cells *in vitro*, depending on the nature of the metal ion (Mn, Fe, Zn, Ni) and on the number and type of functional groups. The manganese pyridiniumyl derivative has shown the highest photodynamic activity, while the nickel analogue is photoinactive.

Another metallo-porphyrin complex, the iron chelate, is more photoactive (towards HIV and simian immunodeficiency virus in MT-4 cells) than the manganese complexes; the zinc derivative is photoinactive.

The hydrophilic sulphonated porphyrins and PCs (AlPorphyrin and AlPC) compounds were tested for photodynamic activity. The disulphonated analogues (with adjacent substituted sulphonated groups) exhibited greater photodynamic activity than their di-(symmetrical), mono-, tri- and tetra-sulphonated counterparts; tumour activity increased with increasing degree of sulphonation.

Third Generation

Many photosensitisers are poorly soluble in aqueous media, particularly at physiological pH, limiting their use.

Alternate delivery strategies range from the use of oil-in-water (o/w) emulsions to carrier vehicles such as liposomes and nanoparticles. Although these systems may increase therapeutic effects, the carrier system may inadvertently decrease the "observed" singlet oxygen quantum yield ($\Phi\Delta$): the singlet oxygen generated by the photosensitiser must diffuse out of the carrier system; and since singlet oxygen is believed to have a narrow radius of action, it may not reach the target cells. The carrier may limit light absorption, reducing singlet oxygen yield.

Another alternative that does not display the scattering problem is the use of moieties. Strategies include directly attaching photosensitisers to biologically active molecules such as antibodies.

Metallation

Various metals form into complexes with photosensitiser macrocycles. Multiple second generation photosensitisers contain a chelated central metal ion. The main candidates are transition metals, although photosensitisers co-ordinated to group 13 (Al, AlPcS4) and group 14 (Si, SiNC and Sn, SnEt2) metals have been synthesised.

The metal ion does not confer definite photoactivity on the complex. Copper (II), cobalt (II), iron (II) and zinc (II) complexes of Hp are all photoinactive in contrast to metal-free porphyrins. However, texaphyrin and PC photosensitisers do not contain metals; only the metallo-complexes have demonstrated efficient photosensitisation.

The central metal ion, bound by a number of photosensitisers, strongly influences the photophysical properties of the photosensitiser. Chelation of paramagnetic metals to a PC chromophore appears to shorten triplet lifetimes (down to nanosecond range), generating variations in the triplet quantum yield and triplet lifetime of the photoexcited triplet state.

Certain heavy metals are known to enhance inter-system crossing (ISC). Generally, diamagnetic metals promote ISC and have a long triplet lifetime. In contrast, paramagnetic species deactivate excited states, reducing the excited-state lifetime and preventing photochemical reactions. However, exceptions to this generalisation include copper octaethylbenzochlorin.

Many metallated paramagnetic texaphyrin species exhibit triplet-state lifetimes in the nanosecond range. These results are mirrored by metallated PCs. PCs metallated with diamagnetic ions, such as Zn^{2+}, Al^{3+} and Ga^{3+}, generally yield photosensitisers with desirable quantum yields and lifetimes (ΦT 0.56, 0.50 and 0.34 and τT 187, 126 and 35 μs, respectively). Photosensitiser $ZnPcS_4$ has a singlet oxygen quantum yield of 0.70; nearly twice that of most other mPCs ($\Phi\Delta$ at least 0.40).

Expanded Metallo-porphyrins

Expanded porphyrins have a larger central binding cavity, increasing the range of potential metals.

Diamagnetic metallo-texaphyrins have shown photophysical properties; high triplet quantum yields and efficient generation of singlet oxygen. In particular, the zinc and cadmium derivatives display triplet quantum yields close to unity. In contrast, the paramagnetic metallo-texaphyrins, Mn-Tex, Sm-Tex and Eu-Tex, have undetectable triplet quantum yields. This behaviour is parallel with that observed for the corresponding metallo-porphyrins.

The cadmium-texaphyrin derivative has shown *in vitro* photodynamic activity against human leukemia cells and Gram positive (*Staphylococcus*) and Gram negative (*Escherichia coli*) bacteria. Although follow-up studies have been limited with this photosensitiser due to the toxicity of the complexed cadmium ion.

A zinc-metallated *seco*-porphyrazine has a high quantum singlet oxygen yield ($\Phi\Delta$ 0.74). This expanded porphyrin-like photosensitiser has shown the best singlet oxygen photosensitising ability of any of the reported *seco*-porphyrazines. Platinum and palladium derivatives have been synthesised with singlet oxygen quantum yields of 0.59 and 0.54, respectively.

Metallochlorins/Bacteriochlorins

The tin (IV) purpurins are more active when compared with analogous zinc (II) purpurins, against human cancers.

Sulphonated benzochlorin derivatives demonstrated a reduced phototherapeutic response against murine leukemia L1210 cells *in vitro* and transplanted urothelial cell carcinoma in rats, whereas the tin (IV) metallated benzochlorins exhibited an increased photodynamic effect in the same tumour model.

Copper octaethylbenzochlorin demonstrated greater photoactivity towards leukemia cells *in vitro* and a rat bladder tumour model. It may derive from interactions between the cationic iminium group and biomolecules. Such interactions may allow electron-transfer reactions to take place via the short-lived excited singlet state and lead to the formation of radicals and radical ions. The copper-free derivative exhibited a tumour response with short intervals between drug administration and photodynamic activity. Increased *in vivo* activity was observed with the zinc benzochlorin analogue.

Metallo-phthalocyanines

PCs properties are strongly influenced by the central metal ion. Co-ordination of transition metal ions gives metallo-complexes with short triplet lifetimes (nanosecond range), resulting in different triplet quantum yields and lifetimes (with respect to the non-metallated analogues). Diamagnetic metals such as zinc, aluminium and gallium, generate metallo-phthalocyanines (MPC) with high triplet quantum yields ($\Phi T \geq 0.4$) and short lifetimes (ZnPCS4 $\tau T = 490$ Fs and AlPcS4 $\tau T = 400$ Fs) and high singlet oxygen quantum yields ($\Phi \Delta \geq 0.7$). As a result, ZnPc and AlPc have been evaluated as second generation photosensitisers active against certain tumours.

Metallo-naphthocyaninesulfobenzo-Porphyrazines (M-NSBP)

Aluminium has been successfully coordinated to M-NSBP. The resulting complex showed photodynamic activity against EMT-6 tumour-bearing Balb/c mice (disulphonated analogue demonstrated greater photoactivity than the mono-derivative).

Metallo-naphthalocyanines

Work with zinc NC with various amido substituents revealed that the best phototherapeutic response (Lewis lung carcinoma in mice) with a tetrabenzamido analogue. Complexes of silicon (IV) NCs with two axial ligands in anticipation the ligands minimise aggregation. Disubstituted analogues as potential photodynamic agents (a siloxane NC substituted with two methoxyethyleneglycol ligands) are an efficient photosensitiser against Lewis lung carcinoma in mice. SiNC[OSi(i-Bu)2-n-C18H37]2 is effective against Balb/c mice MS-2 fibrosarcoma cells.Siloxane NCs may be efficacious photosensitisers against EMT-6 tumours in Balb/c mice. The ability of metallo-NC derivatives (AlNc) to generate singlet oxygen is weaker than the analogous (sulphonated) metallo-PCs (AlPC); reportedly 1.6–3 orders of magnitude less.

In porphyrin systems, the zinc ion appears to hinder the photodynamic activity of the compound. By contrast, in the higher/expanded π-systems, zinc-chelated dyes form complexes with good to high results.

An extensive study of metallated texaphyrins focused on the lanthanide (III) metal ions, Y, In, Lu, Cd, Nd, Sm, Eu, Gd, Tb, Dy, Ho, Er, Tm and Yb found that when diamagnetic Lu (III) was complexed to texaphyrin, an effective photosensitiser (Lutex) was generated. However, using the paramagnetic Gd (III) ion for the Lu metal, exhibited no photodynamic activity. The study found a correlation between the excited-singlet and triplet state lifetimes and the rate of ISC of the diamagnetic texaphyrin complexes, Y(III), In (III) and Lu (III) and the atomic number of the cation.

Paramagnetic metallo-texaphyrins displayed rapid ISC. Triplet lifetimes were strongly affected by the choice of metal ion. The diamagnetic ions (Y, In and Lu) displayed triplet lifetimes ranging from 187, 126 and 35 μs, respectively. Comparable lifetimes for the paramagnetic species (Eu-Tex

6.98 μs, Gd-Tex 1.11, Tb-Tex < 0.2, Dy-Tex $0.44 \times 10-3$, Ho-Tex $0.85 \times 10-3$, Er-Tex $0.76 \times 10-3$, Tm-Tex $0.12 \times 10-3$ and Yb-Tex 0.46) were obtained.

Three measured paramagnetic complexes measured significantly lower than the diamagnetic metallo-texaphyrins.

In general, singlet oxygen quantum yields closely followed the triplet quantum yields.

Various diamagnetic and paramagnetic texaphyrins investigated have independent photophysical behaviour with respect to a complex's magnetism. The diamagnetic complexes were characterised by relatively high fluorescence quantum yields, excited-singlet and triplet lifetimes and singlet oxygen quantum yields; in distinct contrast to the paramagnetic species.

The +2 charged diamagnetic species appeared to exhibit a direct relationship between their fluorescence quantum yields, excited state lifetimes, rate of ISC and the atomic number of the metal ion. The greatest diamagnetic ISC rate was observed for Lu-Tex; a result ascribed to the heavy atom effect. The heavy atom effect also held for the Y-Tex, In-Tex and Lu-Tex triplet quantum yields and lifetimes. The triplet quantum yields and lifetimes both decreased with increasing atomic number. The singlet oxygen quantum yield correlated with this observation.

Photophysical properties displayed by paramagnetic species were more complex. The observed data/behaviour was not correlated with the number of unpaired electrons located on the metal ion. For example:

- ISC rates and the fluorescence lifetimes gradually decreased with increasing atomic number.

- Gd-Tex and Tb-Tex chromophores showed (despite more unpaired electrons) slower rates of ISC and longer lifetimes than Ho-Tex or Dy-Tex.

To achieve selective target cell destruction, while protecting normal tissues, either the photosensitizer can be applied locally to the target area, or targets can be locally illuminated. Skin conditions, including acne, psoriasis and also skin cancers, can be treated topically and locally illuminated. For internal tissues and cancers, intravenously administered photosensitizers can be illuminated using endoscopes and fiber optic catheters.

Photosensitizers can target viral and microbial species, including HIV and MRSA. Using PDT, pathogens present in samples of blood and bone marrow can be decontaminated before the samples are used further for transfusions or transplants. PDT can also eradicate a wide variety of pathogens of the skin and of the oral cavities. Given the seriousness that drug resistant pathogens have now become, there is increasing research into PDT as a new antimicrobial therapy.

Applications

Vascular Targeting

Some photosensitisers naturally accumulate in the endothelial cells of vascular tissue allowing 'vascular targeted' PDT.

Verteporfin was shown to target the neovasculature resulting from macular degeneration in the macula within the first thirty minutes after intravenous administration of the drug.

Compared to normal tissues, most types of cancers are especially active in both the uptake and accumulation of photosensitizers agents, which makes cancers especially vulnerable to PDT. Since photosensitizers can also have a high affinity for vascular endothelial cells.

Acne

PDT is currently in clinical trials as a treatment for severe acne. Initial results have shown for it to be effective as a treatment only for severe acne. The treatment causes severe redness and moderate to severe pain and burning sensation. One phase II trial, while it showed improvement, was not superior to blue/violet light alone.

Ophthalmology

As cited above, verteporfin was widely approved for the treatment of wet AMD beginning in 1999. The drug targets the neovasculature that is caused by the condition.

History

Modern Era

In the late nineteenth century. Finsen successfully demonstrated phototherapy by employing heat-filtered light from a carbon-arc lamp (the "Finsen lamp") in the treatment of a tubercular condition of the skin known as *lupus vulgaris*, for which he won the 1903 Nobel Prize in Physiology or Medicine.

In 1913 another German scientist, Meyer-Betz, described the major stumbling block of photodynamic therapy. After injecting himself with haematoporphyrin (Hp, a photosensitiser), he swiftly experienced a general skin sensitivity upon exposure to sunlight—a recurrent problem with many photosensitisers.

The first evidence that agents, photosensitive synthetic dyes, in combination with a light source and oxygen could have potential therapeutic effect was made at the turn of the 20th century in the laboratory of Hermann von Tappeiner in Munich, Germany. Germany was leading the world in industrial dye synthesis at the time.

While studying the effects of acridine on paramecia cultures, Oscar Raab, a student of von Tappeiner observed a toxic effect. Fortuitously Raab also observed that light was required to kill the paramecia. Subsequent work in von Tappeiner's laboratory showed that oxygen was essential for the 'photodynamic action' – a term coined by von Tappeiner.

Von Tappeiner and colleagues performed the first PDT trial in patients with skin carcinoma using the photosensitizer, eosin. Of 6 patients with a facial basal cell carcinoma, treated with a 1% eosin solution and long-term exposure either to sunlight or arc-lamp light, 4 patients showed total tumour resolution and a relapse-free period of 12 months.

In 1924 Policard revealed the diagnostic capabilities of hematoporphyrin fluorescence when he observed that ultraviolet radiation excited red fluorescence in the sarcomas of laboratory rats. Policard hypothesized that the fluorescence was associated with endogenous hematoporphyrin accumulation.

In 1948 Figge and co-workers showed on laboratory animals that porphyrins exhibit a preferential affinity to rapidly dividing cells, including malignant, embryonic and regenerative cells. They proposed that porphyrins could be used to treat cancer.

Photosensitizer Haematoporphyrin Derivative (HpD), was first characterised in 1960 by Lipson. Lipson sought a diagnostic agent suitable for tumor detection. HpD allowed Lipson to pioneer the use of endoscopes and HpD fluorescence. HpD is a porphyrin species derived from haematoporphyrin, Porphyrins have long been considered as suitable agents for tumour photodiagnosis and tumour PDT because cancerous cells exhibit significantly greater uptake and affinity for porphyrins compared to normal tissues. This had been observed by other researchers prior to Lipson.

Thomas Dougherty and co-workers at Roswell Park Cancer Institute, Buffalo NY, clinically tested PDT in 1978. They treated 113 cutaneous or subcutaneous malignant tumors with HpD and observed total or partial resolution of 111 tumors. Dougherty helped expand clinical trials and formed the International Photodynamic Association, in 1986.

John Toth, product manager for Cooper Medical Devices Corp/Cooper Lasersonics, noticed the "photodynamic chemical effect" of the therapy and wrote the first white paper naming the therapy "Photodynamic Therapy" (PDT) with early clinical argon dye lasers circa 1981. The company set up 10 clinical sites in Japan where the term "radiation" had negative connotations.

HpD, under the brand name Photofrin, was the first PDT agent approved for clinical use in 1993 to treat a form of bladder cancer in Canada. Over the next decade, both PDT and the use of HpD received international attention and greater clinical acceptance and led to the first PDT treatments approved by U.S. Food and Drug Administration Japa and parts of Europe for use against certain cancers of the oesophagus and non-small cell lung cancer.

 Photofrin had the disadvantages of prolonged patient photosensitivity and a weak long-wavelength absorption (630 nm). This led to the development of second generation photosensitisers, including Verteporfin (a benzoporphyrin derivative, also known as Visudyne) and more recently, third generation targetable photosensitisers, such as antibody-directed photosensitisers.

In the 1980s, David Dolphin, Julia Levy and colleagues developed a novel photosensitizer, verteporfin. Verteporfin, a porphyrin derivative, is activated at 690 nm, a much longer wavelength than Photofrin. It has the property of preferential uptake by neovasculature. It has been widely tested for its use in treating skin cancers and received FDA approval in 2000 for the treatment of wet age related macular degeneration. As such it was the first medical treatment ever approved for this condition, which is a major cause of vision loss.

Russia

Russia was the quickest to advance PDT use clinically and made many advances. They pioneered a photosensitizer called Photogem which, like HpD, was derived from haematoporphyrin in 1990 by Mironov and coworkers. Photogem was approved by the Ministry of Health of Russia and tested clinically from February 1992 to 1996. A pronounced therapeutic effect was observed in 91 percent of the 1500 patients. 62 percent had total tumor resolution. A further 29 percent had >50% tumor shrinkage. In early diagnosis patients 92 percent experienced complete resolution.

Russian scientists collaborated with NASA scientists who were looking at the use of LEDs as more suitable light sources, compared to lasers, for PDT applications.

Asia

Since 1990, the Chinese have been developing clinical expertise with PDT, using domestically produced photosensitizers, derived from Haematoporphyrin. China is notable for its expertise in resolving difficult to reach tumours.

Poland

In the 1990s the team of Polish professor Aleksander Sieroń developed the process and devices for PDT treatment, while improving photosensitizer compounds at the University of Bytom in Poland.

Miscellany

PUVA therapy uses psoralen as photosensitiser and UVA ultraviolet as light source, but this form of therapy is usually classified as a separate form of therapy from photodynamic therapy.

To allow treatment of deeper tumours some researchers are using internal chemiluminescence to activate the photosensitiser.

Photothermal Therapy

Photothermal therapy (PTT) refers to efforts to use electromagnetic radiation (most often in infrared wavelengths) for the treatment of various medical conditions, including cancer. This approach is an extension of photodynamic therapy, in which a photosensitizer is excited with specific band light. This activation brings the sensitizer to an excited state where it then releases vibrational energy (heat), which is what kills the targeted cells.

Unlike photodynamic therapy, photothermal therapy does not require oxygen to interact with the target cells or tissues. Current studies also show that photothermal therapy is able to use longer wavelength light, which is less energetic and therefore less harmful to other cells and tissues.

Nanoscale Materials

Most materials of interest currently being investigated for photothermal therapy are on the nanoscale. One of the key reasons behind this is the enhanced permeability and retention effect observed with particles in a certain size range (typically 20 - 300 nm). Molecules in this range have been observed to preferentially accumulate in tumor tissue. When a tumor forms, it requires new blood vessels in order to fuel its growth; these new blood vessels in/near tumors have different properties as compared to regular blood vessels, such as poor lymphatic drainage and a disorganized, leaky vasculature. These factors lead to a significantly higher concentration of certain particles in a tumor as compared to the rest of the body. Coupling this phenomenon with active targeting modalities (e.g., antibodies) has recently been investigated by researchers.

Recent Studies

Gold Nanoparticles

One of the most promising directions in photothermal therapy is the use of gold nanoparticles. Initial efforts with gold nanoparticles, however, were not very effective in vivo because the spherical particles used had peak absorptions limited to 520 to 580 nm for particles ranging from 10 to 100 nm in diameter, respectively. These wavelengths were not effective in vivo because skin, tissues, and hemoglobin have a transmission window from 650 to 900 nm, with peak transmission at approximately 800 nm (known as the Near-Infrared Window). Development of gold nanorods was one solution for the disparity between the wavelengths required to excite spherical gold nanoparticles and the in vivo transmission window. The peak absorption of gold nanorods may be tuned from 550 nm up to 1 micrometre by altering its aspect ratio. Once tuned, the toxic byproducts of CTAB can be removed with non-cytotoxic polyethylene glycol (PEG). PEG not only keeps the nanorods from aggregating in serum once injected, but also increases bloodstream circulation times (leading to better adsorption of nanorods into the cancer tumor). This phenomenon is non-directional (enhanced permeability and retention effect) and has shown to improve tumor accumulation from an intravenous administration (systemic). Several studies report half life circulation times of greater than 15 hours. Once the nanorods have been cleared from the bloodstream, the tumor may be illuminated ex vivo with a diode laser. Nanorods located at distances up to approximately 10 times their diameter absorb roughly 80% of the incident light energy, creating sufficient heat to kill the local (cancer) cells.

Hollow Gold Nanospheres (HGNs)

A novel metal nanostructure, namely hollow gold nanospheres (HGNs or AuNSs), has been recently developed and successfully used for photothermal ablation therapy (PTA) in vitro and in vivo. The HGNs have a unique combination of small size (30-50 nm in outer diameter and 3-6 nm shell thickness), spherical shape, highly uniform shell, and strong, narrow, and tunable near IR absorption. The high optical quality of the HGNs is mainly due to the high uniformity of the Au shell, which is generated using highly uniform Co nanoparticles as a template. The PTA efficiency of HGNs are right times better than solid gold nanoparticles. The absorption of HGNs is also much stronger than solid gold nanoparticles at the SPR region due to the two surfaces (interior and outer) present. Due to the hollow nature, less gold is needed for achieving a certain diameter. The NIR absorbing HGNs require a thin Au shell, which turns out to be most challenging to make. Work is in progress to improve the synthesis of NIR HGNs.

Gold NanoRods (AuNR)

Huang et al. investigated the feasibility of using gold nanorods for both cancer cell imaging as well as photothermal therapy. The authors conjugated antibodies (anti-EGFR monoclonal antibodies) to the surface of gold nanorods, allowing the gold nanorods to bind specifically to certain malignant cancer cells (HSC and HOC malignant cells). After incubating the cells with the gold nanorods, an

800 nm Ti:sapphire laser was used to irradiate the cells at varying powers. The authors reported successful destruction of the malignant cancer cells, while nonmalignant cells were unharmed.

When AuNP are exposed to NIR light, the oscillating electromagnetic field of the light causes the free electrons of the AuNP to collectively coherently oscillate. Changing the size and shape of AuNP, changes the wavelength that gets absorbed. A desired wavelength would be between 700-1000 nm because biological tissue is optically transparent at these wavelengths. While all AuNP properties are sensitive to change in their shape and size, Au nanorods properties are extremely sensitive to any change in any of their dimensions regarding their length and wide or their aspect ratio. When light is shown on a metal NP, the NP forms a dipole oscillation along the direction of the electric field. When the oscillation reaches its maximum, this frequency is called the surface plasmon resonance (SPR). AuNR have two SPR spectrum bands: one in the NIR region caused by its longitudinal oscillation which tends to be stronger with a longer wavelength and one in the visible region caused by the transverse electronic oscillation which tends to be weaker with a shorter wavelength. The SPR characteristics account for the increase in light absorption for the particle. As the AuNR aspect ratio increases, the absorption wavelength is redshifted and light scattering efficiency is increased. The electrons excited by the NIR lose energy quickly after absorption via electron-electron collisions, and as these electrons relax back down, the energy is released as a phonon that then heats the environment of the AuNP which in cancer treatments would be the cancerous cells. This process is observed when a laser has a continuous wave onto the AuNP. Pulsed laser light beams generally results in the AuNP melting or ablation of the particle. Continues wave lasers take minutes rather than a single pulse time for a pulsed laser, continues wave lasers are able to heat larger areas at once.

Gold Nanoshells

Loo et al. investigated gold nanoshells, coating silica nanoparticles with a thin layer of gold. The authors conjugated antibodies (anti-HER2 or anti-IgG) to these nanoshells via PEG linkers. After incubation of SKBr3 cancer cells with the gold nanoshells, an 820 nm laser was used to irradiate the cells. Only the cells incubated with the gold nanoshells conjugated with the specific antibody (anti-HER2) were damaged by the laser.

Graphene and Graphene Oxide

Yang et al. demonstrated the viability of graphene for photothermal therapy in 2010 with in vivo mice models. An 808 nm laser at a power density of 2 W/cm^2 was used to irradiate the tumor sites on mice for 5 minutes. As noted by the authors, the power densities of lasers used to heat gold nanorods range from 2 to 4 W/cm^2. Thus, these nanoscale graphene sheets require a laser power on the lower end of the range used with gold nanoparticles to photothermally ablate tumors.

In 2012, Yang et al. incorporated the promising results regarding nanoscale reduced graphene oxide reported by Robinson et al. into another in vivo mice study. The therapeutic treatment used in this study involved the use of nanoscale reduced graphene oxide sheets, nearly identical to the ones used by Robinson et al. (but without any active targeting sequences attached). Nanoscale reduced graphene oxide sheets were successfully irradiated in order to completely destroy the targeted tumors. Most notably, the required power density of the 808 nm laser was reduced to 0.15 W/cm^2, an order of magnitude lower than previously required power densities. This study demonstrates the higher efficacy of nanoscale reduced graphene oxide sheets as compared to both nanoscale graphene sheets and gold nanorods.

Future Directions

Some research has indicated problems with aggregation of the photosensitizers, local shock waves, hyperthermic effects, but otherwise little phototoxicity. Many potential side effects and complications, as well as other potential applications of photothermal therapy, are yet to be discovered.

Laser Hair Removal

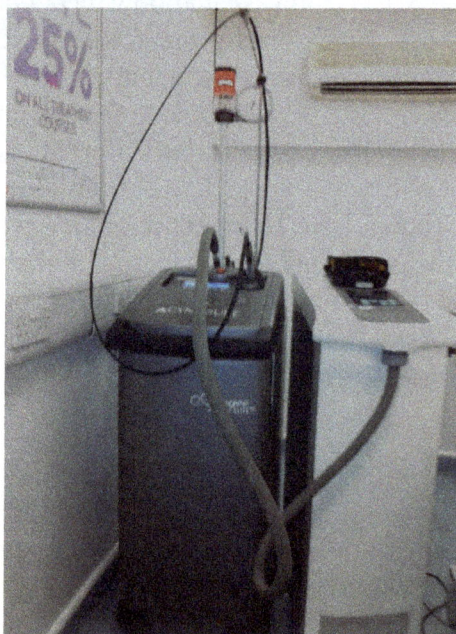

Hair-removal laser working at 755 and 1064 nm. The device to the right provides air cooling.

Laser hair removal is the process of removing unwanted hair by means of exposure to pulses of laser light that destroy the hair follicle. It had been performed experimentally for about twenty years before becoming commercially available in the mid-1990s. One of the first published articles describing laser hair removal was authored by the group at Massachusetts General Hospital in 1998. The efficacy of laser hair removal is now generally accepted in the dermatology community, and laser hair removal is widely practiced in clinics, and even in homes using devices designed and priced for consumer self-treatment. Many reviews of laser hair removal methods, safety, and efficacy have been published in the dermatology literature.

Mechanism of Action

The primary principle behind laser hair removal is *selective photothermolysis* (SPTL), the matching of a specific wavelength of light and pulse duration to obtain optimal effect on a targeted tissue with minimal effect on surrounding tissue. Lasers can cause localized damage by selectively heating dark target matter, melanin, in the area that causes hair growth, the follicle, while not heating the rest of the skin. Light is absorbed by dark objects, so laser energy can be absorbed by dark material in the skin, but with much more speed and intensity. This dark target matter, or *chromophore*, can be naturally-occurring or artificially introduced.

Melanin is considered the primary chromophore for all hair removal lasers currently on the market. Melanin occurs naturally in the skin, and gives skin and hair their color. There are two types of melanin in hair. Eumelanin gives hair brown or black color, while pheomelanin gives hair blonde or red color. Because of the selective absorption of photons of laser light, only black or brown hair can be removed. Laser works best with dark coarse hair. Light skin and dark hair are an ideal combination, being most effective and producing the best results, but new lasers are now able to target black hair in patients with dark skin with some success.

Hair removal lasers have been in use since 1997 and have been approved for "permanent hair reduction" in the United States by the Food and Drug Administration (FDA). Under the FDA's definition, "permanent" hair reduction is the long-term, stable reduction in the number of hairs regrowing after a treatment regime. Indeed, many patients experience complete regrowth of hair on their treated areas in the years following their last treatment. This means that although laser treatments with these devices will permanently reduce the total number of body hairs, they will not result in a permanent removal of all hair.

Laser hair removal has become popular because of its speed and efficacy, although some of the efficacy is dependent upon the skill and experience of the laser operator, and the choice and availability of different laser technologies used for the procedure. Some will need touch-up treatments, especially on large areas, after the initial set of 3-8 treatments.

Comparisons with other Removal Techniques

Comparison with Intense Pulsed Light

A 2006 review article in the journal "Lasers in Medical Science" compared intense pulsed light (IPL) epilators and both alexandrite and diode lasers. The review found no statistical difference in short term effectiveness, but a higher incidence of side effects with diode laser based treatment. Hair reduction after 6 months was reported as 68.75% for alexandrite lasers, 71.71% for diode lasers, and 66.96% for IPL. Side effects were reported as 9.5% for alexandrite lasers, 28.9% for diode lasers, and 15.3% for IPL. All side effects were found to be temporary and even pigmentation changes returned to normal within 6 months.

IPL, though technically not containing a laser, are sometimes incorrectly referred to as "laser hair removal". IPL-based methods, sometimes called "phototricholysis", or "photoepilation", use xenon flash lamps that emit full spectrum light. IPL systems typically output wavelengths between 400 nm and 1200 nm. Filters are applied to block shorter wavelengths, thereby only utilizing the longer, "redder" wavelengths for clinical applications. IPLs offer certain advantages over laser, principally in the pulse duration. While lasers may output trains of short pulses to simulate a longer pulse, IPL systems can generate pulse widths up to 250ms which is useful for larger diameter targets. Some current IPL systems have proven to be more successful in the removal of hair and blood vessels than many lasers,

Comparison with Electrolysis

Electrolysis is another hair removal method that has been used for over 135 years. Unlike laser epilation, electrolysis can be used to remove 100% of the hair from an area and is effective on hair of all colors, if used at an adequate power level with proper technique. More hair may grow in certain areas that are prone to hormone-induced growth (e.g. a woman's chin and neck) based on individual hormone levels or changes therein, and one's genetic predisposition to grow new hair.

A study conducted in 2000 at the ASVAK Laser Center in Ankara, Turkey comparing alexandrite laser and electrolysis for hair removal on 12 patients concluded that laser hair removal was 60 times faster, less painful and more reliable than electrolysis. It is important to note that the type of electrolysis performed in the study was galvanic electrolysis, rather than thermolysis or a blend of the two. Galvanic current requires 30 seconds to more than a minute to release each hair whereas thermolysis or a blend can require much less. This study thus did not test the capability of all forms of modern electrolysis.

Regulation

In some countries—including the U.S.—hair removal is an unregulated procedure that anyone can do. In some places, only doctors and doctor-supervised personnel can do it while in other cases permission extends to licensed professionals, such as regular nurses, physician assistants, estheticians, and/or cosmetologists. The Florida Board of Medicine has determined that the use of lasers, laser-like devices and intense pulsed light devices is considered the practice of medicine, and requires they be used only by a physician (D.O. or M.D.), a physician assistant under the supervision of a physician, or an advanced registered nurse practitioner under a protocol signed by a physician. An electrologist working under the direct supervision and responsibility of a physician is also allowed to perform laser hair removal in the state of Florida.

Laser Parameters that Affect Results

Several wavelengths of laser energy have been used for hair removal, from visible light to near-infrared radiation. These lasers are characterized by their wavelength, measured in nanometers (nm):

Argon: 488 nm (Turquoise/Cyan) or 514.5 nm (Green) (no longer used for hair removal)

Ruby: 694.3 nm (Deep Red) (only safe for patients with very pale skin)

Alexandrite: 755 nm (Near-Infrared) (safe and effective on all skin types, Fitzpatrick I-VI)

Pulsed diode array: 810 nm (Near-Infrared) (for pale to medium type skin)

Nd:YAG: 1064 nm (Near-Infrared) (made for treating darker skin types, though effective on all skin types)

Intense pulsed light (IPL is not a laser): 650 nm (used for hair removal in pale to medium skin types)

Pulse width (or duration) is one of the most important considerations. The length of the heating pulse relates directly to the damage achieved in the follicle. When attempting to destroy hair follicles the main target is the germ cells which live on the surface of the hair shaft. Light energy is absorbed by the melanin within the hair and heat is generated. The heat then conducts out towards the germ cells. As long as a sufficient temperature is maintained for the required time then these cells will be successfully destroyed. This is absolutely critical - attaining the require temperature is not sufficient unless it is kept at that temperature for the corresponding time. This is determined by the Arrhenius Rate Equation. To achieve these conditions the laser/IPL system must be able to generate the required power output. The main reason why hair removal fails is simply because the equipment cannot generate the desired temperature for the correct time.

Spot size, or the width of the laser beam, directly affects the depth of penetration of the light energy due to scattering effects in the dermal layer. Larger beam diameters result in deeper deposition of energy and hence can induce higher temperatures in deeper follicles. Hair removal lasers have a spot size about the size of a fingertip (3-18mm).

Fluence or energy density is another important consideration. Fluence is measured in joules per square centimeter (J/cm^2). It's important to get treated at high enough settings to heat up the follicles enough to disable them from producing hair.

Epidermal cooling has been determined to allow higher fluences and reduce pain and side effects, especially in darker skin. Three types of cooling have been developed:

- Contact cooling: through a window cooled by circulating water or other internal coolant. This type of cooling is by far the most efficient method of keeping the epidermis protected since it provides a constant heat sink at the skin surface. Sapphire windows are much more conductive than quartz.

- Cryogen spray: sprayed directly onto the skin immediately before and/or after the laser pulse

- Air cooling: forced cold air at -34 degrees C

In essence, the important output parameter when treating hair (and other skin conditions) is power density - this is a combination of energy, spot diameter and pulse duration. These three parameters determine what actually happens when the light energy is absorbed by the tissue chromophore be it melanin, haemoglobin or water, with the amount of tissue damaged being determined by the temperature/time combination.

Number of Sessions

Hair grows in several phases (anagen, telogen, catagen) and a laser can only affect the currently active growing hair follicles (early anagen). Hence, several sessions are needed to kill hair in all phases of growth.

Multiple treatments depending on the type of hair and skin color have been shown to provide long-term reduction of hair. Most patients need a minimum of seven treatments. Current parameters differ from device to device but manufacturers and clinicians generally recommend waiting from three to eight weeks between sessions, depending on the area being treated. The number of sessions depends on various parameters, including the area of the body being treated, skin color, coarseness of hair, reason for hirsutism, and sex. Coarse dark hair on light skin is easiest to treat. Certain areas (notably men's faces) may require considerably more treatments to achieve desired results.

Laser does not work well on light-colored hair, red hair, grey hair, white hair, as well as fine hair of any color, such as vellus. For darker skin patients with black hair, the long-pulsed Nd:YAG laser with a cooling tip can be safe and effective when used by an experienced practitioner.

Typically the shedding of the treated hairs takes about two to three weeks. These hairs should be allowed to fall out on their own and should not be manipulated by the patient for certain reasons, chiefly to avoid infections. Pulling hairs after a session can be more painful as well as counteract the effects of the treatment.

Side Effects and Risks

Some normal side effects may occur after laser hair removal treatments, including itching, pink skin, redness, and swelling around the treatment area or swelling of the follicles (follicular edema). These side effects rarely last more than two or three days. The two most common serious side effects are acne and skin discoloration.

Some level of pain should also be expected during treatments. Numbing creams are available at most clinics, sometimes for an additional cost. Some numbing creams are available over the counter. Use of strong numbing creams over large skin areas being treated at one time must be avoided, as this has seriously harmed, and even killed, patients. Typically, the cream should be applied about 30 minutes before the procedure. Icing the area after the treatment helps relieve the side effects faster. Ibrahimi and Kilmer reported a study of a novel device of diode handpiece with a large spot size which used vacuum-assisted suction to reduce the level of pain associated with laser treatment.

Unwanted side effects such as hypo- or hyper-pigmentation or, in extreme cases, burning of the skin call for an adjustment in laser selection or settings. Risks include the chance of burning the skin or discoloration of the skin, hypopigmentation (white spots), flare of acne, swelling around the hair follicle (considered a normal reaction), scab formation, purpura, and infection. These risks can be reduced by treatment with an appropriate laser type used at appropriate settings for the individual's skin type and treatment area.

Some patients may show side effects from an allergy to either the hair removal gel used with certain laser types or to a numbing cream, or to simply shaving the area too soon in relation to the treatment.

References

- Bolton, James; Colton, Christine (2008). The Ultraviolet Disinfection Handbook. American Water Works Association. pp. 3–4. ISBN 978-1-58321-584-5.

- Bouzari N, Elsaie ML, Nouri K (2012). "Laser and Light for Wound Healing Stimulation". In Nouri K. Lasers in Dermatology and Medicine. Springer London. pp. 267–75. doi:10.1007/978-0-85729-281-0_20. ISBN 978-0-85729-281-0.

- Goldman L (1990). "Dye Lasers in Medicine". In Duarte FJ; Hillman LM. Dye Laser Principles. Boston: Academic Press. pp. 419–32. ISBN 0-12-222700-X.

- Messina, Gabriele (October 2015). "A new UV-LED device for automatic disinfection of stethoscope membranes" (PDF). American Journal of Infection Control. Elsevier. Retrieved 2016-08-15.

- "The Nobel Prize in Physiology or Medicine 1903". Nobelprize.org. Nobel Media AB. 2016-11-01. Archived from the original on 2016-10-22. Retrieved 2016-11-01.

- Saini, Rajan; Lee, Nathan; Liu, Kelly; Poh, Catherine (2016). "Prospects in the Application of Photodynamic Therapy in Oral Cancer and Premalignant Lesions". Cancers. 8 (9): 83. doi:10.3390/cancers8090083. ISSN 2072-6694.

- "Single-agent phototherapy system diagnoses and kills cancer cells | KurzweilAI". www.kurzweilai.net. November 2, 2015. Retrieved 2016-04-27.

- "Packaging and Print Media | PACKAGiNG & Print Media | UV LED ... what's it all about?". packagingmag. co.za. Retrieved 2015-12-30.

- Pei S, Inamadar AC, Adya KA, Tsoukas MM (2015). "Light-based therapies in acne treatment". Indian Dermatol Online J. 6 (3): 145–57. doi:10.4103/2229-5178.156379. PMC 4439741. PMID 26009707.

- "How Finsen's light cured lupus vulgaris". Photodermatol Photoimmunol Photomed. 21: 118–24. 2014-11-12. doi:10.1111/j.1600-0781.2005.00159.x. PMID 15888127. Retrieved 2015-02-25.

Permissions

Index

www.ingramcontent.com/pod-product-compliance
Lightning Source LLC
Chambersburg PA
CBHW061316190326
41458CB00011B/3821